Process Development Division 2016

Core Programming Area at the 2016 AIChE Meeting

San Francisco, California, USA
13-18 November 2016

ISBN: 978-1-5108-3446-0

Printed from e-media with permission by:

Curran Associates, Inc.
57 Morehouse Lane
Red Hook, NY 12571

Some format issues inherent in the e-media version may also appear in this print version.

Printed by Curran Associates, Inc. (2017)

For permission requests, please contact AIChE
at the address below.

AIChE
120 Wall Street, FL 23
New York, NY 10005-4020

Phone: (800) 242-4363
Fax: (203) 775-5177

www.aiche.org

Additional copies of this publication are available from:

Curran Associates, Inc.
57 Morehouse Lane
Red Hook, NY 12571 USA
Phone: 845-758-0400
Fax: 845-758-2633
Email: curran@proceedings.com
Web: www.proceedings.com

TABLE OF CONTENTS

Understanding Pressure Regulator Performance: A Discussion of Droop

Jon Kestner

Product Manager, Pressure Regulators

Swagelok Company

Swagelok

Agenda

- Theory of Pressure Regulators
  - Types, Basic Operation
  - Balance of Forces
- Regulator Performance
  - Droop
  - Spring Rates
  - Technology Options, Dome Loading
- Review

Theory of Regulators:
Regulator Types

BACK-PRESSURE Regulators control pressure from the process by sensing the inlet pressure – they control their own upstream pressure

PRESSURE REDUCING Regulators control pressure to the process by sensing the outlet pressure – they control their own downstream pressure

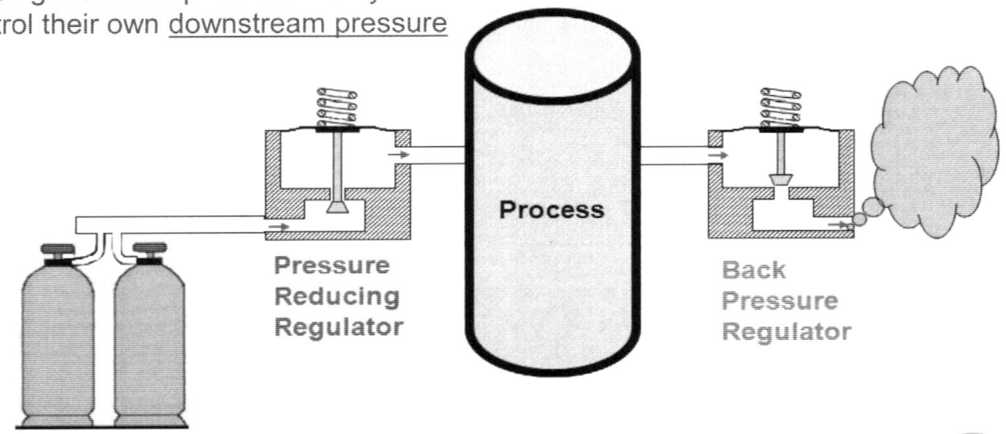

Theory of Regulators:
Regulator Basics

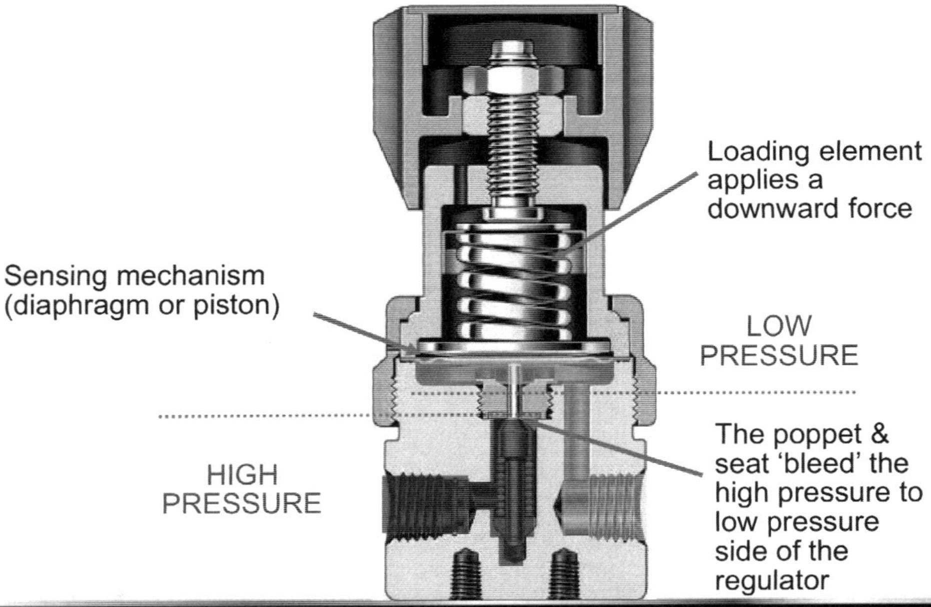

Theory of Regulators:
Balance of Forces

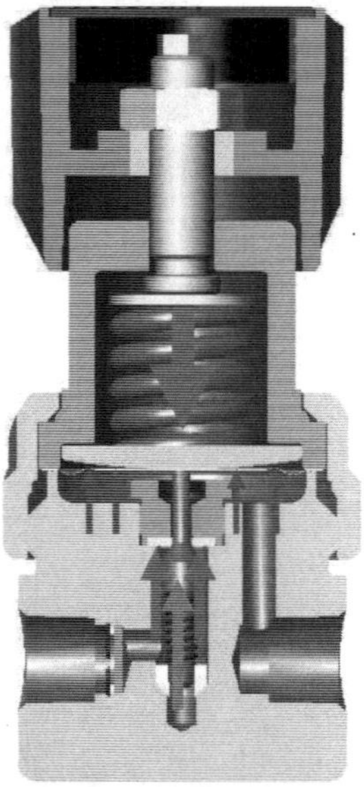

F_1 = **Loading Force**

F_2 = **Inlet Spring Force**

F_3 = **Outlet Pressure Force**

F_4 = **Inlet Pressure Force**

$$F1 = F2 + F3 + F4$$

Regulator Performance:
Droop

- Droop is the reduction of outlet pressure as the flow rate through a regulator increases.

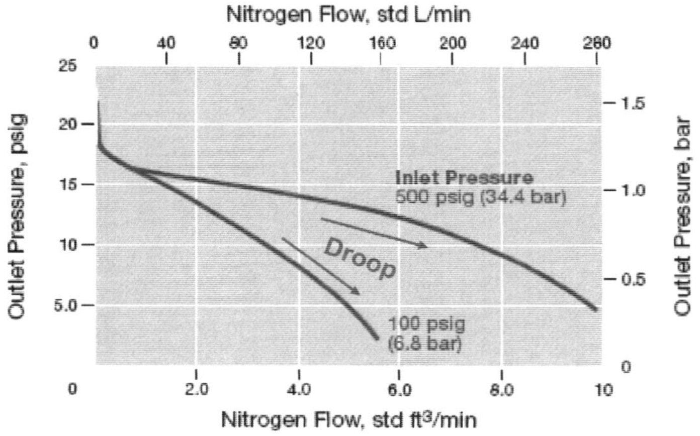

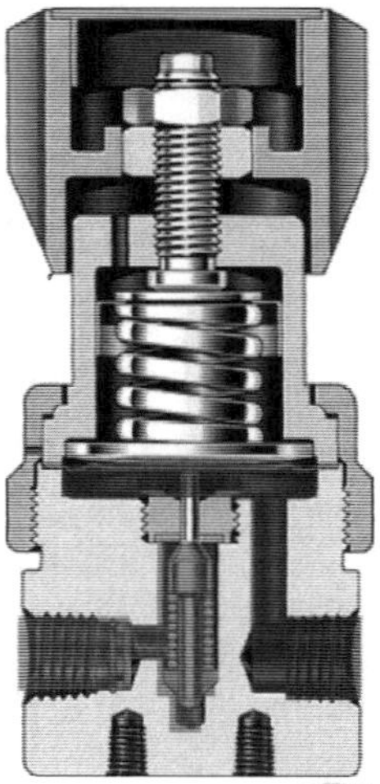

Regulator Performance:
Droop

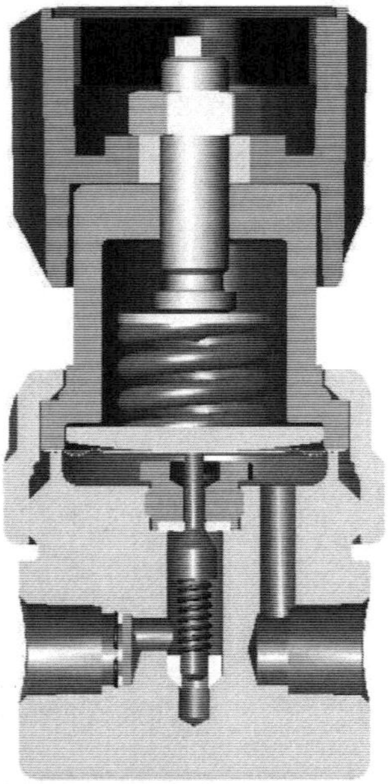

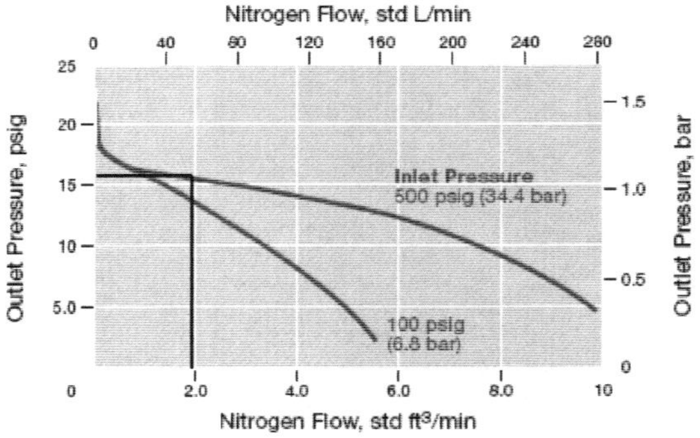

Watch what happens
with loading and
control elements!

Regulator Performance:
Droop

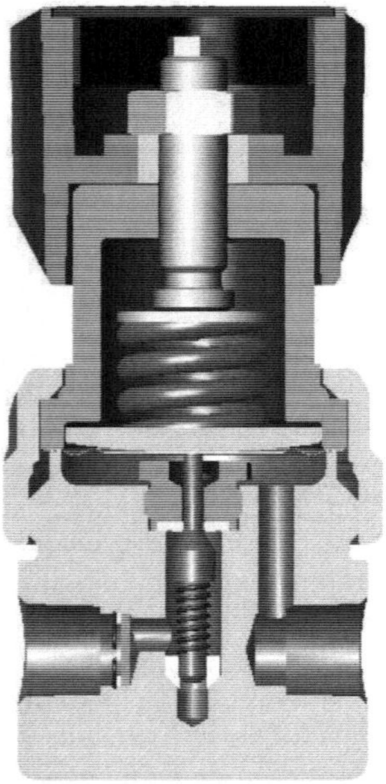

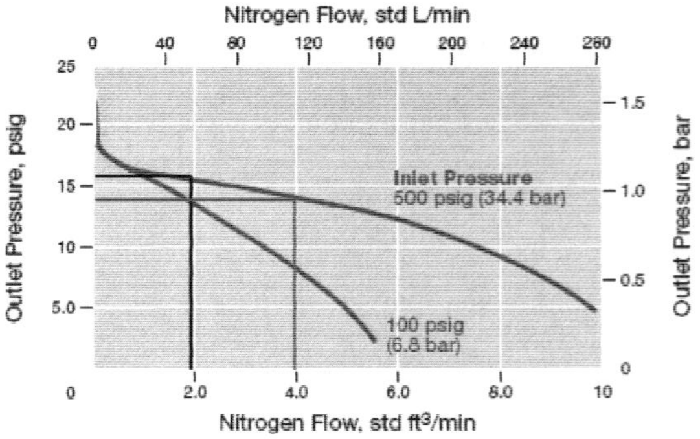

Regulator Performance:
Droop

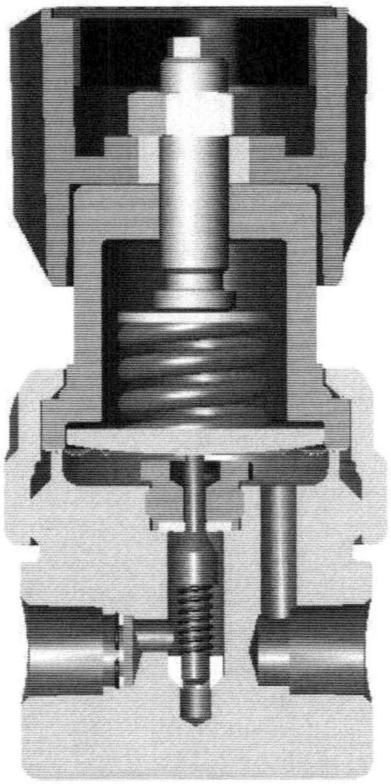

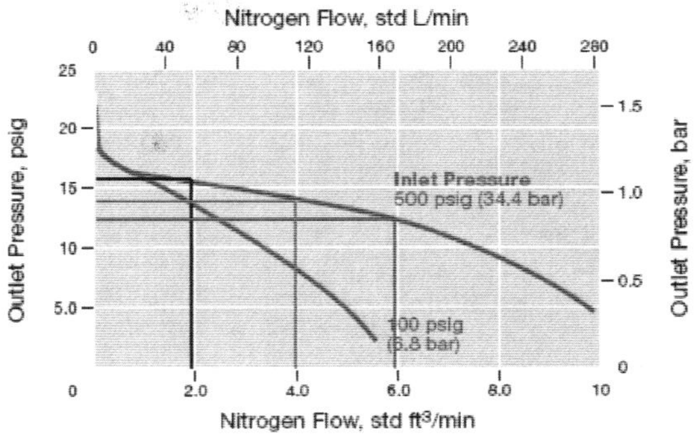

Regulator Performance:
Droop

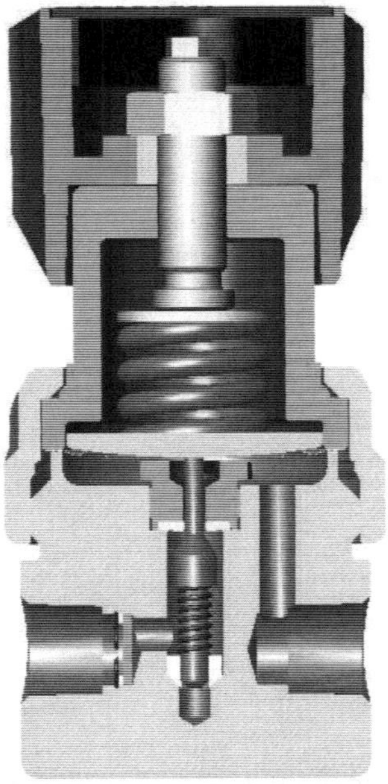

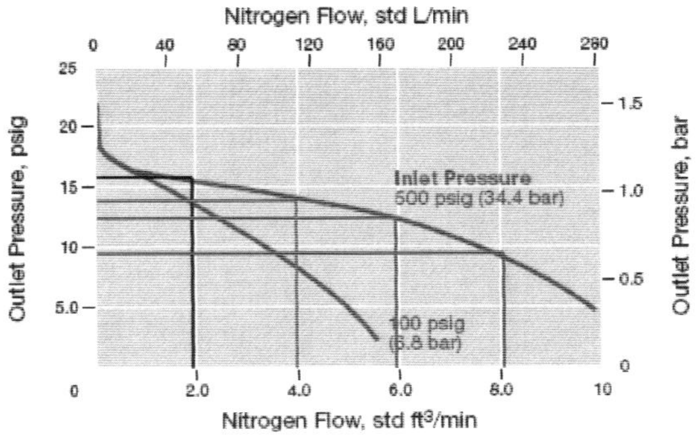

© 2008 Swagelok Company.- Confidential

Regulator Performance:
Droop, Balance of Forces

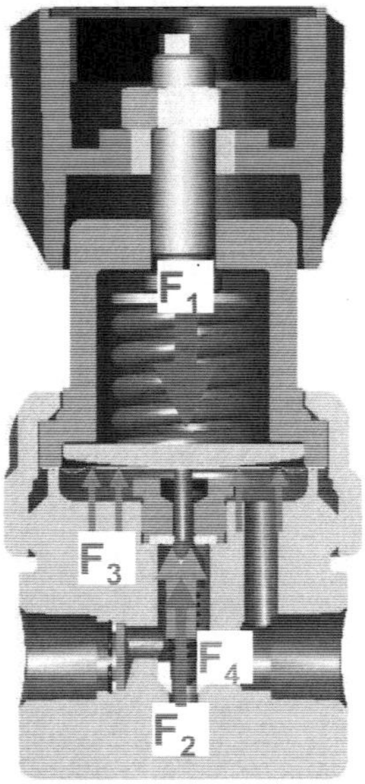

F_1 = **Loading Force**

F_2 = **Poppet Spring Force**

F_3 = **Outlet Pressure Force**

F_4 = **Inlet Pressure Force**

$$F1 = F2 + F3 + F4$$

What happens when more flow is needed?

Regulator Performance:
Droop, Spring Rates

High Spring Rate (400 N / mm)

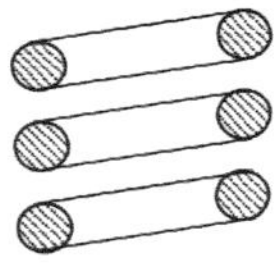

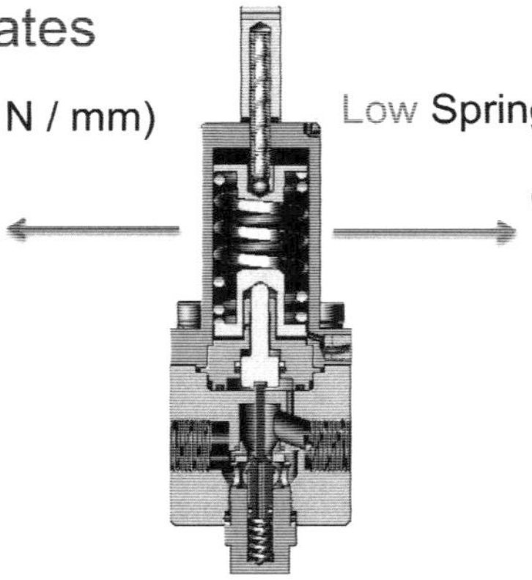

Low Spring Rate (200 N / mm)

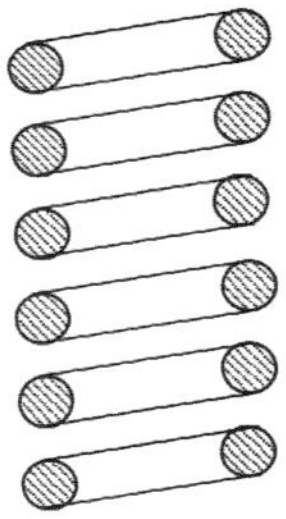

For F1= 3200 N
Spring Compression= 8 mm
If valve stroke= 2.5 mm
ΔF1= -1000 N at full stroke

For a control range of 0 to 14 bar
Droop= 3.2 bar

For F1= 3200 N
Spring Compression=16 mm
If valve stroke= 2.5 mm
ΔF1= -500 N at full stroke

For a control range of 0 to 14 bar
Droop= 1.6 bar

12

Regulator Performance:
Droop, Technology Options

Spring-Loaded

In a spring-loaded regulator, a coil spring is used to generate a load (F_S) against the sensing mechanism. The amount of spring force or load can be adjusted by turning the handle or adjusting screw of the regulator.

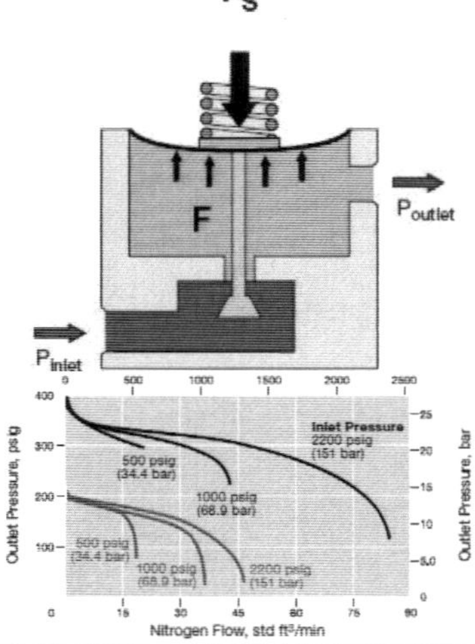

Dome-Loaded

In a dome-loaded regulator, a gas is fed into the dome chamber above the sensing mechanism at a pressure equal to or slightly above the required outlet pressure. This volume of gas is used like a spring. The dome pressure (F_d) is typically supplied by a second regulator called a pilot regulator.

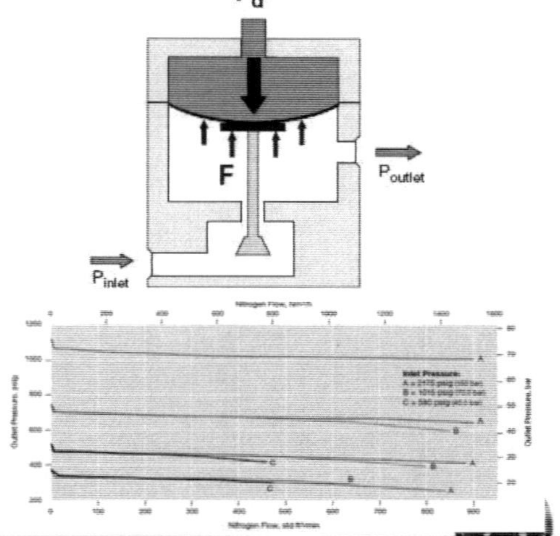

Review

- Two types of pressure regulators
  - Pressure Reducing and Back Pressure
- Control, Loading and Sensing Elements dictate regulator behavior
- Balance of Forces describe regulator performance
- Droop: the reduction of outlet pressure as flow demand increases
- Spring Rate is large driver of droop
- Dome loaded regulators can improve droop performance

Best Practices in Pilot Plant Layout

Richard Palluzi

Richard P Palluzi LLC
Laboratory, pilot plant and research safety, design, training and consulting

72 Summit Drive
Basking Ridge, NJ 07920
rpalluzi@verizon.net
908-285-3782

Overall Goals

- Promote safe operation in all phases
- Maximize operability after construction
  - Crowded units are usually less efficient to operate
- Maximize accessibility during operation
  - For operations, maintenance, and modifications
- Minimize construction costs
  - Crowded units are more expensive and time consuming to build
  - Rework potential due to space issues is high
- Minimize maintenance costs
  - Crowded units require more time and effort to repair
- Minimize space required
  - In order to maximize the use of a given facility

Overall Space Requirements

- All too often, the space "required" for a pilot plant is estimated based on little real analysis
- The result?
  - Inadequate space leading to
    - Higher construction costs,
    - Needlessly difficult operations,
    - Long term operability issues, and
    - Often safety problems
- Few if any of which can ever be corrected regardless of the time, effort and money expended
- Potential solutions

Option 1: A very detailed layout after the final P&ID is developed which addresses all the layout requirements

- Certainly safest approach but usually not practical
- Very tedious and time consuming; hence, very expensive
- Almost always leads to a noticeable delay in starting the project
  - Usually several weeks depending on the size of the unit
- Requires very experienced pilot plant designer to identify all the space needs e.g. supports, piping and wiring runs, accessibility, etc.
- Rarely cost or schedule effective even if one can find an experienced enough pilot plant designer
- Any errors, omissions or modifications can require rework of the layout, often substantial
  - And even the most experienced designer can miss a key element requiring more space or a different layout
  - The effect of even small errors or omissions can result in a series of cascading changes which make reworking the late parts of the layout necessary
- Very inflexible as, once completed, any changes can require rework

Option 2: A much less detailed layout after the P&ID is progressed far enough to minimize the chances for major changes with a conscious decision to allocate a non trivial space contingency

- Next safest approach with the key issue of determining how much contingency is required
- Key factors to evaluate:
  - What are the worst case potential additional space requirements?
  - How confident are you in the major space related portions of the P&ID?
- Common problems:
  - Significant detailed design additions resulting in more piping and wiring (and hence more space needs)
  - Final procurement of larger than expected equipment and/or additional ancillary systems for the equipment selected (and hence more space needs)
  - Significant revisions to the P&ID not considered in the space evaluation
- Determining the appropriate space contingency
  - Note the space contingency needs to be applied somewhat over all of the unit
    - + So simply providing an extra X% at the end of the last skid is not usually adequate
    - + Conversely, assuming a ¼" line will be X% larger is not useful
  - You need to provide the extra space contingency near and around assemblies, e.g. 3" around the reactor or a foot next to the feed system
  - This is a lot harder to do effectively then it sounds and requires a lot of pilot plant experience
- An approach I would strongly recommend

Option 3: Inspired guesswork based on the designer's experience with a very significant increase for space contingency

- Very risky but most common approach
- Totally dependent on
  - The skill and experience of the pilot plant designer
  - How closely their guess as to the final design matches the actual product
- Open to constant arguments that the space requirements are "grossly exaggerated"
  - Difficult to defend and justify except "based on experience"
  - So often results in backing down the space required with attendant increased – but unrecognized - risk
- Miss matches are only identified when it is usually too late to correct resulting in severe layout issues
  - And almost always higher costs and missed schedules

Suggested Best Practices

Be Pessimistic

- Assume the high end space requirements for all equipment
- Diligently try to identify and include all potential growth areas which could be added during the detailed design and include room for them (or a lot of them)
- Allocate extra space for areas
  - Of lower familiarity
  - New or novel approaches
  - New equipment
- Allocate extra, empty space around equipment requiring manual operations
  - Filling
  - Emptying
  - Draining
  - Purging
  - Loading/charging
  - Cleaning
  - Sterilizing
  - Assembling/disassembling
  - Packing
  - Removing/inserting internals
  - Maintenance

Always Allocate Contingency For Space In The Design

Typical Suggested Space Contingencies			
Approach	**Space Contingency Range**	**Typical Probability of Enough Space**	**Comments**
1. Detailed design	10-20%	90-95%	Range is based on level of effort put into the detailed layout
2. Less Detailed Design	50-100%	70-90%	Range is based on the confidence level of the P&ID used for layout and the experience and skill of the person doing the layout
3. Inspired Guesswork	100-300%	25-75%	Range is based on the skill of the pilot plant designer

These assume the proper space contingency has been included

Make Sure You Allow For Room To Fit Up Vessels And Components

- Vessels/Reactors
  - Sight glasses
  - Relief devices
  - Pressure gauges
  - Drain valves
  - Fill valves
  - Block valves
  - Sample ports
  - Instrumentation
  - Purge streams
  - Vents
- Pumps/Compressors
  - Relief devices
  - Pressure gauges
  - Drain valves
  - Fill valves
  - Block valves
  - Calibration ports
- Instruments
  - Drain valves
  - Fill valves
  - Block valves
  - Calibration ports
  - Filters
- Purification units
  - Drain valves
  - Fill valves
  - Block valves
  - Purge streams
  - Vents
- Actuators
- Solenoid valves
- Regulators

Typically these require 100-200% more space than the equipment alone

Plan for Significant Room Around All Equipment

- Assume that the space around the equipment needs to be as least as large as the equipment itself
  - Sometimes it is even larger!

Think About How The Components Will Be Supported And Leave Space

- Strut supports reduce usable room
- Larger/heavier components will require heavier supports or at least more strut closer together
- Brackets take up space
  - Custom brackets can reduce space but at a higher fabrication cost
- Use adjustable supports whenever possible to allow for minor fit up corrections

Piping/Tubing And Wiring Runs Take Up Space Which Adds Up Quickly

- They require room for the piping or conduit, room to make bends, room to make up fittings, room to access valves and covers, etc.
- Ancillary components require room
  - Pressure gauges
  - Filters
- Main runs are only a small part of the piping room required
  - Purges
  - Bypasses
  - Vents
  - Drains
  - Block valves

Use Height When Available

- Put utilities, raceways, racks, marshalling boxes and similar components up high
- Consider low operational need components up out of the way
- Put ancillary equipment out of the way even if it requires a stool or ladder to reach

Panel Boards, Shelves, And Benches Eat Up Space And Block Access

- They requires extra supports
- They block/limit access to around them
- Avoid them whenever possible

Enclosures Require Significant Non-Process Space

- Typical overlooked space needs
  - Supports
  - Insulation
  - Exhaust
  - Access into/out of
- They also make adjacent areas less usable due to restricting access

Accept That Everything Cannot Be Readily Accessible Unless You Accept A Larger Footprint

- Identify components that can be less accessible
  - Utility shut off valves
  - Regulators set once and rarely changed
  - Relief devices changed rarely
  - Reliable utility components
    - Pumps
    - Back pressure regulators
    - Shut off valves
- Identify components that need high accessibility
  - Reactors
  - Sampling
  - Filters
  - Rotating equipment

Make Use Of Depth And Side Walls

- Can often be accessible from both sizes doubling the useable space
- Remember that this will be more costly and time consuming to install

Avoid Space Inefficient Layouts

- Frames, single sided structures, vs. pallets, multiple sided structures, are easy to lay out but very space inefficient

Wiring takes up space

- Routing to each :
  - Instrument
  - Heater
  - Motor
  - TC/RTD
- Junction boxes, splice boxes and similar connection points need both space and access
- Providing code mandated mechanical protection of power feeds can be difficult without adequate space

Hazardous Area Electrical Components Require At Least Twice As Much Space As General Purpose Components

- Conduit and cables often require significantly more space to run in
- Codes require access in front of all panels and disconnects

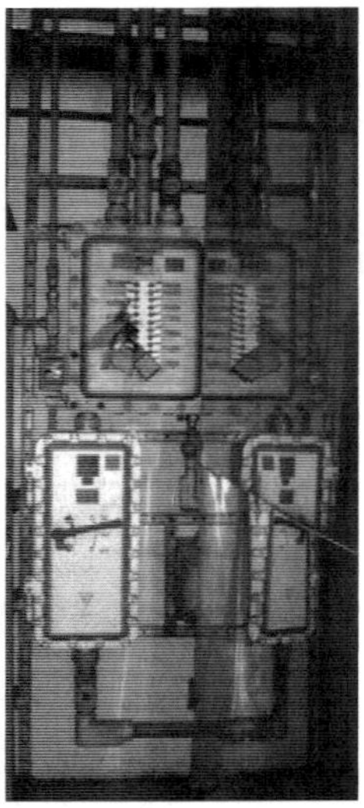

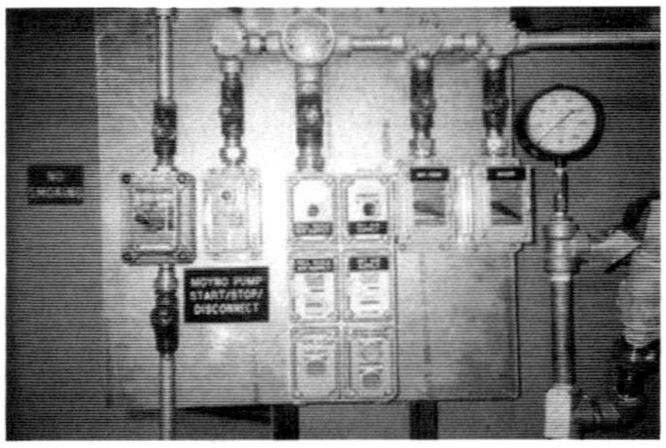

Plan for Insulation

- Larger clamps may be required
- Special supports may be required
- Thermal expansion may limit support options
- Insulation thickness rapidly adds up across a unit
- Installation often requires additional space to install the insulation
  - And if limited, installation costs and time will be greatly increased
  - As will maintenance costs

High Density Layouts Are Expensive

- They require more time and effort to layout and install
- They are less forgiving of any errors, omissions or problems and greatly limit the options available to address
- Potential for unexpected rework is high

Special Provisions May Be Required To Make Restricted Access Spots Viable

- Common areas of concern:
  - Hoods
  - Barricades
  - Enclosures
- Need to be designed in from beginning
- Often require additional access space
- Always more expensive and time consuming

Allow Room For Future Growth

- Usually best if open spaces are scattered around unit rather than additional space allocated in one larger grouping
- This should – ideally - be in addition to space contingency

Summary
The Golden (or Palluzi) Rules

- Pilot plant layout is an art as well as a science so some people will be better at it than others
  - Not a function of engineering or operational expertise
- It <u>always</u> requires more space than you think
  - So you need to estimate high and provide adequate space contingency
- No amount of skilled, experienced, and inspired design work can overcome a lack of adequate space
  - At best it can only somewhat ameliorate the pain
    - \+ And cost
    - \+ And time lost

455075 Best Practices in Pilot Plant Piping

Abstract:

Pilot plants offer unique challenges when it comes to piping. This paper will discuss some of techniques for using piping and tubing on pilot plants. It will help pilot plant users make the right decision when it comes to which system to use to carry fluids throughout their process. The paper will discuss piping, tubing, compression fittings, sanitary fittings, vacuum fittings, threaded pipe, welded pipe, plastic pipe, and other systems.

Best Practices in Design of Pilot Plant Scale Distillation Columns

Richard Palluzi

Richard P Palluzi LLC
Laboratory, pilot plant and research safety, design, training and consulting

72 Summit Drive
Basking Ridge, NJ 07920
rpalluzi@verizon.net
908-285-3782

1

Introduction

- Design of distillation columns for pilot plant service is more difficult than sometimes realized
- Smaller diameter columns experience significant wall and hydraulic effects that can deviate noticeably from predicted performance
- This results in less effective separations and, often, expensive and time consuming rework
- Distillation columns smaller than 12" ID are very difficult to fabricate with tray internals
  - As a result most pilot plant applications tend to be based on packed columns

2

Size vs Effectiveness

- Due to wall effects, smaller diameter distillation columns do not work as effectively as expected
- Based on limited data

ID (In)	Performance	Comments
≥6	Essentially theoretical	
≥2 to< 6	~50% of theoretical effectiveness	Doubling the theoretical height generally works well
≥1 to <2	Varies from 20-50% of theoretical effectiveness	Highly variable
≤1	<20% of theoretical effectiveness; occasionally just barely functional	Difficult to impossible to predict. Best avoided

3

Packing

- Packing smaller ID columns can be very difficult and often impacts the mechanical design
- Column sections longer than 72 inches should have flanges or similar means of disconnect to assist in packing and allow for redistributor plates
- Column sections up to 180 inches can often have the redistributor plate dropped in if done carefully
  - Longer columns or those with inserts may require additional break points
- Inserts into the columns (e.g. TC's, feeds, draw offs, etc.) are significant particularly as the columns get smaller in diameter
  - Suggest that these be minimized and kept as small as possible
    - +Even at the cost of higher pressure drop
- Shaking the column to settle the packing and help close up some of the voids is very useful but progressively harder to do as the column gets larger in diameter and longer in length
  - Using a very high pulsing gas or liquid flow during start up may provide some settling but will require repeated cycles

4

Connections

- At a minimum columns should have a top, middle and bottom connection, typically a flange
  - 1.5 inch ID or less can often use vacuum fittings or compression fittings
- Additional flanges, while often desirable for ease of installation and packing, noticeably increase the potential for leakage and localized cooling
  - Best avoided wherever possible
- Threaded column ports that are frequently opened/modified often have limited lives
  - Consider an additional fitting between the column and the external piping so they can be easily replaced without affecting the column itself
  - Welding tubing is also very effective but does not allow a different fitting later if desired/required
- Vacuum style fittings (e.g. Swagelok VCR and VCO) fittings have proven the most leak free seals
  - High quality compression fittings give satisfactory service but with higher leakage rates
  - Pipe threads are almost become problematical over time

5

Condensers

- Five port heat exchangers are recommended as the extra port allows for balancing the exchanger and minimizing hydraulic problems

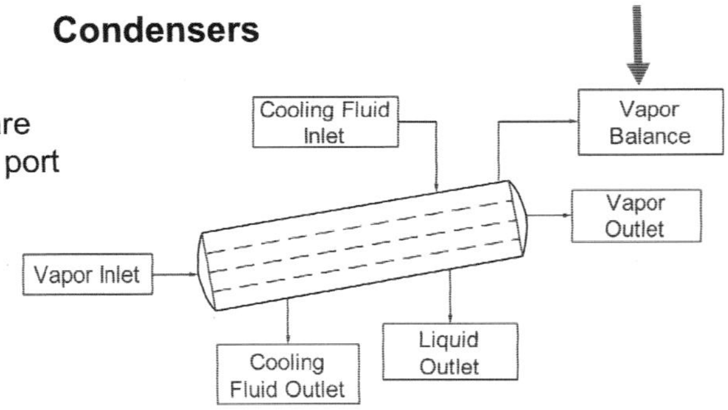

- The product vapor cools rapidly despite insulation so it is prudent to heat trace this line to above the column's vapor outlet temperature until it slopes down to the condenser to avoid unintended reflux
  - This tracing should have it's own control separate from the column so that it can be set clearly higher than the entering vapor

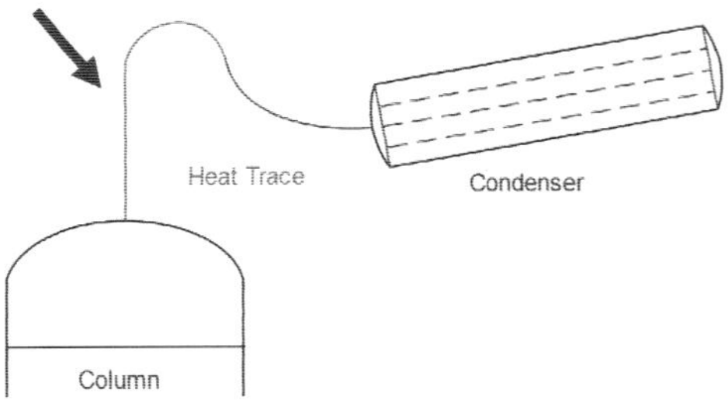

6

7

Condensers (cont'd)

- The minimum condenser height above the column should be at least 12 inch to avoid hydraulic problems
- For research style columns, a timed reflux (i.e. on or off but not modulated) has often proven more stable and always simpler
  - Reflux valve life can be a problem, particularly for systems with reflux rates around 50%, but good quality ball valves have had effective lives

8

Condensers (cont'd)

- The hydraulic head to the overhead product tank, when not refluxing, is usually higher than back to the column
  - Can result in sudden surges that can drain the condenser
  - A small vapor liquid separator between the condenser and the product accumulator is prudent
- Both the line refluxing to the column and the line leading to the vapor liquid separator should have liquid traps to maintain a liquid seal
- Wire mesh, often placed in the top of a column to help in coalescing liquid droplets, produces a higher pressure drop than the reflux return line which can lead to inadvertent bypassing
  - This requires a larger size trap on the reflux line to induce an equivalent higher pressure drop

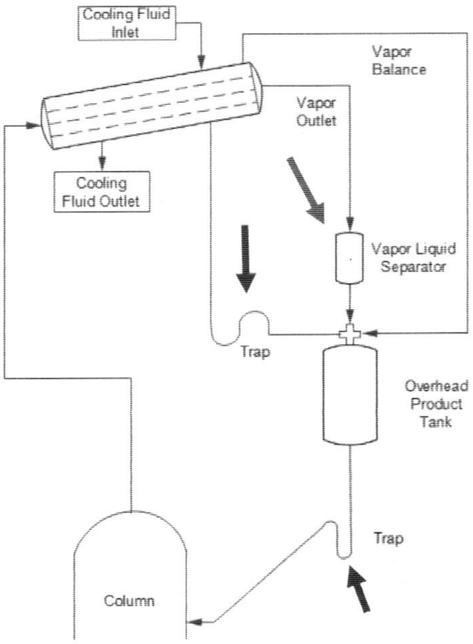

9

Reboiler Design: Thermosyphon Reboilers

- Thermosyphon reboilers are very effective for smaller research columns
  - May be single or multiple "legs" or "arms"
  - Each "leg" should be piped so as to be as close to identical as possible to avoid differences in pressure drop and performance
    - +Includes fittings, bends, and lengths
- In pressurized distillation columns these should be designed to operate 50-80% full
  - They should be heated from the bottom to the 50% full point
- In vacuum distillation columns these should be designed to operate 20-30% full
  - They should be heated from the bottom to the 20% full point
  - The extra head of liquid above the minimum point changes the boiling pressure and makes the thermosyphon less likely to work
  - A conventional reboiler may be a better choice if hold up time is not a concern
- Electrical heaters should have
  - Independent high temperature alarms on each outer wall
    - +So multiple "legs" require multiple alarms
    - +This avoids over temperature when a leg is empty to close to empty
  - A single temperature control point at the bottom
    - +Multiple "legs" do not require more than one control point at the common bottom connection
    - +This avoids each "legs" control loop fighting the others

10

Reboiler Design: Thermosyphon Reboilers (cont'd)

- When multiple "legs" are used, each leg should have as large an ID as possible
  - ID's ≤ 0.75 in are problematical due to hydraulic issues
  - Oversizing the "leg" has few issues but does increase holdup so if that is not an issue go larger
- Minimum reboiler lengths are difficult to determine but appear to be relatively short, typically ≤ 20-24 inch long
  - Oversizing the length can create problems as the height can increase the boiling point and also cause cooling

11

12

Reboiler Design: Conventional Reboilers

- The reboiler vapor cools rapidly despite insulation so it is prudent to heat trace this line to slightly above the reboiler temperature until it reenters the column
  - It may be possible to operate the reboiler at a high enough temperature to account for tis extra heat loss
  - This heat loss, however, is often badly under estimated, resulting in too high a required reboiler temperature
- Electrical heaters should have
  - An independent high temperature alarm on the reboiler outer wall
    - + Again to protect the reboiler when empty or close to empty
  - A temperature control point below the lowest possible reboiler liquid level
- Placing a restriction, such as a regulating valve between the column and the reboiler helps to maintain a liquid level in the bottom of the column for more stable operation
  - Difficult to set unless the tower bottom has a sight glass or level transmitter
  - Either are highly recommended

13

Adiabatic Operation

- True adiabatic operation is very difficult
  - No amount of insulation, even the most carefully applied, can completely prevent non-trivial heat loss
  - Vacuum jacketed research size columns are often problematical
    - +Higher heat losses than expected due to the larger ratio of attachment points to column surface area
    - +Thermal expansion complicates the mechanical design significantly except at low temperatures (< 150-200 °F) often resulting in component failures over time
- Pseudo-adiabatic operation is fairly easy
  - Use of wrap on electric heating blankets controlled on delta temperature (column wall to heater) is relatively straightforward
    - +There must be an insulating layer between the column and the heater for the control to work
  - The more separate zones (heaters), the more the operation approaches adiabatic
    - +At least two (top and bottom) are required
    - +Four to six are usually adequate except for very tall columns

Pseudo-Adiabatic Cross Section

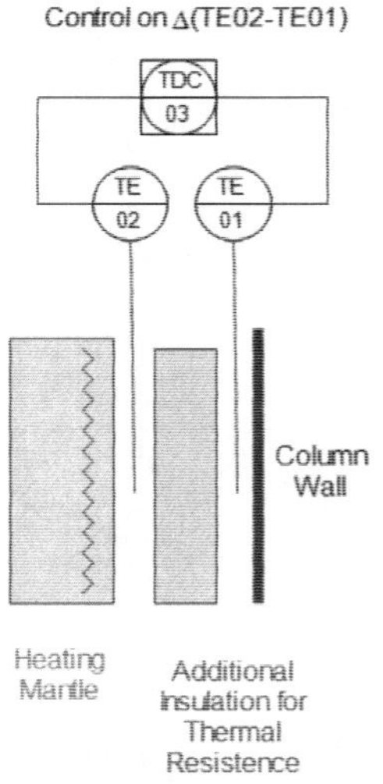

14

15

Level Control

- Level control of distillation columns can be surprisingly difficult due to the routine change in compositions
- Systems that relay on compositional properties often are less effective than normal due to continued bulk property changes
  - Capacitance and resistance probes are often less effective than expected
  - Differential pressure transmitters are often a wiser choice due to their ability to be re-spanned on line
- Sight glasses for visual observations are always prudent
  - Higher temperature systems may require removable insulation to prevent heat loss when the sight glasses are not being actually observed

16

Mounting

- Ideally, distillation columns should be mounted in only one place and allowed to expand and contract in the other direction
- For longer columns, mounting in the middle is often the most practical approach as it limits the column extension above and below the support to more manageable levels
- The top and bottom of most columns often require some limited support, actually guides, to prevent excessive movement and to prevent mechanical damage
- The weight of a packed, operational column is often considerable and a substantial support is required
  - Operating columns usually have some vibration
  - Strut alone is often too flexible and prone to loosening over time
- Inadequately supported columns are prone to increased leakage and, in extreme cases, mechanical damage over time

17

18

19

20

21

22

467135 Pilot Plant Screening Steps, Test Plan Development, Execution, and Results Analysis As Related to Thin-Film Evaporation

Abstract:

Chemical processors invest billions of dollars each year in capital equipment. Many of the processes are new and novel while some are established yet have long running issues that have not been solved. The potential costs of downtime from troubleshooting in the field during production is not justifiable so many companies continue to run their processes at sub-optimal process conditions. Before finalizing a purchase order, engineering, operations, and business leaders want to know that they are signing off on a proven solution.

The solution to vetting new processes or investigating existing ones is utilizing a pilot plant. Information is gathered on pilot scale equipment to scale-up to production scale equipment. The equipment sizing from pilot plant data is used to get ballpark pricing, which can be used to establish a business case for the product you are investigating. Another benefit of pilot plant testing is the small relative quantity of feed material required. It may be difficult to procure mass quantities of feed for new applications, typically using lengthy batch preparation. For existing production evaluation, you can divert a small stream of process material to run the trial without major production losses.

At LCI, we perform testing and piloting services for customers to evaluate new processes and troubleshooting existing processes utilizing our thin-film evaporation and drying technologies. LCI generates processed material samples for its customers for evaluation to determine if the end product will be well received by their customers and profitable enough to invest capital dollars. Pilot testing is an attempt to validate a process through quantitative means via logging of process variables such as mass flow rates for feed, concentrate and distillate streams, heating media type and conditions, operating pressure, and process temperatures and referencing run data versus analytical data from sample analysis. This data is used to draw a correlation between the process conditions and product specifications for quality, whether that be residual solvent content, color, viscosity, reaction completion, et al.

Before pilot testing, new processes are screened using a relatively small sample (~1L) at lab-scale. This qualitative testing determines if there are any product characteristics (such as fouling, foaming, temperature sensitivity, etc) that may present major issues during pilot scale operations. In addition, test conditions parameter ranges can be determined during the sample evaluation. A database of past tests is also utilized to determine any instances of similarity to past tests. These past tests can be used to determine process condition targets or foretell pitfalls that can be avoided with a certain test setup.

Once sample evaluation and database research are completed, a test plan is created with a planned layout. Careful consideration of evaporator style (size, rotor type, heating media, bearing design), material handling needs (pumps, piping, heat tracing, valves, heating media, feed vessel setup), and scope of test objectives determine test time, setup, and quantity of material required.

Execution of pilot plant testing requires a good plan and a skilled operator crew. One of the main goals of pilot testing is to find the boundaries of what the equipment can do and push it to failure to determine the maximum operating conditions related to product temperature, viscosity, concentration, or some other limiting variable.

Once a test is completed, the results logged during the test are compared to analytical tests performed on the collected samples. Optimally, statistical analysis will clarify a relationship between process conditions (feed rate, heating temperature, etc) and product quality specifications (concentration, color, etc). These relationships uncover the optimal test conditions and production-scale operational conditions.

The optimal test conditions/run are the basis for sizing of larger, production scale equipment. For evaporators, the main goal of testing, besides verifying that the equipment type can handle the qualitative restraints of the process, is to determine a U value and subsequently an outside film coefficient. The scale-up requires revised values for the outside film coefficient and conductivity portion of the U value calculation, which will change as the equipment grows in size and the wall thickness changes as per the relevant pressure vessel code.

Pilot plant testing is an essential part of chemical process R&D and production troubleshooting. LCI offers a complete pilot plant test to service the chemical processing industry such as food, pharma, specialty chemical, polymer/plastics, solvent recovery and waste minimization, and others. Alternatively, production facilities with this technology can install a pilot-scale, evaporator setup in their R&D pilot plant to evaluate new product types, batch recipes, or process conditions.

In conclusion, this oral presentation will present an overview of the pilot plant screening steps, test plan development, execution, and results analysis as related to thin-film evaporation. I have a few case studies to illustrate pilot plant importance and lessons learned along the way.

465268 Scheduling and Purchasing Optimization for Olefin Plants with Downstream Constraints

Abstract:

Olefin cracking takes an important role in chemical industry. Efforts are made to develop new technology to increase its production. Multiple cracking furnaces are employed in parallel to convert various hydrocarbon feed stocks to smaller hydrocarbon molecules such as ethylene, propylene, and butadiene. Each furnace is required to be cleaned up periodically because of coke formation in the reaction coils which decays furnace performance. Thus, production scheduling for a furnace system is a complex task, which provides information of feed type, feed flowrate starting and ending time for each batch of each furnace. Due to the large production volume, olefin plants can make hundreds of thousands of profits by optimal scheduling without any investment. The optimal production scheduling can further give direction for raw material purchasing. Currently, most scheduling models focus on maximizing profit without downstream consideration. However, when an ethylene plant is set up, the downstream component separation ability is limited within a certain range. Thus a scheduling and purchasing model should consider not only maximizing its profit but also downstream restrictions. In this paper, a MILP (mixed integer linear programming) model was built for scheduling and purchasing optimization of an olefin plant with consideration of downstream constraints and other constraints, such as non-simultaneous cleaning, recycling ethane relaying cracking. The efficacy of the model is demonstrated by case studies.

466585 a Study of Paraffin Product Distribution in Fischer-Tropsch Synthesis

Abstract:

Natural resources (coal, natural gas, biomass and wastes) can be converted into liquid hydrocarbon fuel via the Fischer–Tropsch Synthesis (FTS) reaction. Although the products of FTS cover an extremely wide range, and include olefins, paraffins and oxygenates with carbon numbers extending from C_1 to more than C_{100}, the major FT products are paraffins and olefins. The Anderson-Schulz-Flory (ASF) distribution model has been used to describe FTS product distribution, where the mole fraction of hydrocarbons is governed by a chain growth probability factor α (Note that in this model $0< \alpha<1$).

Some researchers have chosen to study the ASF distribution of olefins and paraffins separately. The published results of the ASF distribution of paraffin products show deviations from the ideal ASF distribution, such as a high yield of methane; a distinct change in the slope of the line between light hydrocarbons and long chain hydrocarbons; and in some cases, the values of $\alpha >1$ have been observed for light hydrocarbons (C_3-C_5). Unfortunately, none of the kinetic and mechanistic chain growth models can fully explain the paraffin product distribution that have been obtained experimentally. Nowadays, a few researchers try to link the FT products distribution with thermodynamics.

In the current work, a model based on thermodynamic quasi-equilibrium was introduced to describe the paraffin product distribution in FTS. The model was compared to experimental data obtained from two groups of experiments: one using wide ranges of $H_2/CO/CO_2$ mixtures over both cobalt- and iron-based catalysts and the other by co-feeding olefin and paraffin products into the FT reactor with or without feeding CO/H_2. In addition, experimental data from other researchers, covering a wide range of catalysts, operating conditions and reactor types, was also used to test the model. It is shown that the measured data follows the equilibrium curves quite closely. Thereafter, Aspen simulations, incorporating the thermodynamic quasi-equilibrium model, have been used to simulate the production of FTS by varying the reaction conditions (temperature, pressure, feeding different kinds of olefin and paraffin products into the reactor). The comparison between the Aspen simulation results and the data obtained experimentally revealed that most of the experimental results were quite close to those arrived at through Aspen thermodynamic simulations, which indicate that the paraffin product distribution can be described by quasi-thermodynamic equilibrium.

467770 Al-based Coating for Coke Reduction during Ethane Steam Cracking

Abstract:

Surface engineering to suppress coke formation in cracking furnaces is one of the key research topics in the ethylene industry. Alumina containing coatings have been claimed as being a major step to improve furnace run lengths. Therefore, in this work a commercial coating CoatAlloy™ has been compared to the uncoated Fe-Ni-Cr base alloy. First, the effect of the coating pretreatment was evaluated in an electrobalance set-up during ethane steam cracking under industrially relevant conditions ($T_{gasphase}$ = 1173 K, P_{tot} = 0.1 MPa, X_{C2H6} = 70 %, dilution Î′= 0.33 kg_{H2O}/kg_{HC}). Addition of nitrogen proved to be detrimental for the anti-coking performance of the coating. Additionally, the effect of presulfiding with 500 ppmw DMDS/H_2O, continuous addition of 41 ppmw S/HC of DMDS, and combinations thereof were evaluated. The tested samples were examined using online thermogravimetry, scanning electron microscopy and energy diffractive X-ray for surface and cross-section analysis together with X-ray photoelectron spectroscopy and wavelength-dispersive X-ray spectroscopy for surface analysis. The passivating coating performed substantially better than the reference Fe-Ni-Cr alloy while no effect on the product distribution was observed.

470801 Techno-Economic and Process Safety Analysis of Ethane to Ethylene Production Process through Oxidative Dehydrogenation (ODH) Route

Abstract:

Ethylene is one of the most important building blocks in the chemical industry and provides the starting point for many essential consumer products; e.g. paints, plastics, and pharmaceuticals. Conventionally, ethylene is produced from ethane through steam cracking at the refineries. However, harsh reaction condition (850°C reaction temperature) and potential hazardous by-product H_2 formation through steam cracking cause the need to look for alternative safer routes of ethylene production. Idaho National Laboratory (INL) is working on a novel route (ODH) of ethylene production, which produces water instead of hydrogen as a by-product during ethylene production. Our initial techno-economic analysis indicates that the minimum selling price of ethylene has to be around $1.85/kg to make the ODH process to be profitable. This presentation will describe the effects of the presence of steam (as a diluent) on the ODH process economics, and process energy efficiency. Additionally, a detailed inherent process safety analysis result of the ODH process will be presented to indicate the most crucial units of the process.

461302 Synthesis and Adsorption Performance of Zeolites for Separating Alkanes in Naphtha

Abstract:

The utilization efficiency of naphtha can be improved remarkably by separating group compositions in naphtha based on molecular scale management. In order to meeting the raw material requirements of ethylene industry and aromatics industry, zeolites was prepared and used for the adsorption separation of alkanes including n-paraffins and monomethyl iso-paraffins in naphtha, which is a more reasonable strategy than the separation of n-paraffins from naphtha.

MFI zeolites were synthesized for its adsorption performance of monomethyl iso-paraffins based on theoretical analysis and experimental investigation. The adsorption capacity of the synthesized MFI zeolites for 2-methylpentane was 0.0383 g/g at 25 . The optimal conditions for the synthesis of MFI zeolites were: n(SiO_2):n(Al_2O_3): n(Na_2O):n(H_2O) = 80:1:7:2450, aging for 36 h at 25 , crystallization for 48 h at 180 and using n-butylamine as a template. The characterization results showed the synthesized MFI zeolites had a high degree of crystallinity. The average pore size of synthesized MFI zeolites was about 0.58 nm.

The adsorption kinetics of monomethyl iso-paraffins on MFI zeolites was investigated. The diffusion coefficients were gained and the diffusion activation energies of 2-methylpentane, 3-methylpentane, 2-methylheptane and 3-methylheptane were 15.57 kJ/mol, 20.49 kJ/mol, 17.82 kJ/mol and 21.68 kJ/mol. The activation energy of long chain isoparaffin was higher than that of short chain isoparaffin and activation energy of 3-methyl-isoparaffins was higher than that of the corresponding 2-methyl-isoparaffins.

The adsorption separation performance of synthesized MFI zeolites was studied in a fixed bed adsorption column. The adsorption capacities of n-paraffins, monomethyl iso-paraffins and aromatics in MFI zeolites were 0.0529 g/g, 0.0158 g/g and 0.023 g/g. The multimethyl iso-paraffins and cycloalkane were not

adsorbed. N-paraffins and monomethyl iso-paraffins, which were 30.4% and 11.8% of naphtha were separated by MFI zeolites. The potential aromatic content of raffinate oil increased to 31.63 %, which was 10 percent higher than that of naphtha.

467104 a Novel, Efficient Approach to CFD Simulations of Systems with Break-up and Coalescence of Liquid Droplets

Abstract:

A key challenge in CFD simulations of two phase (gas/liquid) systems containing a dispersed liquid phase relates to the breakage and coalescence of droplets. Sophisticated models such as population balance have been used for that purpose, but the transient simulations and small time-steps associated with such approach leads to non-practical turnaround time for simulations of industrial-scale systems.

In this paper, an advanced droplet size model is developed to account for the local effects of turbulent energy dissipation, number density of particles as well as droplet breakup and coalescence. The CFD model is validated with experimental data from V.M. Alipchenkov, et al. (2004) on dispersed-annular flow in a pipe. The model is then applied to simulate industrial scale large pipe (diameter about 0.8 m). This advanced droplet model, compared with the sophisticated population balance model, is found to produce acceptable results with dramatically reduced simulation time. A comparison is also made between the advanced droplet model and the constant droplet size model â€“ to provide guidance when the simplified approach may be appropriate.

472533 Optimizing Overall Energy Consumptions and Dynamic Flexibilities of Air Separation Unit (ASU)

Abstract:

Air Separation Units (ASUs) generate oxygen, nitrogen and argon products from air. ASUs are widely used in industries. Unlike other chemical processes, the raw material of ASUs is free air, thus the cost of raw material is neglectable. Thus, to optimize the operating cost is basically to optimize overall energy consumption of the process. On the other hand, operating flexibilities are also very important for ASUs. The price and demands for oxygen, nitrogen and argon are changing frequently. Flexible productions can meet market requirements more swiftly.

In this paper, multi-objective optimization and dynamic simulation are used to achieve two objectives: 1) reducing production costs; 2) increasing production flexibilities. Multi-objective optimization uses Pareto efficiency to state the relations of the two objectives. Pareto efficiency is not a single value, but a curve showing the frontier of the two objectives.

Dynamic simulation is used to illustrate operating strategies to achieve flexibilities. It tests and validates flexibility scenarios to confirm that they are not only therotically but also practically achievable. Moreover, dynamic simulation gives transition methods between flexibility scenarios.

An industrial scale Air Separation Unit (ASU) is used for case study. Dynamic simulation and multi-objective optimization results show a set of Pareto efficiency pairs for operating costs and flexibilities ranging from 0 to 30%.

November 13-18, 2016 AIChE Annual Meeting Presentation
Paper 88c, Abstract #450363

SIMULATION REQUIREMENTS FOR PROCESS DESIGN PACKAGE OF A COMMERCIAL UNIT

Minghua Ye, Brian Peng, and Sergio Mohedas

November 14, 2016

CHEMISTRY THAT MATTERS™

TYPICAL SCHEDULE FOR A PDP DEVELOPMENT

Pre-PDP: proposal and contract negotiation and finalization

- Definition of technology, design capacity, and performance guarantee,
- Define feedstock, products, vent, and utilities meet the battery limit conditions

Estimation of PDP execution time:

- 20 to 36 weeks, depending on the project size
- Simulations for all MB cases to be finalized within first 4-6 weeks of PDP schedule

Start PDP Development (milestone: Kick-Off Meeting)

- Define a design case (may be different cases for different parts of the plant)
- Specify typical scenarios: start-of-run, end-of-run, and other rating cases
- Construct Simulation Flow Diagram according to PFD

The simulation is a critical milestone for the whole project

(1) Provide heat and material balance (HMB)

(2) Summary of energy and utility consumptions

(3) Summary of waste emissions (vapor, liquid, waste water, and solid);

(4) Generate stream database with detailed physical properties (for equipment and instrument process specifications, pipe sizing, etc.)

(5) Generate process datasheet (operation capacity and conditions) for equipment

No. 1

1 METEOROLOGICAL DATA AT THE PLANT LOCATION

Site location information:

- Extreme weather (the **hottest** and **coldest** temperature)
- Air dry and bulb temperature
- Relative humidity
- Wind Speed and direction
- Atmospheric pressure

$$P_{abs} = P_{gauge} + P_{atm}$$

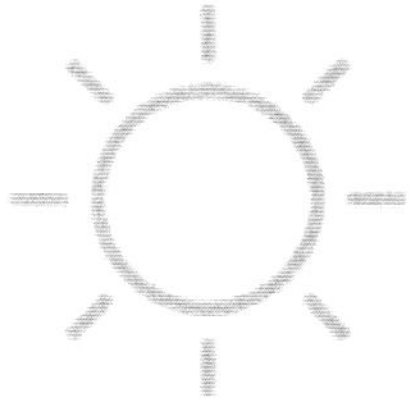

No. 2

1.1 IMPACT OF EXTREME HIGH AND LOW TEMPERATURE

- MDMT, **Minimum Design Metal Temperature**, is the lowest temperature that the vessel can withstand with the chosen material and thickness, as defined in the applicable construction code.
- MDMT is a critical factor in designing a pressure vessel that depends on the environment where the pressure vessel has to be installed, and the minimum temperature likely to be reached in operation.
- Extreme lowest temperature in the area is used to determine or cross check equipment **MDMT**

	Houston, TX		**Denver, CO**	
	T, °C	**Occurred**	**T, °C**	**Occurred**
Coldest Temperature	**-15**	**1/18/1930**	**-33.9**	**1/9/1875**
Hottest Temperature	**42.8**	**9/4/2000 and 8/27/2011**	**40.6**	**8/8/1878, 7/20/2005, 6/25 and 26/2012**

References: (1) ASME Code, Section VIII: Division I, Paragraph UG-20(b).
(2) *API 510, Pressure Vessel Inspection Code: In-Service Inspection, Rating, Repair, and Alteration Downstream Segment, Ninth Edition, June 2006.*
(3) *API RP 17G, Recommended Practice for Completion/Workover Risers, Second Edition, July 2006 (Reaffirmed April 2011).*

No. 3

1.2 IMPACT OF DRY / WET BULB TEMPERATURE AND WIND DATA TO DESIGN OF COOING WATER TOWER

- Use the highest or the design wet bulb temperature during summer (exclude 0.4%, 1%, or 2% certain extreme data in summer for economic consideration).
- Wind speed and direction to be considered for selection of cooling tower type and design.
- Every location has a unique design (worst case) wet bulb temperature that is published by organizations such as ASHRAE and can be obtained easily.
- Economical cooling water temperature approach toward the wet bulb temperature 4 °C.
  - An approach < **2.8** °C are not economical, nor will be certified by CTI (Cooling Technology Institute).
  - An approaches > **4** °C result in higher condensing temperatures (reduced chiller efficiency and performance) without much savings on the cooling tower.

Reference: Stanley M. Walas, "Chemical Process Equipment Selection and Design", copy right © 1990, by Butterworth-Heinemann

No. 4

1.3 ILLUSTRATION OF WATER COOLING TOWER

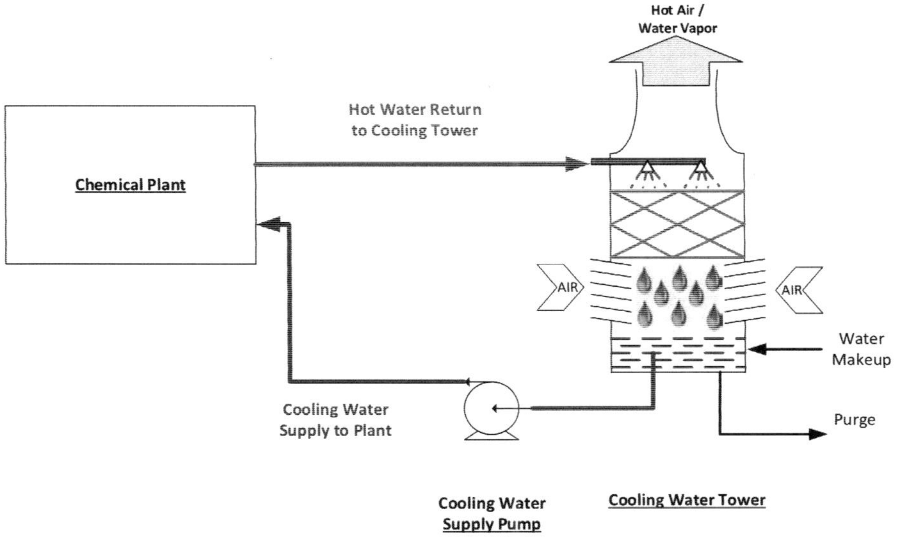

No. 5

سابك
sabic

2 PLANT DEFINITION

Plant Location

- The required geological data such as elevation, earthquake, rainfall, snowfall, solar radiation, extreme weather of hurricane, dust storm, sandstorm are required for design of structure and construction

Plant Design Capacity

- To clarify the basis of the capacity:
- One specific product or multi-products
- May specify feed
- On-stream time (default 8,000 hours/year)

Design Life (to be consistent with technical agreement)

Turndown Requirements (% of Normal Capacity)

Provisions for future expansion (if required)
Integration with existing facilities (if appropriate)

No. 6

سابك
sabic

UNITS TO BE USED IN SIMULATION

United States: English Unit

- Mass: lb
- Length: ft, inch
- Volume: bbl, ft^3
- Flow rate: gpm
- Energy: btu
- Power: mmBtu/hr, hp
- Temperature: °F, R

Most countries: SI or MKL Unit

- Mass: kg
- Length: ft, inch
- Volume: m^3
- Energy: J, kJ, cal
- Power: W, MW
- Temperature: °C, K

➤ Pipes or nozzles: inches or millimeters

No. 7

4 STANDARD CONDITIONS

List of Standard Reference Conditions		
Temperature	Absolute Pressure	Publishing or Establishing Entity
0 °C	101.325 kPa	NIST, ISO10780, formerly IUPAC
15 °C	101.325 kPa	ICAO's ISA, ISO13443, EEA, EGIA
20 °C	101.325 kPa	EPA, NIST
25 °C	101.325 kPa	EPA
60 °F	14.696 psi	SPE, U.S. OSHA, SCAQMD

Density of Some Crude Oil at Different Temperature

Temperature	Density (±0.005 kg /m³)		
°C	Light oil	Heavy oil	Medium oil (50:50 mixture)
20	832.991	952.687	890.499
25	829.500	949.462	887.262
Deviation, kg/m³	3.491	3.225	3.237

Reference: George et., "Equation of State of Crude Oil Samples", J. Pet Environ Biotechnol 2013, 4 (6)

No. 8

5 EQUIPMENT AND STREAM NAME DESIGNATION

- Equipment designation shall be accordance with the code of the client (plant) and consistent with PFD and P&IDs.
- Stream number should be consistent with the number shown on PFD.
- Numbering the equipment and stream according to the unit assignment.
  - P-101 for pump #01 in area #1
  - P-210 for pump #10 in area #2
  - Steam# 100 in area #1
  - Stream #305 in area #3

General Process Equipment Designation Code	
Equipment type	**Code**
Agitators and Mixers	**A, M, MD**
Air Cooled Condensing Unit, Air Coolers	**AC**
Boilers	**BO**
Centrifuges	**CE**
Columns / Towers	**C, T, V**
Compressors, Turbine	**C, K**
Cooling Towers	**CT**
De-superheaters	**DS**
Drums, Coalescers	**D, V**
Dryers	**DR**
Ejectors	**EJ**
Exchangers	**E**
Filters Strainers	**F**
Packaged Unit or Equipment	**PK**
Pumps	**P**
Tanks, Storage Tanks	**TK**

No. 9

سابك
sabic

6 AVAILABLE UTILITY

The utility need to be well defined

LPS	Low-Pressure Steam (3-5 barg)
MPS	Medium-Pressure Steam (10-15 barg)
HPS	High-Pressure steam (40-50 barg)
HTM	Heat Transfer Media (Organic)
CW	Cooling Water
WR	River Water
RW	Refrigerated Water
RB	Refrigerated Brine
CS	Chemical Waste Water with high COD
SS	Sanitary Waste Water **with** high BOD
EL	Electric heat
NG	Natural Gas
FG	Fuel Gas
FO	Fuel Oil
FW	Fire Water
BW	Boiled Water

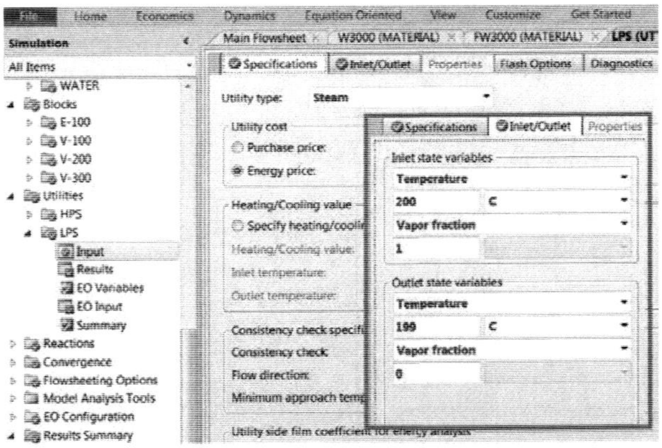

سابك
sabic

7 FEEDSTOCK SPECIFICATION / COMPOSITION

Feedstock specification and composition need to be well defined:
(1) Design case; (2) Start-of Run; (3) End-of Run; (4) Rating cases (with various possible feedstocks or product grades; (5) Emergency startup; (6) Shutdown case

Some important information need to be requested:

- Maximum feed impurities specifications
- Distillation properties such as ASTM D86 (for refining)
  - Key component distribution curve
  - Key component specifications
  - Analytical data of all feedstock compositions
  - Specific gravity and viscosity (for special chemicals)

Other process / technology related properties if applicable, such as:

- Bromine number (to define olefin content)
- PONA analysis (profiles of saturated hydrocarbon, olefin, di-olefin, acetylene, cyclic compounds, aromatics, etc.)
- Octane number (for gasoline, diesel, and other fuel)
- RVP (Reid Vapor Pressure)
- API Density

8 PRODUCT AND INTERMEDIATES SPECIFICATIONS

- Specify required minimums and maximums of parameters or properties, such as flow rate, composition (mole%, wt%, vol%, wppm, etc.)
- Describe when the plant need the specified quality levels, e.g. when and how long high purity product is required and when the specifications can be relaxed to lower level
- Confirm if all products need to meet the (same) design spec or just the combined product has to meet the design specifications.

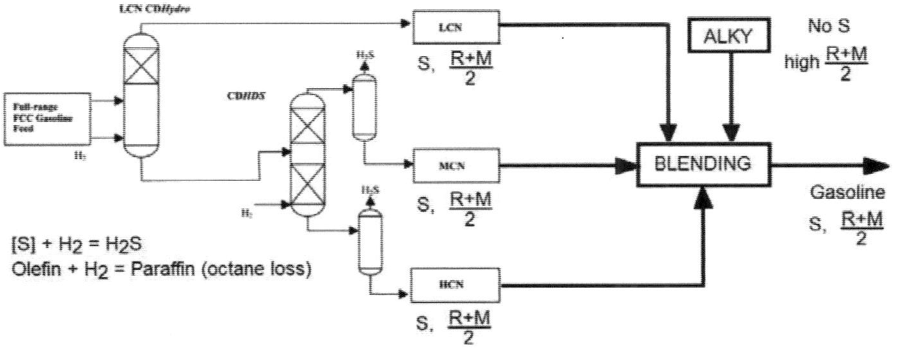

Ref: G. Podrebarac, et al "Doing Wonders for Diesel Production", Hydrocarbon Engineering, Nov. 2009. www.hydrocarbonengineering.com

No. 12

سابك
sabic

9 BATTERY LIMIT CONDITIONS

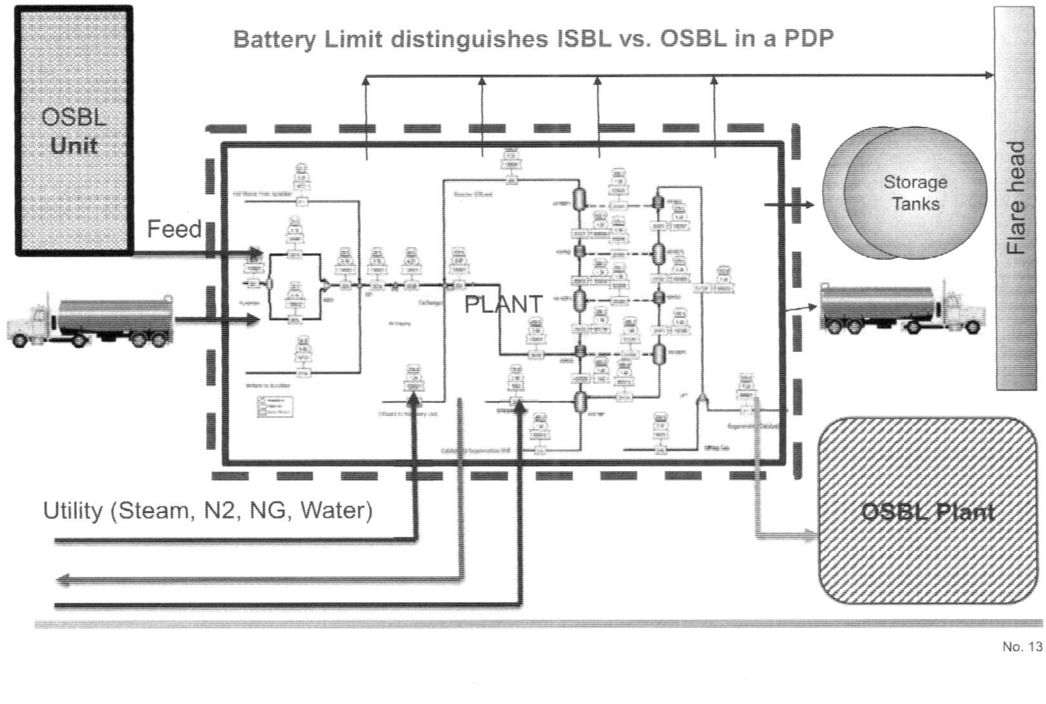

No. 13

10 PROCESS DESIGN CONSIDERATIONS

Type of Cooling
- Air Cooler
- Cooling Water (cooling water, chilled water, sea water, non-recirculating cooling water)
- Coolant

Type of Heating
- Low Pressure (LPS) (P and T, super heated T > Tsat)
- Medium Pressure Steam (MPS)
- High Pressure Steam (HPS)
- Hot Oil
- Molten Salt
- Fired Heater

Integration with Existing Equipment (if applicable – honor the process variable constraints)

Reuse of Existing Equipment (be limited by the size, sometime have to trade between the capital saving and energy consumption)

No. 14

11 THERMODYNAMIC PACKAGE VALIDATION

- Selecting the proper method for estimating properties is one of the most important steps that will affect the rest of the simulation.
  - ✓ VLE, VLLE for separation
  - ✓ Heat and Material Balance
  - ✓ Proper selection of heat and mass transfer equipment
  - ✓ Proper design and sizing of equipment typically distillation, extraction, and CD
- In Aspen Plus, the estimation methods are stored in a "Property Method".

  A property method is a collection of estimation methods to calculate several thermodynamic (fugacity, activity, enthalpy, entropy, Gibbs free energy, and volume) and transport (viscosity, thermal conductivity, diffusion coefficient, and surface tension).
- Aspen Plus stores a large database of interaction parameters that are used with mixing rules to estimate mixtures properties.
- User need to verify the applicability and accuracy of the selected property method against the collected data from plant, public and private literature.
- User may evaluate the parameters for the selected model based on experimental data (Dortmund, NIST, etc.)

No. 15

12 OPERATING CONDITION AND CONSTRAINTS

Recycle Stream:
- Recycle ratio
- Specification of the recycle stream (rate, composition, temperature, pressure)
- Capacity limitation of the equipment (specially compressor for gas recycle)

Equipment
- Size limits, maximum capability and hydraulics (e.g., distillation column, the vapour jet flooding, maximum liquid hold in downcomer, maximum flux in a packed bed, turn-down capability of distributors)
- Available source of energy, heat integration of the whole process.

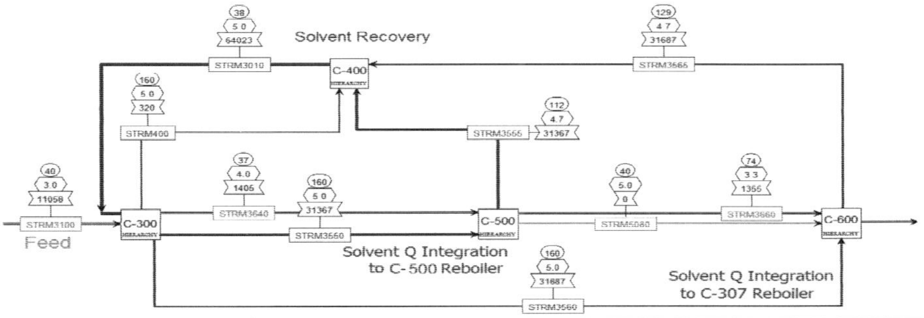

No. 16

13 CONSISTENCY OF PFD, P&IDS, AND SIMULATION FLOW DIAGRAM

Simulation diagram matches process flow diagrams (PFD), piping and instrument diagrams (P&ID)
- In the PFD, the utility such as cooling water and steam may not be shown. Similarly, some equipment may also take a short cut.
- In the P&IDs, all the pipelines will be shown. User need to compare simulation with PFD and P&IDs to verify all the streams are calculated.
- Stream flow rate and properties are used to sizing pipe, connection nozzle, and instrumentations.

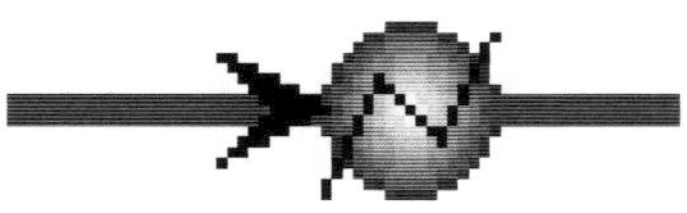

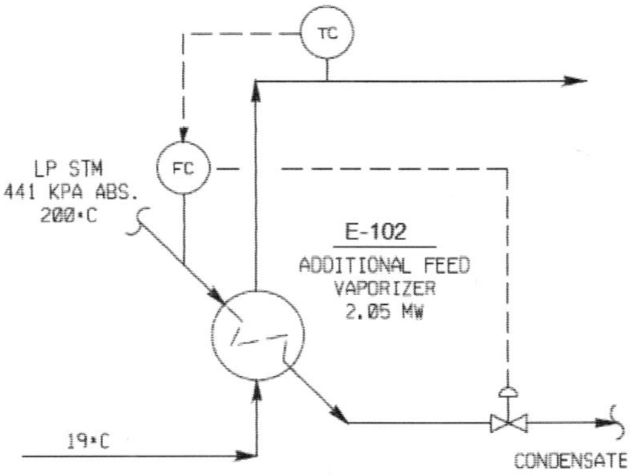

No. 17

13.1 ILLUSTRATION OF A STEAM VAPORIZER P&ID

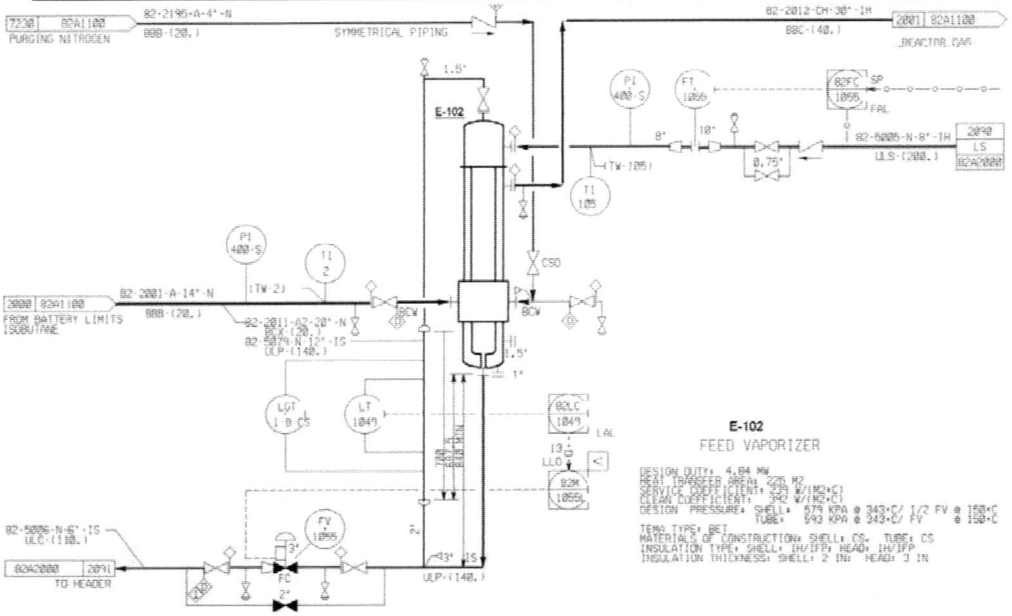

No. 18

SUMMARY

Effective and accurate simulation is the soul of for PDP development. In order to ensure the simulation results can be used reliably to design a plant, following factors must be carefully validated and applied:

1. Meteorological data at the plant location
2. Plant definition and scope
3. Unit to be used in simulation
4. Standard conditions
5. Equipment and stream name designation
6. Available utility
7. Feedstock specification / composition
8. Product and intermediates specifications
9. Battery limit conditions
10. Process design considerations
11. Thermodynamic package validation
12. Operating condition and constraints
13. Consistency of PFD, P&IDs, and Simulation flow diagram

No. 19

سابك
sabic

DISCLAIMER

DISCLAIMER: THE MATERIALS, PRODUCTS AND SERVICES OF SAUDI BASIC INDUSTRIES CORPORATION (SABIC) OR ITS SUBSIDIARIES OR AFFILIATES ("SELLER") ARE SOLD SUBJECT TO SELLER'S STANDARD CONDITIONS OF SALE, WHICH ARE AVAILABLE UPON REQUEST. INFORMATION AND RECOMMENDATIONS CONTAINED IN THIS DOCUMENT ARE GIVEN IN GOOD FAITH. HOWEVER, SELLER MAKES NO EXPRESS OR IMPLIED REPRESENTATION, WARRANTY OR GUARANTEE (i) THAT ANY RESULTS DESCRIBED IN THIS DOCUMENT WILL BE OBTAINED UNDER END-USE CONDITIONS, OR (ii) AS TO THE EFFECTIVENESS OR SAFETY OF ANY DESIGN OR APPLICATION INCORPORATING SELLER'S MATERIALS, PRODUCTS, SERVICES OR RECOMMENDATIONS. UNLESS OTHERWISE PROVIDED IN SELLER'S STANDARD CONDITIONS OF SALE, SELLER SHALL NOT BE RESPONSIBLE FOR ANY LOSS RESULTING FROM ANY USE OF ITS MATERIALS, PRODUCTS, SERVICES OR RECOMMENDATIONS DESCRIBED IN THIS DOCUMENT. Each user is responsible for making its own determination as to the suitability of Seller's materials, products, services or recommendations for the user's particular use through appropriate end-use and other testing and analysis. Nothing in any document or oral statement shall be deemed to alter or waive any provision of Seller's Standard Conditions of Sale or this Disclaimer, unless it is specifically agreed to in a writing signed by Seller. Statements by Seller concerning a possible use of any material, product, service or design do not, are not intended to, and should not be construed to grant any license under any patent or other intellectual property right of Seller or as a recommendation for the use of any material, product, service or design in a manner that infringes any patent or other intellectual property right.

No. 20

Parametric Sensitivity Analysis for the Optimal Design of a Bitumen Upgrading Plant using a Steady-State Model

Paper ID #: 451069

*Jennifer Charry-Sánchez, Alberto Betancourt-Torcat, Ali Almansoori**

Department of Chemical Engineering, The Petroleum Institute, Sas Al Nakhl Campus, P.O. Box 2533, Abu Dhabi, United Arab Emirates.

*Corresponding Author's e-mail: aalmansoori@pi.ac.ae

Keywords: Optimization, Design, Economics, Bitumen Upgrading, Parametric Analysis.

Abstract

The exploitation of non-conventional oil resources have gained global attention in the last decades given their abundance; particularly the Canadian Oil Sands is perceived as a key global energy supplier for the future. The Canadian non-conventional oil resources originate from Oil Sands or Tart Sands deposits; which are technically defined as extra-heavy crude bitumen. This work proposes an optimization model for the design of the most suitable bitumen upgrading pathway in terms of utility operating costs. The optimization approach chooses amongst different available upgrading steps to form the most cost-effective pathway. A sensitivity analysis was performed on the natural gas price for identifying its impact level over the crude bitumen upgrading process. Natural gas is the main fuel feedstock used for the generation of most Oil Sands industry utilities. The optimization results show that thermocracking-based upgraders are most cost-effective than hydrocracking-based pathways. This because hydrocracking-based upgraders are more energy-intensive mainly due to their high hydrogen requirements.

1. Introduction

The search/exploitation of new fuel resources is one of the main drivers of the energy sector worldwide. Non-conventional oil resources such as crude bitumen has been known for decades. However, the expansion of this sector has had to wait for the development of new technologies that allow tapping resources once thought economically infeasible. The Canadian non-conventional oil sector mainly comprises the Oil Sands located in the Athabasca region (Western Canada). The Oil Sands contain the third largest proven oil reserves in the world behind Saudi Arabia and Venezuela [1]. These deposits are constituted by crude bitumen; which is an extra-heavy form of oil very viscous that needs to be mixed with diluent to be transported by pipelines. These resources can be commercialized as raw bitumen or upgraded into synthetic crude oil (SCO) in refineries. SCO is an added-value product obtained from raw bitumen that commercially competes with light petroleum products in the international market. The SCO is processed in refineries to obtain gasolines, jet fuels, and diluents.

The crude bitumen must undergo an energy-intensive upgrading process to be converted into commercial SCO blends. For the upgrading process prime commodities such as: process steam, process heat, electricity, and hydrogen are required. These utility requirements are mainly met using natural gas as primary fuel feedstock given its abundance and relatively low carbon emission rate compared with other fossil fuels [2–5]. This work presents an optimization model for designing the most suitable upgrading pathway in terms of minimum operating costs. Although capital expenditures are an important part of an upgrader plant's annual cost, in general the operating costs are the most significant annual expenses in the lifetime of a bitumen upgrader [6]. Additionally, the accounting treatment of certain capital expenditures may vary significantly between bitumen upgrader operators. For instance, some operators may elect to capitalize or defer and amortize certain expenditures that are recorded as an expense by other producers. Given the nature of capital expenditure accounting treatment, the present study will focus only on utility expenses; which are more tangible. Previous studies on bitumen upgrading have considered the different types of processing pathways as fixed inputs [2,3]; while the proposed model treats each upgrading step and corresponding processing modes as decision variables in the problem's formulation.

The present optimization approach seeks for the optimal bitumen upgrading pathway by selecting the most suitable upgrading steps and corresponding processing modes. This to obtain a specific SCO blend according to composition specifications at minimum operating costs. Moreover, a sensitivity analysis was performed to identify the impact level of natural gas prices over the bitumen upgrading pathway using a steady-state model. The fluctuations in the natural gas price may cause changes in: (1) the design of the optimal upgrading pathway, (2) energy demands, (3) SCO final composition, and (4) greenhouse gas (GHG) emissions level. The proposed steady-state optimization model considers five basic upgrading steps: 1) primary distillation (atmospheric distillation/diluent recovery units), 2) vacuum distillation, 3) cracking (thermocracking/hydrocracking units), 4) hydrotreatment, and 5) blending. Some upgrading steps include various types of upgrading units (e.g., primary distillation and cracking); whereas all steps/units consider distinctive set of processing modes. These processing modes are defined in terms of specific product yields and energy requirements based on reported industrial data [2,3]. The problem was modelled as a mixed-integer nonlinear programming (MINLP) using the GAMS modeling system. The results show that the proposed model is a practical tool to: 1) select the most suitable bitumen upgrading pathway, 2) estimate the upgrading utility costs, 3) identify the most suitable utility supply source based on techno-economics and environmental parameters, and 4) estimate the CO_2 emissions associated to the utility supply.

2. Model Formulation

The proposed mathematical model aims to represent the process of bitumen upgrading into SCO. A sink and source approach was used for determining the most suitable bitumen upgrading pathway consisting of the following steps: 1) primary distillation (atmospheric distillation or diluent recovery unit), 2) vacuum distillation (optional step), 3) cracking (thermocracking or hydrocracking), 4) hydrotreatment, and 5) blending. For each upgrading step there is a set of processing modes defining: *i*) energy requirements, *ii*) products, intermediate-products, and by products yields, and *iii*) destination of the products, intermediate-products, and by products.

2.1 Formulation Indices

The mathematical formulation can be simplified denoting key model's properties as indices. For instance, the upgrading steps can be represented by the index s (e.g., primary distillation, vacuum distillation, cracking, hydrotreatment, and blending), the energy utilities by u (e.g., process steam, process heat, electricity, and hydrogen), the processing modes by p (p_1, p_2, …, p_n), and h the types of hydrocarbons (atmospheric tower bottoms (ATB), vacuum tower bottoms (VTB), naphtha, light gas oil (LGO), heavy gas oil (HGO), and vacuum gas oil (VGO)) processed and/or produced during the upgrading pathway.

2.2 Upgrading Pathway Utility Demands

The utility demands associated to each step of the upgrading pathway follow this general form:

$$UD_{u,s,p} = \sum_{c} HC_{h,s,p}\ \mathrm{UR}_{u,h,s,p} \quad (1)$$

where $UD_{u,s,p}$ (e.g., tonne steam/h, MMBTU/h, MW, tonne H_2/h for process steam, process heat, electricity, and hydrogen consumption, respectively) denotes the type of utility u consumed in the sth upgrading step when following processing mode p, $HC_{h,s,p}$ (e.g., tonne/h) represents the mass flowrate of the hydrocarbon compound h processed/produced in the sth upgrading step when following processing mode p, and $\mathrm{UR}_{u,h,s,p}$ (e.g., tonne steam/tonne, MMBTU/tonne, MWh/tonne, tonne H_2/tonne) indicates the type of utility u required by hydrocarbon compound h processed/produced in the sth upgrading step when following processing mode p.

Likewise, the total amount of utility u demanded by the upgrading pathway is given as follows:

$$TUD_{u} = \sum_{s,p} UD_{u,s,p} \quad (2)$$

where TUD_u (e.g., tonne steam/h, MMBTU/h, MW, tonne H_2/h) is the total amount of utility u demanded by the upgrading pathway determined by means of the optimization process.

2.3 Utility Cost Function

The model's utility cost function is given in terms of the energy forms demanded in the upgrading process as follows:

$$UCOST = \sum_{u} TUD_{u}\ \mathrm{CF}_{u}\ \mathrm{OT} \quad (3)$$

where $UCOST$ (\$/yr) denotes the model's utility cost function, CF_u is the cost factor associated to utility type u given in monetary terms per unit of utility consumed (e.g., \$/tonne steam, \$/MMBTU, \$/MWh, \$/tonne H_2), and OT is a parameter indicating the upgrader's annual operating time in hours (h/yr). The cost of each type of utility depends on the generation technology and supply infrastructure, and whether it employs or not carbon capture and storage (CCS) methods.

2.4 Carbon Dioxide (CO_2) Emissions and Tax

The carbon dioxide (CO_2) emissions generated in the production of the utilities u supplied to meet the energy demands of the bitumen upgrading operations depend on the technology employed and whether it includes or not CCS methods. Accordingly, the CO_2 emissions can be estimated as follows:

$$CO2E_{u} = TUD_{u}\ \mathrm{CEF}_{u} \quad (4)$$

where $CO2E_u$ (tonne/h) denotes the CO_2 emissions associated to the uth utility, and CEF_u (e.g., tonne/tonne steam, tonne/MMBTU, tonne/MWh, tonne/tonne H_2) represents the average CO_2 emission factor for the uth utility supply to the bitumen upgrading plant. Moreover, the total CO_2 emission associated to the bitumen upgrading utility supply is given as follows:

$$TCO2E = \sum_{u} CO2E_{u} \quad (5)$$

where $TCO2E$ (tonne/h) denotes the total CO_2 emissions associated to the utilities supply of the bitumen upgrading process.

In Alberta-Canada a CO_2 tax is applied to large industrial emitters exceeding their annual emission allowance. Accordingly, those producers must paid a levied per tonne of CO_2 over their specified allowance. This can be expressed as follows:

$$TAXCOST = \left[\left(TCO2E\ \mathrm{OT}\right) - \mathrm{CO2ALLOWANCE}\right]\ \mathrm{CTAX} \quad (6)$$

where $TAXCOST$ (\$/yr) is the total levied cost on the bitumen upgrading operations due to CO_2 emissions exceeding the annual allowance, CO2ALLOWANCE (tonnes/yr) is a parameter denoting the annual carbon dioxide emissions allowance assigned to the upgrading plant, and CTAX ($/tonne) is the CO_2 tax rate applied per unit of CO_2 exceeding the allowance.

2.5 Model Constraints

The proposed design optimization model determines the most suitable bitumen upgrading pathway according to economics objectives subject to techno-economic and environmental constraints. The upgrading pathway can be composed of either four or five basic steps *s*. The mathematical formulation constrains that all steps (except for vacuum distillation which is optional) must be chosen to form an upgrading pathway design. This is constrained in the model as follows:

$$U_s = 1, \; s \in \{d, c, ht, b\} \tag{7a}$$

$$U_s \leq 1, \; s \in \{v\} \tag{7b}$$

$$U_s = \begin{cases} 1 & \text{if step } s \text{ is selectedto be part of the upgrading pathway} \\ 0 & \text{otherwise} \end{cases}$$

where U_s is a binary variable indicating the selection of the sth upgrading step. The upgrading step *s* is defined as a set that contains the following elements: primary distillation (*d*), vacuum distillation (*v*), cracking (*c*), hydrotreatment (*ht*), and blending (*b*).

Additionally, only one processing mode *p* per upgrading step *s* can be chosen, this is included in the model as follows:

$$\sum_{p,s'} Y_{p,s,s'} \leq 1, \; \forall s; s \neq s' \tag{8}$$

$$Y_{p,s,s'} = \begin{cases} 1 & \text{if processingmode } p \text{ is chosenfor the } s^{th} \text{ upgrading step} \\ 0 & \text{otherwise} \end{cases}$$

where $Y_{p,s,s'}$ is a binary variable indicating the flow of material from the sth upgrading step under processing mode *p* to step *s'*.

The fractions of the hydrocarbon components *h* (making up the blended SCO) determines the quality of the final upgraded product. Accordingly, the SCO composition can be specified as follows:

$$X_h^{MAX} \geq \frac{HC_{h,s,p}}{BSCO} \geq X_h^{MIN}, \; s \in b; \;\; h \in \{naphtha, LGO, HGO, VGO\} \tag{9}$$

where X_h^{MAX} and X_h^{MIN} denote upper and lower bounds for the fractions of hydrocarbon components *h* making up the blended SCO product, and *BSCO* (tonne/h) is the total amount of blended SCO produced by the upgrading pathway design.

The total amount of blended SCO produced by the upgrading plant can be estimated as follows:

$$BSCO = \sum_{h,s,p} HC_{h,s,p}, \; s \in b; \;\; h \in \{naphtha, LGO, HGO, VGO\} \tag{10}$$

2.6 Problem Summary

The problem's conceptual formulation can be given as follows:

$$\min_{\eta} \quad COST = UCOST + TAXCOST \tag{11}$$

subject to:

Utility supply source (3)
Utility unit prices (3)
CO_2 levies (6)
Upgrading steps (7)
Steps processing modes (8)
Blended SCO specifications (9)

The optimization model set of decision variables η includes: 1) the type of upgrading steps, 2) steps' processing modes, and 3) utility supply sources. The proposed optimization model was formulated as a Mixed Integer Nonlinear Programming (MINLP) in the General Algebraic Modeling System (GAMS). The optimization approach aims to determine the most suitable bitumen upgrading pathway design subject to techno-economic and environmental constraints. The most suitable upgrading plant design was determined taking into account only the operating utility costs. Although the capital costs are generally considered in this type of design analysis, for the present study the capital costs were disregarded since they generally fall within the same order of magnitude. Moreover, when they are amortized over long period of times such is the case of bitumen upgrading plants (e.g., +35 years), their contribution to the total annualized cost is generally a fraction of the operating costs. This can be supported by the fact that most of the existing literature on crude bitumen extraction and upgrading through process modeling and simulation uses the operating utility cost as the core of their analyses [2–5,7–10]. Accordingly, a sensitivity analysis was performed on the natural gas price, which is the key fuel feedstock for utility production in crude bitumen upgrading (e.g., Canadian Oil Sands industry). This to determine the most suitable bitumen upgrading pathway based on the operating utility costs of the process.

3. Case Study

The proposed design optimization model was used to estimate the most suitable upgrading pathway for producing SCO. The case study considers an upgrader plant with a net crude bitumen processing capacity of 1000 tonne/h. The case study consists of a sensitivity analysis performed on the price of natural gas. Natural gas is the dominating fuel feedstock used to produce the main utilities consumed by the Oil Sands industry [2,4,11]. Furthermore, the bitumen upgrading utility demands can be met by means of importing energy from conventional sources or sources featuring reduced carbon emissions (i.e., including CCS methods) [2]. However, only electricity and hydrogen can be imported from low emission suppliers. The utility costs considered in this study correspond to average unit supply costs. Accordingly, the average unit costs were estimated considering that natural gas is the dominating fuel feedstock for utility generation in Western Canada. For instance, most of the process steam is generated by natural gas-fired boilers, process fuel is generally natural gas, hydrogen is produced via steam methane reforming (SMR), and the electricity mix is set to increase its gas-based share while coal is being phased out [12].

The SCO product was specified according to the following composition: naphtha $\geq$14%, LGO $\geq$ 33%, HGO $\geq$ 32%, and VGO $\geq$ 0%. The optimization process minimizes the operating utility costs associated to bitumen upgrading. The input data and models considered in the proposed mathematical framework such as upgrading steps and steps' processing modes correspond to information available in the literature on typical units/plants used to upgrade bitumen in Western Canada [2,3]. For the present study, the following number of processing modes were considered by upgrading step: 1) primary distillation step: three processing modes for the atmospheric distillation units (ADU_p), and three for the diluent recovery units (DRU_p); 2) three processing modes for the vacuum distillation step (VDU_p); 3) cracking step: three processing modes for the thermocracking units (delayed cokers) (TCU_p), and two for the hydrocracking units (LC-Finers) (HCU_p); 4) six processing modes for the hydrotreatment step (HTU_p), and 5) one standard blending step (BU). The sub-index *p* indicates the chosen processing mode by upgrading step. The processing modes data for the upgrading steps was taken from previous studies available in the literature [2,3].

For the present sensitivity analysis 13 different scenarios were studied considering a minimum and maximum natural gas price of $2/ MMBTU and $14/ MMBTU, respectively. A step of $1/ MMBTU was used to generate the different studied scenarios. Table 1 shows the various scenarios considered in the present analysis and their corresponding natural gas price. Also, the table shows the average unit cost per utility as a function of the gas price.

Table 1. Input utility costs as a function of the natural gas price.

Scenario	Natural Gas Prices ($/MM BTU)	Average Unit Costs per Type of Utility					
		Process Steam ($/tonne)	Process Heat ($/MM BTU)	Electricity ($/MWh)	Electricity with CCS ($/MWh)	Hydrogen ($/tonne)	Hydrogen with CCS ($/tonne)
S1	2	7.4	2	48	64	2750	3250
S2	3	11	3	55	74	2910	3430
S3	4	14.7	4	61	82	3080	3630
S4	5	18.4	5	68	90	3240	3820
S5	6	22.1	6	74	97	3410	4020
S6	7	25.7	7	80	104	3570	4210
S7	8	29.4	8	87	113	3740	4410
S8	9	33.1	9	94	121	3900	4600
S9	10	36.8	10	100	128	4070	4800
S10	11	40.5	11	107	136	4230	5000
S11	12	44.1	12	113	144	4400	5200
S12	13	47.8	13	120	152	4560	5380
S13	14	51.5	14	126	159	4730	5580

The optimization results show that only thermocracking-based upgrading pathways are selected regardless the natural gas price level. This is explained by the fact that thermocracking units are significantly less energy intensive and therefore generate lower CO_2 emissions compared with hydrocracking units. This findings are in agreement with previous reported studies [4,5]. Delayed cokers mainly consume process fuel that is directly converted into thermal energy in the cokers' furnaces to break the heavy bitumen feed into lighter hydrocarbon components such as those making up the SCO product. Delayed cokers also require electricity. On the other hand, hydrocracking units such as LC-Finers mainly depend on hydrogen to break the heavy bitumen into lighter hydrocarbons by hydrogen addition. Additionally, they also require process heat and electricity during their operations. LC-Finers intensively consume hydrogen, while the hydrogen is one of the most expensive utilities consumed in the bitumen upgrading process. This because hydrogen production is still in a transient stage, and has not reached maturity. Thus, it is still relatively expensive to produce. One of the advantages of hydrocracking-based upgraders are their higher products yield. However, these conversion gains are offset by their significantly high energy requirements.

For the first three scenarios (S1-S3) the most suitable upgrading pathway was found to produce approximately 127,500 bbl SCO/d. The pathway consists of the following upgrading steps and processing modes: diluent recovery unit (DRU_2), delayed coker unit (TCU_3), hydrotreatment unit (HTU_4), and blending unit (BU). This pathway was chosen for the scenarios located at the low-end of the gas prices (see Table 1, S1-S3). The chosen processing mode for the diluent recovery unit yields the highest quantity of LGO (which is sent to the hydrotreatment step) from the available options at relatively low energy expenses.

Regarding the delayed coker, its chosen processing mode features the lowest electricity requirements, but highest process heat requirements. Additionally, this type of delayed coker includes the lowest products yield amongst the available options. This outcome allows maintaining relatively low the amount of hydrogen consumed in the hydrotreatment step. This because the hydrogen consumption directly depends on the amount of hydrocarbon products processed by the hydrotreater units; which mainly proceeds from the cracking step. Moreover, as shown in Table 2 the hydrogen supply is by far the dominating expense among the annual utility costs. Also, at this point the process heat does not represent a major burden on the total operating cost given the low natural gas price range.

Concerning the hydrotreatment step, the chosen processing mode set includes the overall lowest hydrogen requirements when accounting for all the different hydrocarbon products (naphtha, LGO, HGO) treated at this step. Moreover, the blending step corresponds to a standard process.

Table 2. Parametric analysis cost results considering fluctuations in the natural gas price.

Scenario	Natural Gas Prices ($/MM BTU)	Annual Energy Costs (MM $/yr)						Unit Cost ($/bbl SCO)
		H_2	Electricity	Process Heat	Process Steam	Carbon Tax	Total	
S1	2	315	22	29	3	30	399	9.49
S2	3	334	25	44	4	30	437	10.39
S3	4	353	28	59	6	30	476	11.30
S4	5	392	32	53	7	27	511	11.90
S5	6	413	34	63	9	27	546	12.72
S6	7	432	37	74	10	27	580	13.52
S7	8	453	40	84	11	27	615	14.35
S8	9	472	44	95	13	27	651	15.15
S9	10	493	47	105	14	27	686	15.98
S10	11	512	50	116	16	27	721	16.78
S11	12	533	53	126	17	27	756	17.60
S12	13	552	56	137	19	27	791	18.41
S13	14	573	59	147	20	27	826	19.23

However, when the natural gas price increases and reaches $5/ MMBTU and beyond (see Table 2, S4-S13), the thermocracking unit switches to a delayed coker with a different processing mode (TCU_1); whereas the rest of the upgrading steps and processing modes remained unchanged. The new delayed coker features the lowest utility requirements, but considers the highest hydrocarbon products yield amongst the available options. Which are two of the most desirable characteristics sought in a cracking unit. The new upgrading pathway could produce approximately 130,000 bbl SCO/d. Although the new processing mode involves higher hydrogen consumption rates in the hydrotreatment step compared with the previous pathway (given the higher conversion rates in the coker), the higher expected hydrogen expenses are offset by significantly lower process heat (natural gas) requirements in the coker. For instance, in the fourth scenario (S4) the process fuel unit cost increases by 150% compared with the first scenario (S1); whereas the unit cost for hydrogen only grew by 18%. This is, the price of process fuel is equal to that of natural gas because they are equivalent; whereas the unit cost of hydrogen increases at a much lower rate as a function of gas price. This because the total hydrogen supply cost involves additional significant factors besides the feedstock (gas) price such as: capital costs, transportation costs, storage costs, and energy conversion efficiency. On the other hand, natural gas is directly used as process fuel. Therefore, from S4 and forward it results more economic to keep the process fuel demands as low as possible. This is accomplished by switching the thermocracking unit to delayed coker TCU_1.

Regarding the utility supply, for all studied scenarios none of the supply sources included any CO_2 mitigation strategy (e.g., CCS methods). This is, the carbon levies paid by the upgrader operator due to exceeding emissions over the cap of 100,000 tonnes of CO_2 per year do not represent a major burden on the total annual operating cost (see Table 2). Accordingly, it is more economical for the upgrader operator to import utility from conventional sources (and pay the carbon penalties) than importing more costly utilities from supplier employing CO_2 mitigation methods.

4. Conclusion

This work presents a design optimization model for estimating the most suitable bitumen upgrading pathway based on the operating utility costs. A sensitivity analysis was carried out to determine the influence of the natural gas price over the optimal upgrading pathway in terms of operating costs. The optimization results show that a thermocracking-based pathway is the most economical type of upgrading plant for converting bitumen into SCO because it is less energy-intensive. Moreover, for low gas price levels (lower or equal than $4/MMBTU) it is more economical to maintain the hydrogen cost as low as possible since it represents the dominating operating cost by far. On the other hand, for medium to high gas prices levels (higher or equal than $5/MMBTU) it is more economical to opt for maintaining the process heat requirements low because its price increases at the same rate that gas; whereas the hydrogen cost increases at a much lower rate. Thus, the process heat cost becomes increasingly important in the total operating cost. Additionally, for all studied scenarios the utility supply correspond to conventional sources.

5. References

[1] Betancourt-Torcat A, Almansoori A, Elkamel A, Ricardez-Sandoval L. Stochastic modeling of the oil sands operations under greenhouse gas emission restrictions and water management. Energy & Fuels 2013;27:5559–78.

[2] Charry-Sanchez J, Betancourt-Torcat A, Ricardez-Sandoval L. An Optimization Energy Model for the Upgrading Processes of Canadian Unconventional Oil. Energy 2014;68:629–43. doi:10.1016/j.energy.2014.03.016.

[3] Gomes U, Patil B, Betancourt-Torcat A, Ricardez-Sandoval L. Optimal Infrastructure of the Upgrading Operations in the Oil Sands under Uncertainty: A Multiscenario MINLP Approach. Ind Eng Chem Res 2014;53:16406–24.

[4] Nimana B, Canter C, Kumar A. Energy consumption and greenhouse gas emissions in upgrading and refining of Canada's oil sands products. Energy 2015;83:65–79. doi:10.1016/j.energy.2015.01.085.

[5] Alvarez-Majmutov A, Chen J. Analyzing the energy intensity and greenhouse gas emission of Canadian oil sands crude upgrading through process modeling and simulation. Front Chem Sci Eng 2014;8:212–8. doi:10.1007/s11705-014-1424-z.

[6] Syncrude Canada. Syncrude 2014 Sustainability Report. Fort McMurray, Alberta, Canada: 2015.

[7] Betancourt-Torcat A, Gutierrez G, Elkamel A, Ricardez-Sandoval L. Integrated Energy Optimization Model for Oil Sands Operations. Ind Eng Chem Res 2011:12641–63.

[8] Betancourt-Torcat A, Elkamel A, Ricardez-Sandoval L. Optimal integration of nuclear energy and water management into the oil sands operations. AIChE J 2012;58:3433–53. doi:10.1002/aic.13736.

[9] Ordorica-Garcia G, Croiset E, Douglas P, Elkamel A, Gupta M. Modeling the Energy Demands

and Greenhouse Gas Emissions of the Canadian Oil Sands Industry. Energy & Fuels 2007;21:2098–111. doi:10.1021/ef0700984.
[10] Nimana B, Canter C, Kumar A. Energy consumption and greenhouse gas emissions in the recovery and extraction of crude bitumen from Canada's oil sands. Appl Energy 2015;143:189–99. doi:10.1016/j.apenergy.2015.01.024.
[11] Ordorica-Garcia G, Elkamel A, Douglas PL, Croiset E, Gupta M. Energy Optimization Model with CO2 Emission Constraints for the Canadian Oil Sands Industry. Energy & Fuels 2008;22:2660–70. doi:10.1021/ef700770n.
[12] Government of Alberta. Phase-out of coal-fired emissions in Alberta. Edmonton, Canada: 2016.

454770 Model ComparisonsÂ of Cooled Tubular Multi-Tube Fixed Bed Reactor - Analysis and Control of Highly Exothermal Fast Equilibrium Reactions

Abstract:

Many industrially important vapor phase reactions are equilibrium controlled and highly exothermal in nature. Very often a cascaded multi-staged adiabatic reaction system is adopted in the industry where the heat of reaction is transferred by inter-stage cooling and equilibrium limitation is relieved by partial removal of product through condensation. Re-initiation of the reaction for the leftover reactants requires re-heating, and total installation and operating costs to carry out these conversions are generally high, especially if high conversion is expected for concentrated inlet reactants. One possible improvement in operation mode is to design and build a thermally cooled multi-tube tubular reactor system where both heat of reaction and equilibrium limitation can be alleviated in a one piece stand-alone reactor.

However, this operation mode is more complex than the above stated one in terms of system design and control. We found that pronounced heat transfer effect in this cooled, highly exothermal reactor created a need to model detailed temperature and conversion profiles within each tube of the tubular reactor. Radial heat removal is strongly influenced by the effective radial thermal conductivity which decreases significantly near the tube wall. Considering the possibility of instability and runaway in this type of reactor, a two-dimensional model serves a much better purpose in design and control than a corresponding one-dimensional model. The type of equilibrium reactions considered is very fast and highly exothermic, therefore effective thermal conduction should be distinguished between the gas and solid catalysts because solid temperature profile is generally higher than that of gas, and temperatures predicted by the two-dimensional heterogeneous models are generally very different from the one-dimensional homogeneous models. The development of the modern-day computer allows efficient computation in a two-dimensional model for most steady-state and some dynamic calculations. The one-dimensional model, backed by its corresponding two-dimensional model, can be used for on-line process control applications more effectively.

A two-dimensional heterogeneous tubular reactor model is proposed. Validation of the model requires experimental work to determine key process parameters. Because most of the physical properties and reaction kinetics of the system are already known, conceptual modeling of the process option could allow early visualization of certain aspects of the proposed design and help develop necessary control strategies for the operation of such reactors. The model can be run dynamically and can also be used to study transient behaviors of startup and shut down operations.

References:

1. Lordanidis, A., A. Mathematical Modeling of Catalytic Fixed Bed Reactors. Ph.D. Thesis, University of Twente, 2002.
2. Froment, G., F. and Bischoff, K., B. Chemical Reactor Analysis and Design. 2nd Edition, John Wiley & Sons, Inc. 1990.

463156 Predicting Nitrosamine Formation in Amine-Based Carbon Dioxide Capture Systems with Aspen Plus Simulation

Abstract:

Carbon dioxide capture, the separation of carbon dioxide from fossil fuel-fired power plants, is a critical endeavor to mitigate global climate change. Currently, the most mature technology for carbon dioxide capture is amine-based absorption. However, this technology can also generate nitrosamines, a group of carcinogenic compounds, from the reactions between amines and nitrogen oxides present in the power plant flue gas. Public concerns for nitrosamines have impeded the implementation of carbon dioxide capture. The investigation of nitrosamine formation usually relies on experimental measurement of routine samples from carbon dioxide capture plants, which is difficult to test â €œwhat ifâ€ conditions.

In this study, a simulation module was developed to predict nitrosamine formation in industrial-scale carbon dioxide capture systems. Aspen Plus is currently the most widely used simulation software for predicting the carbon dioxide capture performance of industrial-scale systems. However, there is no available simulation module in Aspen Plus to assess the formation of nitrosamines. To develop a rigorous simulation module, potential chemical reactions associated with nitrosamine formation were integrated. A rate-based multi-stage simulation was performed with RadFrac block, incorporating heat and mass transfer correlations accounting for column characteristics and hydraulics. The simulation indicated that our module is able to estimate nitrosamine concentration as low as ppm level in the bulk solution. This simulation module offers a robust and efficient strategy to evaluate the risk of nitrosamine formation from carbon dioxide capture systems.

458557 Hydroformylation of 1-Dodecene in Microemulsions: Proof of Concept and Long-Term Operability on a Mini-Plant Scale

Abstract:

Within the chemical industry hydroformylation is one of the main reaction steps forming aldehydes. For state of the art industrial processes, it is typically realized as a homogeneously catalyzed process. This offers major advantages regarding product selectivity and mild reaction conditions [1]. Moreover, water soluble, ligand-modified transition metal catalysts offer efficient catalyst recycling by simple means of phase separation. However, with an increasing miscibility gap between reacting olefins and catalyst solution, these processes are limited to short-chained olefins. If long-chained olefins are to be converted, severe conditions regarding pressure (300 bar) and temperature (200 °C) are still necessary for industrial applications [1]. One approach to overcome this issue is the application of microemulsion systems, which is currently being investigated within the Collaborative Research Centre InPROMPT / TRR 63 [2]. Here, microemulsions offer a promising and efficient possibility to enable a homogeneous catalytic reaction at mild conditions as the interfacial area between olefin and catalyst solution is increased. Simultaneously, the thermomorphic behavior of microemulsions with several miscibility gaps and phase separation states can be exploited. Thus, efficient catalyst recycling and product separation becomes feasible [3].

To analyze the applicability of this novel concept, a mini-plant has been built at the Chair of Process Dynamics and Operations of the Technische Universität Berlin [4]. This offers the possibility to verify obtained results from preliminary lab-scale experiments, regarding reaction performance and phase separation behaviour. Also, the identification of effects appearing during long-term continuous operation is possible, which then can be further investigated on lab-scale. Consequently, arising challenges can be tackled at an early stage, reducing the overall time needed for process development.

Therefore, this contribution firstly discusses the installed fully automated mini-plant, which is modularized into a feed, mixer-settler, and recycle section. It can be operated at temperatures of up to 478 K, pressures of up to 100 bar and process streams of a maximum at 1.1 kg/h. Regarding catalyst and reactants cost, the total process volume is kept small at 1.5 l. Given that synthesis gas is a main reactant, the plant and installed equipment are ATEX Zone II-conform and a safety concept, based on a detailed HAZOP analysis has been applied. Using Siemens PCS 7 the plant is fully automated with more than 50 sensors and actuators. Additionally, concentration measurements are carried out with an online micro GC to analyze gas composition, as well as an online Raman spectrometer and an offline GC for liquid composition sampling. The component system consists of 1-dodecene (CAS: 112-41-4) as a model olefin, the non-ionic surfactant Marlipal 24/70 (CAS: 68439-50-9), a rhodium catalyst precursor (CAS: 14874-82-9), and the water soluble ligand SulfoXantPhos (CAS: 161265-03-8). With this set-up, mini-plant operations can be carried out continuously in a three-shift system.

Subsequently, gathered operation data from several long-term mini-plant campaign of around 200 h operation time is presented. Here, two major aspects are discussed. The reaction performance at mild conditions of 95 °C and 15 bar and the phase separation performances with the corresponding catalyst recycling efficiency. Regarding the first, a high reaction yield of 40 % and an overall selectivity of 95 % was achieved. The latter one is highlighted by a total oil phase purity of more than 96 % (total amount of oily components in the oil phase), whereas 99.9 % of the rhodium catalyst was recovered and recycled. Thus, a proof of concept for the hydroformylation of long-chained olefins at mild process conditions is presented for a long-term mini-plant operation. In order to evaluate the reproducibility of the reaction and separation performance in the technical system, these results are then compared to preliminary lab-scale findings.

Additionally, several effects are addressed, which arose from the aforementioned long-term plant operation, such as surfactant loss via the product flow, component fractionation, and accumulation in the settler unit. These effects are not present at the lab-scale, but have a major impact on the operability of the system. Especially an increased byproduct formation of undesired hydrogenates was observed with fractions of up to 30 % of the total conversion. To overcome this obstacle and to enable a successful long-term process operation, several counter measures have been tested and successfully implemented in the mini-plant. Resulting in new mini-plant runs with exceptionally high selectivity and further improved phase separation, as stated above. To outline the applied strategies, the manipulation of recycle streams inside the plant and the variation of residence times in reactor and settler unit are discussed.

Concluding, this contribution constitutes the applicability of microemulsion systems for the hydroformylation of long chained olefins. Based on long-term mini-plant campaigns, a proof of concept for such an application is shown. Here, the benefits of homogenous catalytic systems are exploited at considerably low reaction conditions, whereas also an efficient catalyst recycling was possible. Therefore, opportunities for larger scale processes, but also remaining challenges concerning this novel process concept are pointed out.

Acknowledgements

This work is part of the Collaborative Research Center "Integrated Chemical Processes in Liquid Multiphase Systems" (subproject B4) coordinated by the Technical University of Berlin. Financial support by the German Research Foundation (Deutsche Forschungsgemeinschaft, DFG) is gratefully acknowledged (TRR 63). Furthermore, the authors gratefully acknowledge the support of Umicore N.V. for sponsoring the rhodium catalyst precursor “Acetylacetonatodicarbonylrhodium(I) (CAS: 14874-82-9)”, Sasol Ltd. for the surfactant used in the described experiments, the support of SIEMENS AG for sponsoring the entire process control system SIMATIC PCS7 for the automation of the mini-plant, and Rhodius GmbH for sponsoring the knitted fabrics. Finally, the support of the Federal Institute for Materials Research and Testing (BAM) is gratefully acknowledged.

References

[1] R. Franke, D. Selent, A. Börner, *Applied Hydroformylation*, Chem. Rev, 2012, 112 (11), 5675-5732.

[2] M.-J. Schwuger, K. Stickdorn, R. Schomäcker, *Microemulsions in Technical Processes*, Chem. Rev , 1995, 95 (4), pp 849–864.

[3] T. Hamerla, A. Rost, Y. Kasaka, R. Schomäcker, *Hydroformylation of 1-Dodecene with Water-Soluble Rhodium Catalysts with Bidentate Ligands in Multiphase Systems*, ChemCatChem, 2013, vol. 5 (7), 1854-1862.

[4] M. Müller, Y. Kasaka, D. Müller, R. Schomäcker, G. Wozny, *Process Design for the Separation of Three Liquid Phases for a Continuous Hydroformylation Process in a Miniplant Scale*. Ind. Eng. Chem. Res., 2013, 52, 7259 – 7264.

466031 From Lab to Miniplant: Effective Process Development for Deep Desulfurization of Industrial Coke Oven Gas

Abstract:

As a consequence of the energy revolution and the ambition to reduce emissions, known and in chemical engineering established processes are adapted to convert off-gas of steel plants to basic chemicals. An initial and therefore crucial role in this plan is the development of a process that removes organosulfur compounds from coke oven gas. This is necessary to treat this valuable gas, that consists of approx. 60 vol.% hydrogen, 20 vol.% methane, and 6 vol.% carbon monoxide, with catalysts that are highly sensitive to sulfur (Twigg and Spencer 2001).
In accordance with other desulfurization processes (e.g., natural gas) heterogeneous catalysis in a fixed bed is chosen to fulfil this task since it shows the highest conversion rates compared to other processes. Here, an industrial catalyst is analyzed and investigated for its potential to convert organosulfur compounds like COS or CS2 to H2S which can consequently be removed from the gas by a scrubber or a guard-bed (Rhodes et al. 2000):

COS + H_2Â â†” H_2S + CO (1)
CS_2 + $2H_2$Â â†” $2H_2S$ + C (2)

However, due to multiple additional components in the gas stream (mainly olefins, benzene, and oxygen) the direct transfer from established processes is not trivial and process modifications are required. Therefore, a full process development is necessary from catalyst screening to the crucial step of tests under real conditions to investigate the influence of side reactions.
In this work, the development of a new catalytic process in a fixed bed reactor will be presented. The two-stage development consists of studies on a lab-scale that are conducted as a first step to effectively screen industrial catalysts. The lab reactor has a diameter of 3 cm and is operated with the main components only. The most promising

catalyst candidates are subsequently used for tests with a mobile and modular miniplant on the coke plant ArcelorMittal Bottrop GmbH in Germany. The miniplant is operated with industrial coke oven gas in by-pass to the real plant, therefore, the gas stream contains the full range of components with their fluctuations in concentration. The plant was designed to enable a fast testing of catalysts under industrial conditions for proof of concept and to gain information on (exothermic) side reactions. The miniplant reactor has a diameter of 10 cm and can be filled with a maximum of 5 beds with each 10 cm height. The results of both stages in the process development are used for the development of a model that can be used for simulations with Aspen Plus® to design the complete process.
The results of lab- and miniplant tests will be presented and discussed. It was shown in lab experiments and confirmed in a first measurement campaign with industrial gas, that the catalyst is able to remove up to 98% of the organic sulfur in the gas stream. The influence of side components will be considered and resulting challenges for the transfer from lab to miniplant scale will be elucidated.

References
Rhodes, C., Riddel, S. A., West, J., Williams, B., & Hutchings, G. J. (2000). The low-temperature hydrolysis of carbonyl sulfide and carbon disulfide: a review. Catalysis Today, doi: 10.1016/S0920-5861(00)00309-6.
Twigg, M. V., & Spencer, M. S. (2001). Deactivation of supported copper metal catalysts for hydrogenation reactions. Applied Catalysis A: General, doi: 10.1016/S0926-860X(00)00854-1

453171 Tributyl Citrate Production Via Reactive Distillation

Abstract:

Over the last decades, a market shift towards the consumption of biobased plasticizers has been observed. Such trend obeys a generalized concern about toxicity, biodegradability and sustainability of traditional petroleum derived plasticizers (mainly phthalates). Tributyl citrate (TBC) and Acetyl Tributyl citrate are two well-known bioderived plasticizer which has been used as substitutes for various phthalic esters. TBC production is generally accomplished by direct esterification of citric acid and butanol in a semi-batch process. This reaction is limited by chemical equilibrium, so excess of butanol and water removal are common strategies to improve reaction conversion at the industrial scale. Such operating conditions makes the process energy and mass intensive, with low efficiency and costly.

Taking into account the above, this work focuses on the evaluation of a reactive distillation system for the production of TBC. A pilot scale reactive distillation system was used to evaluate the process performance under different conditions. A 7m tall 3" diameter stainless steel reactive distillation unit with structured packing similar to KATAPAK SP-11 with Amberlyst 70 as catalyst in the reactive sections was used. Pall rings were used as internals in the rectifying and striping sections. The column was coupled with a top decanter to allow removing produced water when need it. The column performance was studied at local atmospheric pressure (74 kPa), using unreacted and pre-reacted mixtures of different molar ratios of butanol and citric acid, under different feed loadings and at different reflux ratios.

Under different operating condition citric acid conversion was found to be close to 100% and a selectivity towards TBC above 90% was reached. A mixture of TBC and Butanol was withdrawn as the bottoms product to avoid thermal degradation of the citric species. With this butanol content, the temperature at the reboiler was kept below 426 K. Experimental results were compared to the simulated operation using a Radfrac module included in Aspen Plus. The necessary reaction kinetics and thermodynamics for the reactive system were obtained experimentally and correlated with suitable models (Power law kinetics and UNIQUAC activity based model). An equilibrium stage model was used

during simulations, and it was enhanced by including Murphree efficiency, catalysts effectiveness factor and heat losses. A good agreement between experimental and simulation data were observed, indicating that the model can be used for up-scaling the operation.

464161 Design Alternatives and Control Performance in the Pilot Scale Production of Isoamyl Acetate Via Reactive Distillation

Abstract:

In the present work, two different Reactive Distillation (RD) designs for the pilot scale production of isoamyl acetate are studied by means of steady state simulations. In the first design near stoichiometric amounts of acetic acid and isoamyl alcohol are fed directly to the reactive and rectifying sections of the RD column. On the second design, acetic acid in excess and isoamyl alcohol are first fed to a continuously stirred tank reactor (CSTR) and its product stream is fed to the reactive section of the column. Among different optimization variables, the overhead vapor flowrate appears to be the most influencing variable in the energy requirements of the processes, and its value and composition are responsible for the two liquid phase formation in the condensate. This variable also determines the global performance of the process schemes yielding a >99% conversion and >99% isoamyl acetate mass purity of the bottoms product. Having a CSTR upstream to the RD column allows an operation with a lower overhead vapor flowrate, thus a lower energy consumption is achieved for this design.

Indirect composition control of the bottoms product in the direct feed design has been implemented by controlling the temperature of one of the reactive trays. This control strategy shows a good disturbance rejection in the dynamic behavior of the main process variables to variations in the total feed rate and reactant feed ratio.

465885 Pilot Plant Study on Pyrolysis of Pulverized Coal in Circulating Fluidized Bed with Heat Char Carrier

Abstract:

Pilot plant study on pyrolysis of pulverized coal in circulating fluidized bed with heat char carrier

Pengfei Dong, Zhiping Zhu*, Guangjun Meng, Qinggang Lu, Kuangshi Yu

Institue of Engineering Thermophysics, Chinese Academy of Sciences, Beijing 100190, China

Abstract:

The high dust content and poor quality of tar are the main restriction for industrialization development of pulverized coal pyrolysis. A hierarchical dedusting process, mainly consisting of dust removal of pulverized coal and heat char carrier, dust removal during pyrolysis reaction and dust removal of pyrolysis products, has been proposed by institute of engineering thermophysics (IET), Chinese Academy of Sciences (CAS) to reduce the dust content of tar in the pyrolysis of pulverized coal. Using heat char as solid heat carrier promotes the heavy components of tar cracking, which is beneficial to upgrade the quality of the tar so as to get more high-value light tar boiling points below 360 ¡æ. Experiments on pyrolysis of Chinese Shenmu pulverized coal (with the range of particle size: 0¡«10mm) were carried out in a pilot plant with a treating capacity of 10 t/h. The effects of hierarchical dedusting on dust content and tar composition were investigated. It was found that the dust content decreases by hierarchical dedusting. When temperature is 700 ¡æ with coal feeding rate of 10 t/h, the dust content of tar reaches the minimum value of 4.12 wt%, and the light tar content of tar reaches the highest value of 86 wt%.

Keywords: circulating fluidized bed; pulverized coal pyrolysis; hierarchical dedusting

Acknowledgements:

This work was financially supported by the Strategic Priority Research Program of the Chinese Academy of Sciences (No.XDA07010300).

447399 Continuous Biocompatible Reactive Extraction Concept for the Purification of Di-Carboxylic Acids from Dilute Aqueous Solutions

Abstract:

Product concentrations for the biocatalytic production of dicarboxylic acids achievable by whole-cell biocatalysis have recently been shown to reach industrial levels. Caused by their high reactivity, dicarboxylic acids are an attractive substance class for the chemical industry and are even described to serve as future platform chemicals. The development of downstream processing (DSP) concepts still is the bottleneck towards an industrial application of dicarboxylic acids from biocatalytic origin.

Reactive extraction (RE) has shown to be applicable as selective and efficient purification step for the direct recovery of dicarboxylic acids from a dilute aqueous solution, with the mechanism of the RE recently described in our previous work [1]. A hydrophobic complex is formed between hydrophilic dicarboxylic acid (*cis,cis* muconic acid and itaconic acid) and the extractant (tri-n-octylamine) and is extracted into the organic phase. For the recovery of the dicarboxylic acid from the organic RE phase two different strategies have been developed recently: *ph*-shift and addition of a water soluble amine lead to a full recovery of the acid and allow for a recycle of the organic RE phase containing the extractant.

Within this work batch extraction experiments published previously [1] were transferred into a continuous DSP concept using a 15 stage lab-scale Mixer-Settler setup. The transfer to continuous operation and the scale up (factor 20) in the Mixer-Settler experiments marks a first step to fulfill the requirements for an industrial application. Different process parameters (e.g. mass flow, number of stages for RE and REEX) were investigated with respect to the overall DSP yield achievable. Additionally a recycle of the organic phase (including amines for RE) and a combined RE and REEX was investigated in order to prove the applicability of the developed DSP concept. Stability investigations over the course of up to 12 h at equilibrium state within the Mixer-Settler set up of the system

demonstrate the robustness of the presented DSP concept. Aqueous phase after REEX was shown to enable final purification and thus allow for a further development of an in situ DSP concept for dicarboxylic acids from biocatalytic origin.

[1] J. Gorden, T. Zeiner, C. Brandenbusch, Fluid Phase Equilibria, 393 (2015) 78-84

462006 Combining PAT with Data Visualization and Machine Learning to Understand and Control Batch Cooling Crystallization

Abstract:

With the rapid development of PAT, online crystallization monitoring tools have become commonplace. Spectroscopic monitoring tools, such as ATR-FTIR, and particle-detecting tools, such as FBRM, are especially popular in that regard [1].

Using these tools, large data sets can be collected, which presumably contain detailed information on crystallization dynamics. But this is where the promise of big data often falls short. Simply collecting raw data does not automatically lead to understanding. Instead, the data most be processed, analyzed, and presented in a coherent way.

In this talk, we will describe how data from ATR-FTIR, FBRM, and temperature measurements can be brought together to visualize batch cooling crystallization as movement through a two-dimensional space [2]. The data visualization, we will show, fosters an intuitive understanding of the crystallization dynamics. It also sets the stage for a low-dimensional mathematical model of crystallization and dissolution dynamics that can be learned directly from data.

Combined, the data visualization and application of machine learning creates a framework for establishing control. In particular, we demonstrateâ€"on two experimental case systemsâ€"the use of the framework to accurately control the mean size of crystals produced by unseeded batch cooling crystallization [2, 3].

[1] L. L. Simon, H. Pataki, G. Marosi, F. Meemken, K. HungerbÃ¼hler, A. Baiker, *et al.Â* 2015. Assessment of Recent Process Analytical Technology (PAT) Trends: A Multiauthor Review. *Organic Process Research & Development* 19, 3-62.

[2] Griffin, D. J., Grover, M. A., Kawajiri, Y. and Rousseau, R. W. 2015. Mass-Count Plots for Crystal Size Control. *Chemical Engineering Science* 137, 338-351.

[3] Griffin, D. J., Grover, M. A., Kawajiri, Y. and Rousseau, R. W. 2016. Data-Driven Modeling and Dynamic Programming Applied to Batch Cooling Crystallization. *Industrial & Engineering Chemistry Research* 55, 1361-1372.

467965 Stability Analysis of Solar Silicon Wafer Manufacturing By the Horizontal Ribbon Growth Process

Abstract:

The increasing demand for energy and the prospect of global warming urges the world community to seek cleaner and cheaper energy resources to replace non-renewable fossil fuels. The silicon based photo-voltaic system is one of the most promising renewable energy resources. Its application has grown rapidly during the past 2-3 decades as production cost has been reduced and the potential for application is virtually without limit.

This promotes the project of cost reduction for silicon wafer via different technology development. Traditional silicon growth processes are either expensive or compromise the material efficiency for faster production speed. The horizontal ribbon growth (HRG) process has the potential to overcome these limitations. In the HRG process, heat is removed through the top surface of the molten silicon pool, meanwhile, a thin silicon solid sheet is produced and extracted continuously. However, despite the theoretical and experimental efforts to develop and commercialize this technology for more than 60 years, many technical challenges have been reported to achieve a stable and fast ribbon production.

One of the principal challenges encountered is the instability at the wafer-melt interface due to various coupled effects. Theoretical studies have been developed for the instabilities during unidirectional solidification. In 1964, Mullins and Sekerka developed the foundation for the thermal and solutal conditions to ensure the stable crystallization front under an arbitrary sinusoidal perturbation of infinitesimal amplitude to the interface [1]. However, the flow pattern in the melt during the crystallization also plays an important role on the impurity distribution, which alters the micro-structure and macro material property of the grown crystal. This flow pattern can be induced by the rotation of the crucible or crystal, withdrawal of the material, density change during crystallization, etc. Coriell and Sekerka performed extensive studies in the coupled convective and morphological instabilities. [2][3]

In this work, we developed a linear stability theory to study the crystallization interface of the Horizontal Ribbon Growth (HRG) process. The conditions for the onset of stability are investigated theoretically and numerically on the basis of diffusion-convection equations for the thermal, solutal fields coupled to the Navier-Stokes equation describing the flow field in the molten pool near the interface. In order to understand the unexplained "facet lines" observed in the experiments, the interface stability conditions of the system under different operating conditions are tested.

1. **Mullins, W. W., & Sekerka, R. F**, "Stability of a planar interface during solidification of a dilute binary alloy," *Journal of applied physics*, **35**(2), 444-451(1964).
2. **Coriell, S. R., Cordes, M. R., Boettinger, W. J., & Sekerka, R. F**, "Convective and interfacial instabilities during unidirectional solidification of a binary alloy," *Journal of Crystal Growth*, **49**(1), 13-28 (1980).
3. **Coriell, S. R., McFadden, G. B., Boisvert, R. F., & Sekerka, R. F**, "Effect of a forced Couette flow on coupled convective and morphological instabilities during unidirectional solidification," *Journal of crystal growth*,**69**(1), 15-22 (1984).

458953 Protein Enzymatic Hydrolysis Integrated with Ultrafiltration

Abstract:

Thermolysin was used to hydrolyse serum bovine albumin. The molecular mass and concentration of hydrolysates were identified by SE-HPLC. The substrate pre-incubation at high temperatures (>46°C) contributed to the increase of the hydrolysis rate and final efficiency, but at 70°C a negative effect caused by the aggregation of albumin molecules was noticed. Pre-incubation time is significant up to 0.5 hour.

Strong product inhibition was detected, thus the values of hydrolysis degree obtained in classical reactors were low. The kinetic equation was elaborated and its constant values were estimated.

The data yielded in an integrated bioreactor-ultrafiltration process provides strong evidence that considered enzymatic process could be significantly intensified. The ultrafiltration membrane with a small pores size was selected to take off the products (peptides), including enzyme inhibitors.

The utility of a previously elaborated kinetic model of biopolymer hydrolysis was checked and a qualitative and quantitative description of the proteolysis products was performed. The data gathered suggest that process efficiency and permeate composition depend on the proper selection of dilution rate and the separation properties of the membrane.

469298 The Dynamics of Biodiesel Profile and Fungal Diversity of Activated Sludge during Lipid Accumulation on Acetic Acid

Abstract:

Recent studies indicated that the microbial consortia of wastewater treatment plant (WWTP) activated sludge feeding on acetic acid can accumulate lipids, which can be transesterified into biodiesel. The composition of the raw activated sludge, however changes significantly with the fluctuations of operational and environmental conditions in a WWTP. The variations in raw activated sludge composition may pose additional processing cost after extraction and transesterification of the lipids if there is significant variation of the biodiesel profile (measured as fatty acid methyl esters or FAMEs) among raw activated sludges. This study describes the changes on biodiesel profile during lipid accumulation on acetic acid. The changes on the diversity of fungal community are also correlated to the changes in biodiesel profile. The findings suggest that even with variations in the biodiesel profiles from lipids of raw activated sludge, the enhancement of lipid content by feeding on acetic acid induces speciation of microbial community, specifically fungi, resulting in the homogenization of biodiesel profiles.

464945 A Novel and Systematic Method for Process Intensification

Abstract:

Process intensification combines multiple operations in a single unit, and leads to substantially smaller, cleaner, safer, and more energy-efficient technologies [1]. It has been shown that intensification represents a limit case of tight integration through significant material recycling [2]. Although there are established methods for process synthesis and integration, few methods exist for the systematic identification and selection of intensification opportunities. Current superstructure-based process synthesis relies on pre-specified equipment configurations and is unable to automatically construct novel equipment configurations for intensification. Phenomena based process intensification [3-5] allows a bottom-up approach for the identification, screening and incorporation of intensification opportunities, but this approach often requires sequential or decomposition based solution strategies to solve large models, which can lead to suboptimal solutions.

In this work, we propose a unified process synthesis and intensification approach that leverages on an original representation of the flowsheet superstructure. Unlike a classical flowsheet superstructure which encompasses all feasible sequences of unit operations (e.g., reactors, separators, movers, etc.), our proposed superstructure is an ensemble of building blocks, where each block represents a unit use of a material and nexus between the blocks are achieved by intra-block mass and energy transfer. An intensified unit is realized by selecting and assembling multiple neighboring blocks with different functionalities. While reaction, mixing, heating and cooling are achieved within the block, separation related phenomena (e.g., phase separation, phase-contact, etc.,) are achieved via the boundaries between two neighboring blocks. This enables to include many intensification alternatives, including divided wall column distillation, reactive distillation, reactive absorption and membrane reactor, within the same superstructure without any *a priori* postulation.

The overall process intensification model is formulated as a single mixed-integer nonlinear optimization (MINLP)

problem and is solved to optimality. Through a range of applications, we demonstrate that the proposed intensification method is capable of generating innovative and intensified configurations and obtaining optimal process conditions for given feed and product specifications.

References:

[1] Reay, D.; Ramshaw, C.; Harvey, A. Process intensification: engineering for efficiency, sustainability and flexibility. Waltham, MA: Butterworth-Heinemann, 2013.

[2] Baldea, M. From process integration to process intensification. *Computers & Chemical Engineering*2015, 81:104â €“114.

[3] Arizmendi-Sanchez, J. A., and Sharratt, P. Phenomena-based modularisation of chemical process models to approach intensive options. *Chemical Engineering Journal* 2008, 135(1):83-94.

[4] Lutze P, Babi DK, Woodley JM, Gani R. Phenomena based methodology for process synthesis incorporating process intensification. *Industrial & Engineering Chemistry Research* 2013, 52(22):7127â€“44.

[5] Babi, D. K., Holtbruegge, J., Lutze, P., Gorak, A., Woodley, J. M., and Gani, R. Sustainable process synthesisâ €“intensification. *Computers & Chemical Engineering* 2015, doi:10.1016/j. compchemeng.2015.04.030.

469723 High-Resolution Process Design: Flowsheet Optimization with Embedded High Fidelity Unit Models

Abstract:

High-resolution process design: Flowsheet optimization with embedded high fidelity unit models

R.C. Pattison and M. Baldea

McKetta Department of Chemical Engineering, The University of Texas at Austin, Austin, TX 78712

Optimal process design is key to maximizing the profitability of chemical plants. Due to the large upfront capital investments required to construct new processes or retrofit existing plants, it is imperative that the design is optimized using modern gradient-based optimization tools coupled with detailed models of the processing units present in the flowsheet.

Equation-based simulation and optimization have witnessed substantial progress, and most current simulators offer equation-oriented simulation options as well. In this presentation, we illustrate a novel equation-oriented simulation tool which allows for high-fidelity models of select unit operations (including reactors, multi-stream heat exchangers) to be easily embedded into the process flowsheet model. These models capture phenomena occuring over multiple length scales, including smaller dimensions (e.g., diffusion and reaction in a catalyst pellet) which are typically not considered in flowsheet simulation or optimal design calculations.

Our framework is based on a pseudo-transient representation of the unit operation models developed in our previous work [1-3]. The pseudo-transient modeling framework relies on a systematic reformulation of the material and energy balances into a differential and algebraic equation model having the same steady state solution to that of the original algebraic equation model. The (typically) partial differential equations associated with detailed unit models lend themselves naturally to a pseudo-transient reformulation which can then be incorporated into the process flowsheet model. This, in turn, enables the detailed steady state unit designs to be optimized simultaneously with the design of the entire process flowsheet. To this end, we use a time relaxation-based optimization algorithm previously developed in our group [1].

The framework will be illustrated with industrial case studies, including the design optimization of an intensified dimethyl ether production process featuring a dividing wall distillation column and the optimal design of a natural gas liquefaction process.

References

[1] Pattison, R.P.; Baldea, M. Equation-oriented flowsheet simulation and optimization using pseudo-transient models. *AIChE J.*, **2014**, 60, 4104-4123.

[2] Pattison, R.P.; Baldea, M. Multistream heat exchangers: Equation-oriented modeling and flowsheet optimization. *AIChE J.*, **2015**, 61, 1856-1866.

[3] Pattison, R.P.; Gupta, A.M.; Baldea, M. Equation-oriented optimization of process flowsheets with dividing-wall columns. *AIChE J.*, **2016**, 62, 704-716.

456412 Process Intensification of Multicomponent Distillation Configurations Using Minimum Additional Number of Heat and Mass Integration Sections

Abstract:

We present a powerful and simple-to-use fact about heat and mass integration to consolidate distillation columns in a multicomponent distillation configuration. First, thermal couplings are introduced to all intermediate transfer streams in the distillation column. And then, distillation columns associated with a lighter final product reboiler and a heavier final product condenser are consolidated to produce a heat and mass integrated configuration. We call this novel set of configurations the HMP configurations.

Coupled with the already developed techniques[1] that enumerate all basic and thermally coupled distillation configurations, a systematic way of synthesizing all HMP configurations for *n*-component separations is introduced here. Also, we compare the energy savings of HMP configurations with the well-known fully thermally coupled (FTC) configuration using the state-of-the-art Global Minimization Algorithm (GMA)[2] developed based on Underwood equations. GMA provides a global optimization based rank-list of basic and thermally coupled configurations according to minimum total vapor duty requirement under minimum reflux conditions for separating any ideal or near-ideal multicomponent mixture. We demonstrate through case studies that HMP configurations can have very similar and sometimes even the same minimum total vapor duty requirement as the FTC configuration, while using far less number of column sections, intermediate transfer streams, and thermal couplings than the FTC configuration.

These findings not only intensify the search space of distillation configurations, but also provide industrial practitioners with insights in designing energy efficient distillation configurations that are easy to operate and control.

[1] Shah VH, Agrawal R. A matrix method for multicomponent distillation sequences. AIChE J. 2010;56(7):1759â€"1775.
[2] Nallasivam U, Shah VH, Shenvi AA, Huff J, Tawarmalani M, Agrawal R. Global optimization of multicomponent distillation configurations: 2. Enumeration based global optimization algorithm. AIChE J. 2016;62(6):2071â€"2086.

469729 Novel Optimization-Based Approach for Process Design and Intensification of High-Dimensional Modular Systems

Abstract:

In the recent years, due to the advent of the shale gas revolution, the US has incremented the annual production of processed natural gas by approximately 21% from 16.3 to 19.8 trillion cubic feet[1]. To accelerate the potential benefits brought by the abundant amounts of shale gas in the country, approaches are needed for the optimization and techno-economic viability analysis of novel natural gas utilization processes that produce heat, power, fuels and chemicals. However, the design of such processes is a challenging task as these systems are typically operated in a highly constrained and integrated environment that is represented by complex large-scale and nonlinear models. In this presentation, a novel optimization-based operability approach for process design and intensification of high-dimensional and modular energy systems is proposed. These modular units if enabled have the potential to transform the US economy by utilizing local stranded gas without the need of building expensive pipelines.

Process operability was originally developed as an approach for the design and control of complex chemical processes[2]. This approach has been applied to processes that are described by high-dimensional linear models[3,4]. For nonlinear systems, an operability-based approach for design and intensification of low-dimensional membrane reactor systems has also been proposed[5,6]. However, operability methods for nonlinear systems are currently limited by the problem size that they can address. The proposed approach bridges this gap in the literature by addressing the challenges of process nonlinearity and model size. The proposed method also broadens the scope of the traditional path of the operability approaches for design and control, mainly oriented to obtain the achievable output set (AOS) from the available input set (AIS), and compare the computed AOS to a desired output set (DOS). In particular, an optimization algorithm based on nonlinear programming tools is formulated for the high-dimensional calculations of the desired input set (DIS) that is feasible considering process constraints, performance levels and intensification

targets.

To illustrate the effectiveness of the developed method, two modular systems are addressed: i) a membrane reactor (MR) for the direct methane aromatization (DMA) conversion to benzene and hydrogen; and ii) a natural gas combined cycle (NGCC) system for heat and power generation. Results on the application of this novel method as a tool for process intensification show reduction of the DMA-MR footprint (â‰ˆ77% reactor volume and 80% membrane area reduction) for an equivalent level of performance, when compared to the base case. These results indicate that the novel approach can be a powerful tool for enabling the realization of the concept of the modular economy.

References

1. EIA (2016). U.S. Energy Information Administration. Natural Gas Report. Available at: http://www.eia.gov/dnav/ng/ng_sum_lsum_dcu_nus_m.htm
2. Vinson D. R., Georgakis C. A new measure of process output controllability. J. Proc. Cont. 2000, 10(2-3), 185-194.
3. Lima F. V., Georgakis C. Design of output constraints for model-based non-square controllers using interval operability. J. Proc. Cont. 2008, 18(6):610-620.
4. Lima F. V., Georgakis C. Input-output operability of control systems: the steady-state case. J. Proc. Cont. 2010, 20(6):769-776.
5. Carrasco J. C. and Lima F. V. Nonlinear operability of a membrane reactor for direct methane aromatization. In Proceedings of the 2015 IFAC ADCHEM Symposium, Whistler, Canada, June 2015.
6. Carrasco J. C. and Lima F. V. Novel operability-based approach for process design and intensification: application to a membrane reactor for DMA. Submitted for publication, 2016.

470719 Towards Optimal and Sustainable Operation of Separation Processes: The Computational Approach

Abstract:

Separation processes (such as Simulating Moving Bed, Pressure Swing Adsorption Systems and chromatographic processes) are widely applied in the chemical/biochemical industry for the separation and/or purification of various components [1-3]. Commonly, chromatographic separations are assessed by their product recovery rate, as well as the purity of the end product, with the latter being of vital importance. Consequently it is essential to guarantee that the process runs under optimal operating conditions, achieving high recovery rates of products that meet the desired specifications. However, downstream processing is coupled with high capital and operational costs that are significantly increased as a function of the feed stream produced by the upstream [4]. Current industrial needs suggest the development of sustainable processes, characterized by lower investment and/or operation costs. To answer to global demand and competition arising from novel solutions such as single-use technologies [5, 6], separation processes and particularly preparative chromatography, requires significant advancements.

Along this line, in this work we focus on the development of advanced control strategies for a semi-continuous, chromatographic separation process [7] aiming to increase both product purity and yield. In this framework we examine a highly nonlinear, periodic system and we design novel control strategies to ensure that optimal operation is maintained. The controllers are designed and tested following our recently presented PAROC framework that comprises four steps: (i) development of a high-fidelity process model, (ii) approximation of the complex, process model, (iii) design of a multi-parametric controller and (iv) â€˜closed-loopâ€™, in-silico validation of the controller against the process model. The controllers are tested in-silico against the high-fidelity process model and demonstrate an overall satisfactory performance, tracking the predefined setpoints and providing input profiles of reduced utility consumption. In addition, the proposed control scheme accounts for variable feeding composition, thus enabling the downstream processing to operate independently from the mixture composition resulting from the upstream.

Acknowledgements

The authors would like to thank Mr R. Oberdieck and Miss A. Quiroga-Campano for their contribution in the assessment of the controller validation. The authors would also like to thank Mr. Fabian Steinebach & Prof. M. Morbidelli from ETH Zurich, as well as Dr. Thomas Mueller-Spaeth from ChomaCon AG for their valuable input on the understanding of the MCSGP process. Financial support from the European Commission (OPTICO/G.A. No.280813) & Texas A&M University are also gratefully acknowledged.

References

1. Subramani, H.J., K. Hidajat, and A.K. Ray, *Optimization of Simulated Moving Bed and Varicol Processes for Glucoseâ€"Fructose Separation.* Chemical Engineering Research and Design, 2003. **81**(5): p. 549-567.

2. Khajuria, H. and E.N. Pistikopoulos, *Dynamic modeling and explicit/multi-parametric MPC control of pressure swing adsorption systems.* Journal of Process Control, 2011. **21**(1): p. 151-163.

3. Müller-Späth, T., et al., *Chromatographic separation of three monoclonal antibody variants using multicolumn countercurrent solvent gradient purification (MCSGP).* Biotechnology and Bioengineering, 2008. **100**(6): p. 1166-1177.

4. Gronemeyer, P., R. Ditz, and J. Strube, *Trends in Upstream and Downstream Process Development for Antibody Manufacturing.* Bioengineering, 2014. **1**(4): p. 188-212.

5. Shukla, A.A. and U. Gottschalk, *Single-use disposable technologies for biopharmaceutical manufacturing.* Trends in Biotechnology, 2013. **31**(3): p. 147-154.

6. Xenopoulos, A., *A new, integrated, continuous purification process template for monoclonal antibodies: Process modeling and cost of goods studies.* J Biotechnol, 2015. **213**(0): p. 42-53.

7. Krättli, M., F. Steinebach, and M. Morbidelli, *Online control of the twin-column countercurrent solvent gradient process for biochromatography.* Journal of Chromatography A, 2013. **1293**: p. 51-59.

454543 Multi-Scale Computer Aided Synthesis–Design–Intensification Method for Sustainable Hybrid Solutions

Abstract:

Process synthesis-design refers to identification of the processing route to reach a desired product from a specified set of raw materials, design of the operations involved in the processing route along with calculations of utility requirements, waste generations and emissions to the environment. Process intensification refers to strategies by which new processes can be designed or existing processes can be redesigned to be more efficient, compact and sustainable.

Therefore, in order to achieve sustainable process design it would be logical to perform process synthesis-design and process intensification together.

In this work, a four stage hierarchical decomposition solution strategy is proposed to generate sustainable hybrid solutions. The framework operates at different scales, the unit operations scale, the task scale and the phenomena scale. In stage1, process synthesis is performed (at the unit operations scale) using a newly developed computer-aided flowsheet design method (CAFD) (Tula et al, 2015), which applies principles of computer-aided molecular design to the synthesis and design of process flowsheets. That is, use process-groups representing different unit operations (reactor, distillation, flash, crystallization, etc.), bonds representing streams and/or recycles, rules for chemical feasibility also representing process flowsheet feasibility and sum of group contributions representing the performance of the flowsheet that can later be used to quickly screen the alternatives and to generate the optimal processing route. In stage 2, the base case design is first established based on the generated processing route. This base case design is further analyzed to identify process hotspots (process bottlenecks) using a comprehensive indicator based analysis incorporating process economics, sustainability measures (Carvalho et al., 2013) and life cycle assessment factors (Kalakul et al., 2014). These indicators corresponding to the hotspots are translated to design targets that are targeted in the next stage to achieve more sustainable and intensified designs. In stage 3,

integrated task phenomena based synthesis-intensification method is applied (Babi et al., 2015). The main idea here is to generate all the phenomena involved in the process by breaking the tasks involved into corresponding phenomena. Later these phenomena scale units are combined using combination rules in order to generate new and/or existing unit operations that constitute the (more sustainable) flowsheet alternatives which satisfy, the process design targets. In this way, truly predictive and innovative hybrid solutions are generated much in the same way as atoms are combined to form molecules with desired properties (that is, analogous to computer-aided molecular design) which are otherwise could not be found from the higher scales. In the final stage, detailed model based simulations and/or experiments are performed to verify and compare the new solutions.

In this presentation, a multi-scale computer-aided framework will be presented together with the corresponding databases, computer-aided methods and tools needed to implement and apply the synthesis-design-intensification method. The application of this framework will be highlighted through case studies generating hybrid solutions for energy efficient processing schemes involving downstream separations and intensified reaction-separations.

References:

[1] Babi, D. K., Holtbruegge, J., Lutze, P., Gorak, A., Woodley, J. M., & Gani, R. 2015. Computers & Chemical Engineering.

[2] Carvalho, A., Matos, H. A., & Gani, R. 2013. Computers & Chemical Engineering, 50, 8–27

[3] Kalakul, S., Malakul, P., Siemanond, K., & Gani, R. 2014. Journal of Cleaner Production

[4] Tula, A. K., Eden M. R., Gani, R. 2015. Computers & Chemical Engineering.

458409 Microwave Assisted Crystallization: Intensification of Direct Nucleation Control

Abstract:

Direct nucleation control (DNC) is a feedback control strategy used to control the size distribution of crystals (one of the many possible control objectives) during a crystallization process. DNC has been shown for maintaining a narrow size distribution by removal of unwanted fine particles. Application of DNC for obtaining desired size distribution is based on alternating dissolution and growth cycles, achieved via temperature cycling and triggered as a result of deviations in the particle count set point (measured by in-situ measurement PAT tool).

However, a major draw-back of the DNC implemented in conventionally heated/cooled reactors is that the temperature cycles needed to dissolve the nuclei considerably increases the process/batch time. However recently it has been demonstrated that by application of microwave fields for heating, the same control performance can be achieved with a 50% reduction in process/batch time. The study showing the intensification of the DNC in terms of process time by integrating microwave fields into a crystallizer was performed in a 1 litre jacket vessel in which the slurry was circulated via an external loop through a microwave cavity. However a better application would be when microwave fields are truly introduced inside the crystallizer thus eliminating external loops for slurry circulation which can be problematic.

In this study we present DNC experiments using a microwave integrated crystallizer. The crystallizer utilizes internal transmission line technology made available by Sairem SAS which enables direct introduction of microwave fields into the slurry. In this study we demonstrate cooling crystallization with DNC control and address key challenges in process integration and implementation, namely: Evaluation and optimization of the microwave heating to ensure homogeneous temperature distribution in the system, Optimization of the reactor configuration for safe introduction of PAT tools essential for DNC feedback (e.g FBRM), by avoiding the exposure to high intensity MW field and

evaluation of the performance of microwave assisted DNC in a cooling crystallization process.

467819 A Systems Approach for the Development of Osn Membrane Cascades

Abstract:

The processes involved in separating the components of chemical mixtures into pure or purer forms, such as distillation, account for 10-15% of the worldâ€™s energy consumption. Methods to purify chemicals that are more energy efficient could, if applied to the U.S. petroleum, chemical and paper manufacturing sectors alone, save 100 million tonnes of CO2 emissions and US$4 billion in energy costs annually [1]. Membrane-based separation methods, or other non-thermal ones, can be an order of magnitude more energy efficient than heat-driven separations that use distillation, but these methods are underdeveloped [2].

Given the recent advances in organic solvent nanofiltration (OSN) membranes [3], the focus of this presentation is on OSN membrane cascades for separating organic/hydrocarbon mixtures. As a case study, we investigate the design of a lab-scale membrane cascade for purifying a binary mixture of heptane and hexadecane, with a 75-25 weight fraction. The main objective is to maximize heptane purity in the permeate stream by using a maximum of 10 identical membrane units, each having an effective area of 14 cm.

In a first step, we use a superstructure optimization approach based on mixed-integer nonlinear programming (MINLP) to determine the most promising cascade configurations, in terms of the number of stages in the cascade, the number of membrane units in each stage, and the interconnections thereof, along with certain operating parameters [4,5]. Constraints are imposed for limiting the transmembrane pressure drop and the stage-cut in each unit. The permeate and retentate flux and composition for each membrane unit are predicted by a classical solution-diffusion model, where the permeability coefficients for heptane and hexadecane are estimated based on dedicated experiments for the membranes at hand, and the activity coefficients of heptane and hexadecane are approximated using the UNIFAC method for different mixture compositions. An optimal 3-stage cascade obtained with this approach is shown on the figure below, with GAMS 24.5.6 and the global optimizer BARON 15.9 used to solve the MINLP problem.

In a second step, we implement the optimised OSN membrane cascades experimentally in order to verify the predictions. For the 3-stage cascade shown on the figure, the mismatch between the model predictions and the measurements remains small, within a few percents, thereby validating the model-based optimization approach.

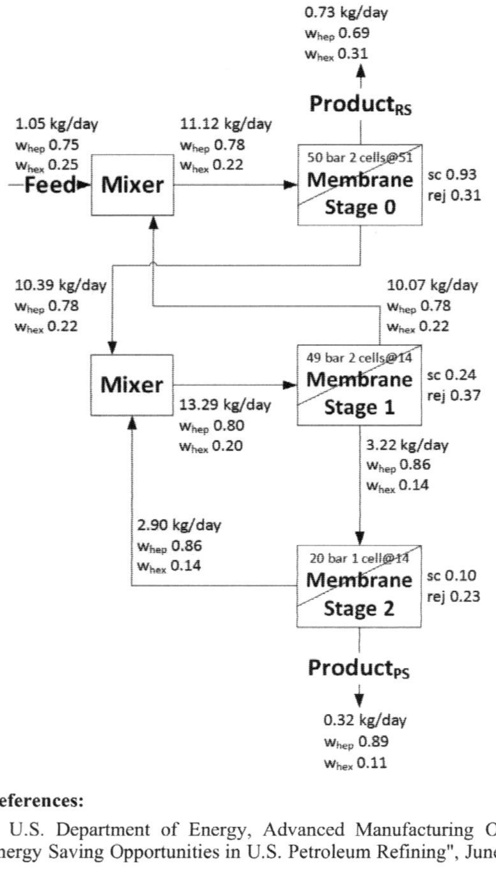

References:

1. U.S. Department of Energy, Advanced Manufacturing Office, "Bandwidth Study on Energy Use and Potential Energy Saving Opportunities in U.S. Petroleum Refining", June 2015.

2. D.S. Sholl, R.P. Lively, "Seven chemical separations to change the world," Nature 532:435-43, 28 April 2016. U.S. Petroleum Refining (US Dept. Energy, 2015).

3. P. Marchetti, M.F.J. Solomon, G. Szekely, A.G. Livingston, "Molecular separation with organic solvent nanofiltration: A critical review," Chemical Reviews 114:10735-10806, 2014.

4. R.H. Qi, M.A. Henson, "Membrane system design for multicomponent gas mixtures via mixed-integer nonlinear programming," Computers & Chemical Engineering 24:2719-2737, 2000.

5. V.S.K. Adi, M. Cook, L.G. Peeva, A.G. Livingston, B. Chachuat, "Optimization of OSN membrane cascades for separating organic mixtures," Proc. 26th European Symposium on Computer Aided Process Engineering (ESCAPE), 12-15 June 2016, PortoroÅ¾, Slovenia.

475310 Introduction to Altsep: Sustainable Separation Processes Roadmap Review

Abstract:

Creating a sustainable future requires sustainable manufacturing processes. This is especially true for the separation processes central to chemical manufacturing. Conventional fluid separations account for over a third of energy use and over half of the capital investment of chemical plants. To pave the way toward transforming the energy efficiency of fluid separations, the ACS Green Chemistry Institute (GCI) has partnered with AIChE to coordinate the development of a NIST-sponsored innovation roadmap. Based on a series of workshops, the draft AltSep roadmap has identified a number of research, development, and demonstration initiatives. These initiatives are critical for establishing a molecular property-based system for the industrial selection, simulation, and design of energy-efficient separation processes and accelerating the industrial application of such alternative separation processes.

A series of speakers will present on the following topics:

- Background and need for the roadmap,
- System to synthesize and design alternative separation processes, and
- Challenges to overcome barriers to adoption of sustainable separation solutions.

Following these presentations, feedback will be solicited from the audience to improve the quality of the roadmap findings.

460002 Computational Solid State Chemistry Applications in Pharmaceutical Industry

Abstract:

Given the complexity of the pharmaceutical solid state landscape and challenges facing the pharmaceutical industry, an accelerated Drug Development greatly benefits from guidance provided by computational methods. The emerging field of Computational Pharmaceutical Solid State Chemistry covers the whole spectrum of state of the art computational approaches, which are used to support all steps related to the development of solid state pharmaceuticals.[1] This work will consider virtual screening approaches allowing solvent selection for impurity purge, coformer selection for cocrystallization and solid solvate/hydrate formation screening.

1. Abramov, Y.A., Ed. *Computational Pharmaceutical Solid State Chemistry*, John Wiley & Sons, 2016.

461286 Process Optimization Using a Distillation Model Coupled with a Solubility Model for Increased Process Yield Following Multiple Distillations and Solvent Swap

Abstract:

Yield variability and extended distillation times lead to the need for process modeling. The combined distillation and solubility model predicts both the acetonitrile concentration and resulting yield loss in the mother liquors. The model was used to reduce yield loss and cycle time while providing additional potential for solvent recovery of acetonitrile. The distillation model was generated using Aspen properties with the system at 0.1 bar absolute pressure and Euler integration for a stepwise mass balance approach while the solubility model was generated using data regression from mother liquors. The solubility data was necessary to be used from mother liquor rather than in a pure system due to increased solubility of the intermediate while in the presence of the byproducts of the desired reaction. The model has predicted the acetonitrile composition to be within 0.03 mass fraction of the experimental data. The yield loss using the mother liquors model with modeled acetonitrile mass fraction resulted in a difference of no more than 1.30% yield over the 0.001 to 0.114 acetonitrile mass fraction range. The model was used to maximize yield and was verified on pilot plant scale.

464090 Large-Scale Crystallization of a Pure Metastable Polymorph By Reaction Coupling

Abstract:

A maximum of more than 40 g of acetaminophen Form II crystals with a yield of 65 mol % is made reproducibly in a 500 mL reactor by coupling the acetylation of *p*-aminophenol with the neutralization of acetic acid before the crystallization of acetaminophen Form II crystals. This novel working principle involves a sudden drop of the solubility of acetaminophen from the acetic acid–water environment to the acetate–water system in addition to temperature cooling but without agitation. This particular processing pathway is capable of maintaining a large supersaturation for quite some time for the system to enter the metastable zone of Form II with respect to the acetate–water system. The large amount of Form II crystals so produced are then isolated by filtration and oven drying to prevent Form II crystals in the mother liquor from transforming to the thermodynamically stable Form I crystals by the Ostwald's rule of stages.

461590 PAT Enabled Drying Platform for API Process Development â€“ Investigation of Solvent Trapping Associated with Form Transition and Crystal Collapse during Drying

Abstract:

An active pharmaceutical ingredient was isolated as a crystalline hydrate which converts to an amorphous solid during drying. Challenges associated with material handling and incomplete drying were identified at the factory scale. It was suspected that differences in processing equipment and conditions were leading to rapid loss of water, collapse of the crystalline lattice, and trapping of the process solvent at levels above specification.

A PAT enabled drying platform was used and advanced laboratory methodologies were developed to capture detailed process data and gain a fundamental understanding of the drying process, test various theories for solvent trapping, and evaluate drying from an alternative solvent system. Experiments performed led to increased process understanding, identification of conditions needed for successful drying, and ultimate process selection to improve process efficiency, robustness, and portability. The newly developed process was successfully demonstrated at pilot plant scale with performance in line with model projections. Results of this investigation and the methods applied will be described in detail during this presentation.

454769 Flow through Packed Bedsâ€¦an Inconvenient Truth

Abstract:

Flow Through Packed Bedsâ€¦An Inconvenient Truth

Jacob Arredondo, Jonathan Worstell, and Meng Zhang

Packed beds are a staple of the chemical industry, and as such, are a critical topic in undergraduate education. Students encounter packed beds in assignments for separations, kinetics, and laboratory courses; and they are usually drawn directly to the Ergun equation for answers.

The Ergun equation, while well known, generally gives poor results, especially for non-spherical packing. Furthermore, the Ergun equation provides no available mechanism for scaling up or scaling down of chemical processes.

In this presentation, we outline a simple undergraduate experiment that demonstrates to students the ineffectiveness of the Ergun equation and facilitates the development of a better solution using dimensional analysis. This assignment is designed to promote critical thinking and shows students that textbooks and academic papers are not always the definitive authority. Additionally, with no moving parts, the experimental apparatus is very robust, i.e. student resistant, in high volume teaching laboratories.

459994 Steal This Project: Entrepreneurial-Based Learning By Building Your Own Reactor

Abstract:

Quite often in reaction engineering education, students are rigorously trained in applying material and energy balances and reaction kinetics for the purposes of reactor selection, sizing, and operating parameters. In chemical engineering laboratory instruction, students are often given the chance to operate and test various reactors or are given the opportunity to experimentally measure reaction kinetics. However, due to a variety of factors, students are rarely afforded the opportunity to build a chemical reactor, which can be an exercise in creativity and safety. At Kettering University, a Keen partner, the reaction engineering laboratory curriculum was revised to incorporate an entrepreneurial-based learning project where student teams selected a chemical reaction of industrial significance and through a guided process designed and demonstrated the reactor. Keen provided grant funds to develop the curriculum and to procure kits of parts which the students can use to build tubular reactors. By the end of 2016 this project will be offered for four quarters. The purpose of this project is open-ended and an exercise in creativity and safe practices as a reaction engineer.

The student teams for this project are expected to research and identify a chemical reaction of potential commercial relevance they wish to evaluate. Due to safety, time, and cost certain limitations such as operating temperature, flammability, toxicity, and material compatibility were be considered while researching chemical reactions. Once a reaction is identified, the student team will propose it to the professor for a safety review and approval. The team will then design a chemical prototype system to evaluate the chemical reaction for possible production to market. The team will propose the design to the class and the professor for approval. The team will then construct the chemical reactor system and evaluate their chemical reaction. It also is required that the team determines a method to determine conversion of the reactants and quantify the products. The final deliverable for this project is an operating chemical reactor system and a PowerPoint presentation. Stretch goals for this project is the determination of the conversion and rate law.

How the project was implemented in the curriculum, rubrics, sample student work, lessons learned, and student survey data will be presented.

462100 Steal This Project: Solar-Thermal Conversion of Biomass to Methanol As an Introduction to Chemical Engineering Design

Abstract:

This project introduces the fundamentals of engineering design over the course of a semester by asking students to design a solar-thermal plant for the production of methanol from biomass. The project involves the gasification of lignocellulosic biomass at high-temperature to produce synthesis gas, which is converted and purified downstream. The high-temperatures are achieved via solarthermal heating during the day, and natural gas combustion at night, requiring students to design a hybrid reactor system.

This project introduces several topics in chemical engineering design that students have likely not encountered before such as life-cycle analysis and solids handling, in addition to common concepts such as reactor design and process heat integration. Students individually perform work on the project in three phases throughout the semester in the form of take-home exams as pertinent topics are introduced. During the final 6 weeks of the semester students break off into teams of 4 or 5 and complete the project. Teams are required to build three different simulations, two in Aspen HYSYS™ and one in Aspen Plus™, thus familiarizing the students with both software packages. Students must also incorporate textbook design calculations, kinetic modeling and cash flow analysis.

This presentation will describe the project, the unique way in which it was taught, and show sample results from students.

465960 The Virtual Sandbox: Particle Flow Physics Taught with Interactive Tools

Abstract:

Processes in the chemical industry involving particles and powders are (i) difficult to describe, and (ii) often pose a risk for process stability and safety. To guarantee process and product safety, process simulation, as well as the simulation of particle flow, became a main trend over the past decades. Accordingly, there is a strong need to cover the simulation of particulate systems in basic and advanced courses of particle technology and multiphase reactors. The virtual sandbox project focuses on the adoption of the augmented reality sandbox presented by Reed et.al. (2014) for teaching the basics of particle flow physics. It comprises a sandbox, a Kinect camera, and a simulation program. The sand surface is scanned, and loaded into the particle simulator LIGGGHTS®. The simulation results can be projected onto the sand to close to loop from reality to simulation, and back to reality.

The deeper scientific meaning of the project can be summarized as follows: the first key element is to teach the students the importance and comfort to simulate particle based processes, rather than performing experiments which suffer from the opaque nature of sand. The second element is to introduce the complexity and the effect of liquid bridges that exist between individual grains. Our idea is to guide the students through a set of models that are needed to describe these bridges: starting with models that describe the genesis of a pendular bridge (Wu et al., 2016), over the seminal work of Halsey and Levine (1998) on roughness effects, to the most recent developments in force models. Third, basics concepts of 3D vision are introduced, in order to make students aware of the enormous capabilities of modern camera systems. Last but not least, new teaching methods, e.g., inquiry-based learning, are developed and applied to the above topics: via additional small-scale and easy to reproduce experiments we make an attempt to illustrate the above scientific concepts even to high school students (age 16 to 18), as well as primary school pupils (age 6 to 10).

Acknowledgement
The project is founded by the Austrian Science Fund FWF WKP67.

References

T. Halsey, A. Levine, "How Sandcastles Fall", Physical Review Letters 1998, 80, 3141-3144.

S.E. Reed, L. Kreylos, S. Hsi, L.H. Kellogg, G. Schladow, M.B. Yikilmaz, H. Segale, J. Silverman, S. Yalowitz, and E. Sato, "Shaping Watersheds Exhibit: An Interactive, Augmented Reality Sandbox for Advancing Earth Science Education", Americal Geophysical Union, Fall Meeting 2014.

M. Wu, S. Radl, J.G. Khinast, A model to predict liquid bridge formation between wet particles based on direct numerical simulations, AIChE Journal 2016,62, 1877-1897.

471828 Process Controls Final Projects Inspired By Real Unit Operations Laboratory Modules

Abstract:

Northeastern University specializes in opportunities for experiential learning, both through a co-op program as well as through direct coursework. The style of experiential and inquiry-based learning have been particularly reinforced through two intermediate-level Unit Operations courses, in which students have been tasked with developing experiments, selecting measurement parameters, and designing analysis. Studentsâ€™ previous experiences have been utilized in the development of final projects for their Process Controls course.

In order to provide a culmination of the entire Process Controls course and to challenge students to design a complete controls system relying on some combination of feedback, feedforward, and/or cascade control, a final project was designed to connect the concept of Controls to a real-world system. Student teams were given the option of selecting from six different modules they had observed and analyzed in a previous laboratory course, including distillation, liquid-liquid extraction, heat exchangers, reverse osmosis, and water remediation. These modules had required complete manual control, and had a range of parameters that could be adjusted during operation. For their final project, students were challenged to design a complete control system that required them to select control variables, handle multiple disturbances, and operate within acceptable limits. Students derived overall balances, designed block diagrams, calculated tuning parameters both by hand and through simulation software, and analyzed their proposed control system for stability and economic viability.

The project culminated in both a technical written report and a presentation. Across three semesters and multiple sections, every team has designed a different system resulting in different quality of control. Requiring the students to defend their design decisions and forcing them to account for a range of disturbances has helped students develop a much more complete grasp of Process Controls and given them greater confidence exiting the course.

This presentation will provide examples of the modules and assessment of the project to date.

472364 Developing Professional Skills through a Service Learning Project on Aluminum Recycling

Abstract:

Given the volume of scientific and technical material that must be covered and assessed in most standard core chemical engineering courses, addressing the softer educational objectives required for ABET accreditation often poses a challenge for instructors. These objectives include teamwork experience, presentation skills, and an awareness of social responsibility and the global implications of engineering decisions. In this project, students were provided an opportunity to develop those professional skills by applying basic technical knowledge to understand and influence a lifestyle choice: recycling aluminum. The students were enrolled in a sophomore-level material and energy balances course. In a semester-long project, the students worked on teams to learn about aluminum refining and recycling technology from a chemical engineering perspective. Using skills learned in the course material, they were able to perform basic calculations that gave them a grasp of the difference in energy requirements between the two processes. They also participated in a service project collecting and recycling cans during tailgate events at football games, which provided them with practical experience of local recycling infrastructure. Finally, they implemented their learning to prepare a presentation for lay audiences regarding the energy requirements and environmental effects of aluminum refining. Each group delivered their presentation to a different external audience. Student surveys reveal that their teamwork, management, and communication skills were enhanced through the project, as was their cognizance of how engineering can relate to societal challenges. Students also reported that the project helped them better connect to the technical course material by relating it to a real process whose broader implications they could understand.

472416 Case Studies for Vertically Integrated Curriculum Projects

Abstract:

Often, students are unable or unwilling to â€œconnect the dotsâ€ when it comes to various courses in the chemical engineering curriculum. To combat courses being interpreted as isolated or unrelated to one another, U.C. San Diego has begun experimenting with a curriculum-wide course project. In this project, students are exposed to a single industrial-scale chemical process and repeatedly analyze it from different perspectives in each class. Starting with material balances, expanding to complex energy/entropy/exergy balances in thermodynamics, to designing the reactors, and so on. In this discussion, several case studies already implemented as UCSD are discussed, including: methanol synthesis, the Haber-Bosch process, vinyl chloride production, and hydrodealkylation (HDA) of toluene. Specific focus will be attached to how the projects progressed and expanded from course to course.

472951 It's Not Stealing If It's Free: Practical Process Instrumentation, Data Analysis, and Control with Open Source Hardware and Software

Abstract:

Data acquisition, analysis, interpretation, and visualization skills aren't just for "data scientists"—all chemical engineers benefit from a strong practical foundation in data. Unfortunately, many undergraduates have a hard time connecting the abstract concepts from their computer science course requirements back to the core chemical engineering curriculum. I'll present a hands-on project-based course concept that reinforces chemical engineering fundamentals while emphasizing the incredible utility of computational tools in the practical context of PHYSICAL COMPUTING—coding motivated by the task of working with sensors and actuators and the challenge of gleaning insights from the messy world of real data.

467494 Cyclic Polyethylene Furanoate As a Monomer from Renewable Resources for Ring Opening Polymerization

Abstract:

Introduction

The replacement of oil based chemicals by renewable resource based chemicals is one of the major issues chemical industry is facing right now. The alternative polymer poly (ethylene furanoate), PEF, is a possible substitute for poly (ethylene terephthalate), PET. Among different process alternatives, PEF can be produced at high polymerization rates and molecular weights via ring opening polymerization of cyclic oligo (ethylene furanoate), cyOEF. On the other hand, the effective production of these cyclic oligomers at high purity is still the major obstacle towards the process industrialization. In this work, different strategies aimed to produce these cyclic molecules from the two monomers Furan dicarboxylic acid (FDCA) and ethylene glycol (EG) are explored. And a model is presented to describe the equilibrium behavior of said cyclic oligomers.

Three pathways towards cyOEF

The three synthetic strategies shown in figure 1 have been investigated. In all cases, high dilutions in organic solvents are required to achieve high yields; the final product is a distribution of cyclic oligomers with different number of repeating units.

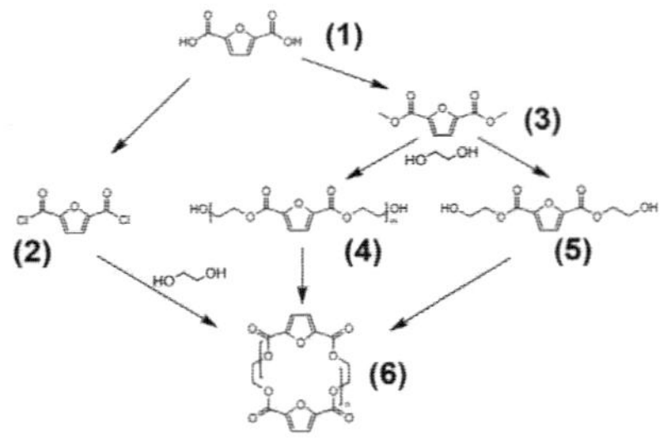

Figure 1 Reaction pathways for the production of cyOEF.

The first (scheme 1-2-6) is a fast cycle formation using the rapid esterification of acid chlorides with alcohols. Moderate selectivity towards cycles of around 50% are in contrast to a complex reaction setup and toxic chemicals involved for the chlorination. The product size distribution is kinetically controlled and consists mainly of C3, with C2 and C4 being the second most abundant species.

The depolymerization (scheme 1-3-4-6) and reactive distillation (scheme 1-3-5-6) are the two more promising routes towards cycles. For depolymerization FDCA and EG are converted to oligomeric chains of around 12 repeating units. This prepolymer is subsequently diluted in a high boiling organic solvent and converted to a mixture of cyclics and linears at 180-200 °C. In case of reactive distillation the FDCA and EG (in excess) are converted to very short chains (1-2 repeating units). The chains are then diluted in a high boiling organic solvent and finally converted to cyOEF by distilling a portion of the solvent together with the excess EG.

Both routes yield a similar distribution of cyclic species. Possible purification routes include precipitation and adsorption processes, which have both been applied to yield cycles with high purity starting from linear species (>99% by HPLC area). The yield of both reactions lies in the range of 60 to 80 % depending on the process conditions. One striking feature of this type of reaction is that it is possible to recycle the solvent and the linear side products in order to increase the yield of the process and furthermore make it more sustainable.

Modeling of the ring chain equilibrium

The equilibrium composition of cyclics and linears in 2-Methylnaphthalene and Dichlorobenzene (DCB) could be modeled using the Jacobson-Stockmayer model (Figure 2). Under high dilutions this equilibrium shifts towards the formation of cyclic oligomers. It is based on computing the probability of ring closure using the assumption of randomly coiled hydrocarbon chains for linear polymers. Two fitting parameters are involved: one is the equilibrium constant for the step growth polymerization, which is in equilibrium with the cyclisation reactions, and the second one is the effective length of the bonds in the polymer chains.

As shown in the figure below, our own results from depolymerization and reactive distillation experiments were fitted using the model. These results suggest that it is indeed possible to use this approach to describe the ring chain equilibrium. The model is currently under refinement in order to describe system-specific effects such as the complete absence of cycles with six repeating units.

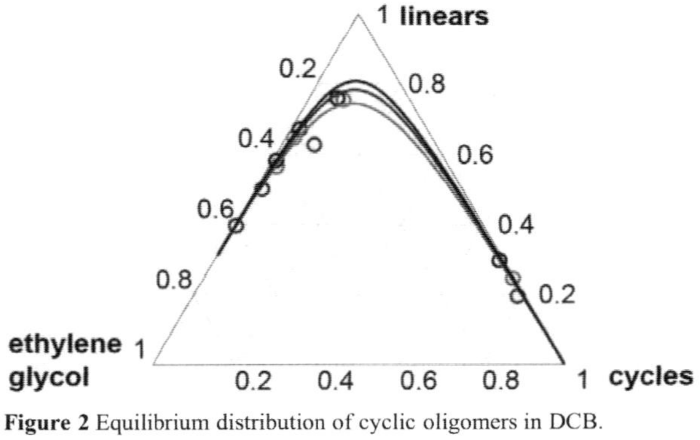

Figure 2 Equilibrium distribution of cyclic oligomers in DCB.

References

[1] Gomes M, Gandini A, *J. Polym. Sci., Part A: Polym. Chem.*, 49, 3759-3768 (2011).

[2] Burch, RR, Lustig, SR, Spinu, M *Macromolecules*, 33, 5053-5064 (2000).

[3] Stockmayer, WH, Jacobson H, *J. Chem. Phys.*, 18, 1600-1606 (1950).

453695 Polymeric Nano-Metal Composite Membranes for Water Remediation

Abstract:

Responsive polymers react to external stimuli like temperature, pH, radiation, ionic strength or chemical composition. These polymers in combination with nanomaterials provide novel hybrid properties such as bactericidal, electrical, hydrophilic, catalytic, selectivity, etc., and they can be self-supported or in a convenient matrix such as a membrane.

These membranes, functionalized with polymers having pH responsiveness (hydrogels), is a promising research field for numerous applications. The functional groups of the hydrogels gain or lose protons based on the pH, collapsing or extending their network, making them swell or un-swell solutions. These nanostructured materials can be exploited to synthesize metal nanoparticles and as ion exchangers. They also prevent the loss of nanomaterials to the environment and improve reactivity due to larger surface areas, expanding its range of applications to selective catalysis, remediation and mining.

The present work describes "green" techniques used to create nanocomposites based on polyvinilidene fluoride (PVDF) hollow fiber (HF) and flat sheet (FS) membranes, both sponge-like and thin. Hydrophobic HF membranes were synthesized with 57.2% porosity and 0.72 μm mean pore size. These membranes were then hydrophilized by a water-based process of cross-linking polyvinylpyrrolidone (PVP) by persulfate (PS). Though, hydrophilization of PVDF membranes using PVP is known in industry, as a post-process has only been implemented at lab-scale.

The PVP cross-linking is confirmed but also a ring-opening is present due to etherification. Hydrophilization was measured by contact angle changes (CA) (76.3° to 60.3°) and surface free energy calculations (40.3 to 49.8 $mJ{\cdot}m^{-2}$). HF and FS were subsequently functionalized with a poly(acrylic acid) (PAA) hydrogel through free radical polymerizations (thermal or redox) using PS as initiator. This work advanced FS membrane

functionalization from lab-scale to full-scale by modifications of the polymerization procedures.

Swelling tests were performed to hydrogels/functionalized membranes finding responsiveness up to 1676% swelling at pH 9.0. Also, mesh size values (ξ) were calculated showing an inverse relationship with membrane permeability, tuning the flux through the membrane. An increase of ξ from 5.89 to 8.33 nm (pH 3.0 to 9.0) represented a 93% permeability decrease (53.9 to 3.4 $L{\cdot}m^{-2}{\cdot}h^{-1}$) at pressures between 1.0 and 4.0 bar.

Throughout a double ion exchange of sodium/iron and a subsequent reduction, bimetallic Fe/Pd nanoparticles (60 to 300 nm) were synthesized. The Fe exchanged was almost 500 mg Fe/g PAA. Similarly, it was possible to use the exhausted accelerants of Fe ions of the redox polymerization to synthesize nanoparticles, but having only 9.2 mg Fe/g PAA of loading.

Finally, depending on the Pd concentration in the membrane, the catalytic activity in the reduction of trichloroethylene (TCE) was studied, having enhanced reactivities (up to ten times) in sponge-like membranes compared to previous studies. Using hydrogels alone, similar results to nanoparticles in solution phase were found without nanoparticle loss. 80% of reduction was achieved within 2 hours with chloride production near to stoichiometric values (3/1), demonstrating absence of intermediates.

This research is supported by the National Science Foundation EPSCoR, and by the National Institute of Environmental Health Sciences (P42ES007380). HF membranes were synthesized at the Singapore Membrane Technology Center, NTU (Singapore). Nanostone Water (Oceanside, CA) provided support and collaboration in the development of full-scale flat sheet membranes.

454787 Thermodynamic and Economic Assessment of the Production of Ethylene and Propylene from Bioethanol

Abstract:

Introduction

The increasing worldwide pollution and the environmental problems associated with the petroleum extraction and use, have led to the development of new energy resources and sustainable raw materials [1]. The production of bioethanol from biomass offers a sustainable biotechnology solution at competitive costs with high availability. In Colombia (South America), bioethanol production has increased because the local policies encourage its production, being the third producer in America [2]. However, the focus on bioethanol production has been related to its use as biofuel. Alternatively, there are some high-commercial-value compounds —mostly produced from petroleum— that can also be obtained from renewable sources [3]. Recently, bioethanol has been proposed as a source for producing commodities such as light olefins in the framework of the biorefinery concept [4]. Among these biomaterials, ethylene stands out as one of the most produced chemicals worldwide, having a projected production of 400 million ton in 2020 [1]. In addition, the demand for propylene is growing faster, having a projected production of 100 million ton in 2021[5]. Therefore, this study aims to propose a conceptual design for producing ethylene and propylene from bioethanol samples procured in Colombia, integrating the reaction and separation stages and assessing its technical and economic feasibility. Aspen Plus and Aspen Hysys were used as process simulators.

Methodology

Aspen Plus V7.3 and Aspen Hysys V7.3 (Bedford, USA) were used as process simulators throughout the study. In both, the Soave-Redlich-Kwong (SRK) equation of state was used as thermodynamic model. First, the bioethanol dehydration reaction was analyzed in the RGibbs reactor module in Aspen Plus. Water/ethanol (S/E) feed ratio and reaction temperature were varied in the module in order to assess their effect on products selectivity. Ethylene, propylene and *n*-butene were selected as possible products, and a selectivity response surface was obtained in Matlab R 2014 (Salt Lake City, USA). Aspen Hysys was used to perform the sizing of the process units, heat integration, and process yields. Products purification was modeled in the Distillation Column module and the energetic analysis was performed using Heat Exchanger and Heaters/Coolers modules in Aspen Hysys. Finally, a class-V-order economic evaluation was done to the final flowsheets by determining the capital expenses for all units in the processes [6].

Results and analysis

Figure 1a shows the surface response for ethylene selectivity during bioethanol dehydration simulation, when varying S/E ratio and temperature. The higher the temperature and S/E ratio, the higher the ethylene yield, with a selectivity exceeding 60% beyond S/E = 5:1 and 600 °C. On the other hand, Figure 1b shows a maximum peak of 45% selectivity for propylene at 400 °C, and a slightly positive effect of the S/E ratio. *n*-butene formation also presented a similar trend (Figure 1c), showing a maximum selectivity of 14% at 300 °C. Heavier olefins are formed by ethylene condensation, which is thermodynamically favored at lower temperatures [7]. However, the absence of ethylene at these lower temperatures reduces the amount of this compounds.

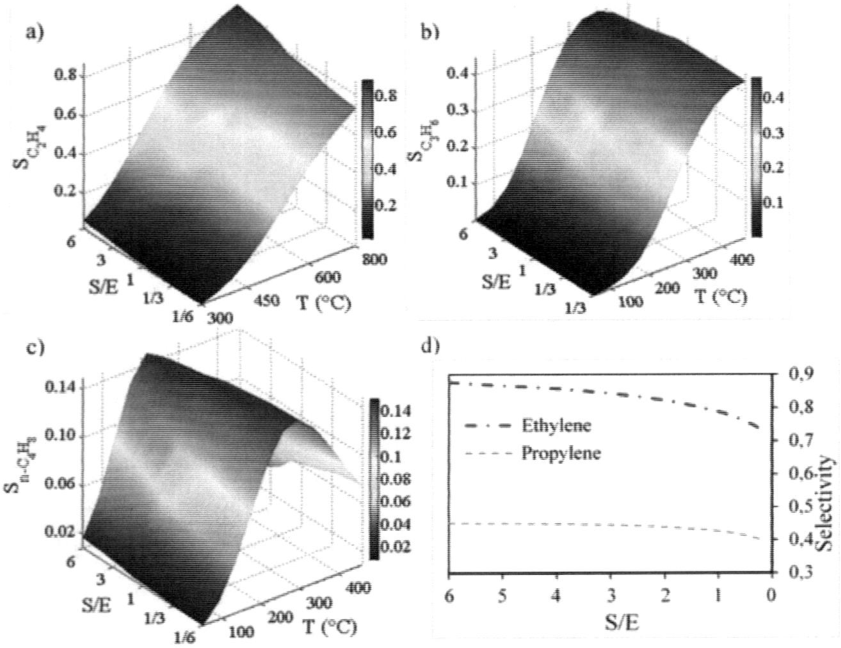

Figure 1. Matlab response surface from Aspen Plus simulated data for the production of (a) ethylene, (b) propylene, and (c) *n*-butene using different water:ethanol ratios (S/E) in the temperature range of 200–800 °C; and (d) the effect of the S/E ratio on the maximum ethylene and propylene yields.

Figure 1d shows the effect of the S/E on the maximum both ethylene and propylene selectivity. The excess of water is slightly positive for the ethylene formation because its acts as a quasi-inert diluent from a thermodynamic point of view [7]. In this way, a bioethanol dehydration reactor feed with a stream of S/E = 5 and operating at 615 °C was selected as the proper design to produce ethylene. In addition, a dehydration reactor operating at 400 °C was selected to produce propylene. These reactors were included in two different processes to analyze the conceptual design of two plants, one producing ethylene and another producing propylene from bioethanol.

Figure 2 shows the designed processes for producing (a) ethylene and (b) propylene from bioethanol. The processes begin with the feed stream of 300,000 ton/year of raw, fermented bioethanol (5 mol% ethanol, S1), which corresponds to the 10% of the national bioethanol production. This stream is heated to 40 °C before its further introduction to a flash separator V-100, which operates at 95 °C and 1 atm. These conditions were established according to the ethanol-water equilibrium to obtain an S/E = 5:1 (16.4 mol% ethanol). After V-100, the product stream rich in ethanol (S3) is heated to either 615 °C or 400 °C (reaction temperatures for ethylene or propylene production, respectively) by a recycle of the flash bottoms stream (S4). S4 is rich in water and is used to the process heat integration. The product streams (S7 in Figure 2a and S6 in Figure 2b) are pressurized to 19 atm in the compressor K-100. The pressurized streams are cooled to 18 °C by fresh water streams. These streams would deliver the energy duty for heating streams (S20 in Figure 2a and S19 in Figure 2b). Subsequently, the cooled streams are used as an inlet in a three-phases flash separator V-101 with the purpose of removing the residual water after the reaction. Finally, the product streams are introduced into distillation columns to separate the olefins.

The ethylene process delivered a stream of 99.7 mol% ethylene with an almost 100% recovery (S16 in Figure 2a), producing 0.60 mol ethylene/mol ethanol feed. In addition, propylene process delivers a stream with 92.0 mol% propylene in S15 in Figure 2b. This corresponds to the 70% of the propylene entering to the distillation tower, producing 0.43 mol propylene/mol ethanol feed.

A class-V economic assessment delivered investments of USD 2 million for producing ethylene and USD 2.23 million for propylene. These costs represent 159 USD/ton-year of ethylene and 327 USD/ton-year for propylene. True [8] reported investment costs of 1300 USD/ton-year for ethylene and 1450 USD/ton-year for propyylene produced from petroleum for a plant constructed in 2016. However, these costs include lumpsum, engineering, procurement, construction and commissioning turnkey contract; which not were included in the current analysis. Nevertheless, this class-V economic assessment shows that olefins production from bioethanol could be a profitable and sustainable alternative to petrochemical processes.

a)

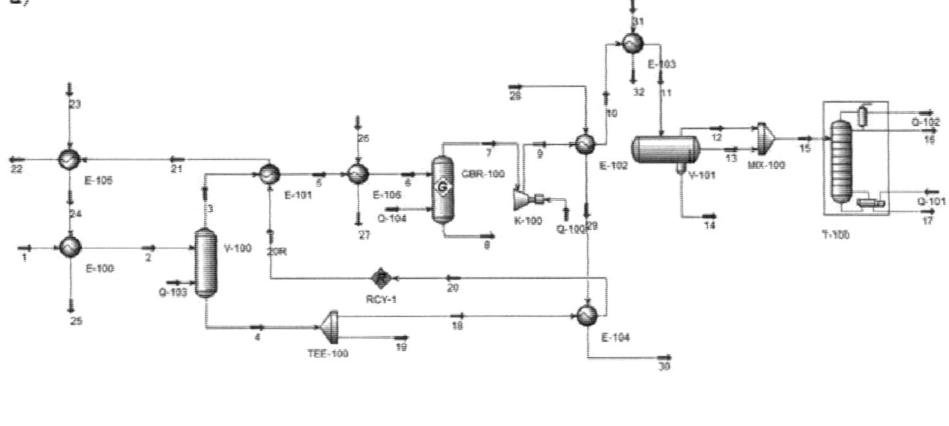

b)

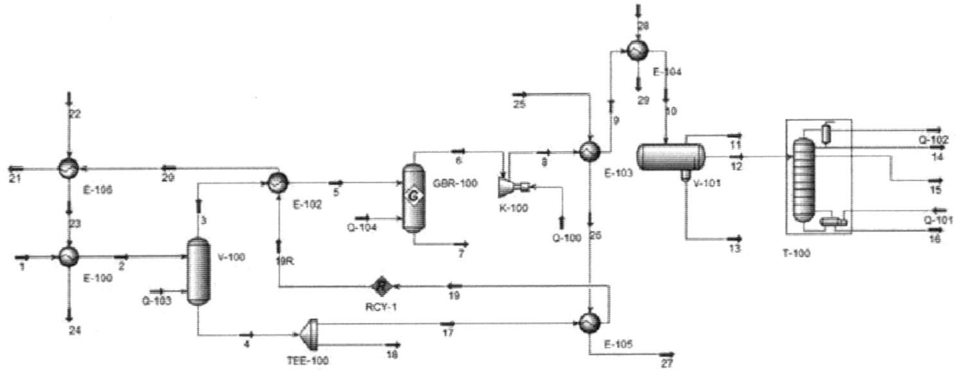

Figure 2. Process flowsheets designed in Aspen Hysys for producing (a) ethylene and (b) propylene from bioethanol. Blue lines represent mass streams and red lines represent energy streams. Equipment notation: E: Heat exchanger, V: Flash separator, K: Compressor, MIX: Mixer, T: Distillation tower, TEE: Splitter, and GBR: Gibbs reactor.

Conclusion

This study evaluated a conceptual design for producing light olefins from bioethanol base on thermodynamic predictions and using commercial software. To the best of our knowledge, it is the first time that reaction and separation stages are integrated and economically evaluated in two different processes, one to produce ethylene and another to produce propylene from bioethanol. Due to water showed a positive effect on the ethanol dehydration reaction, both processes start with a simple flash separator operating at 95 °C and 1 atm to reach the ethanol initial concentration needed in the reactors (S/E=5, 17.6 mol% ethanol). Two different reactors were designed, one operating at 615 °C for producing ethylene and another at 450 °C for producing propylene. The designed ethylene process delivered a stream of 99.7 mol% ethylene with a 60% yield and an investment cost of USD 2 million, equivalent to 159 USD/ton-year of ethylene. On the other hand, propylene production from bioethanol delivers a stream with 92 mol% propylene and 43% yield. Propylene process has an investment cost of USD 2.23 million (327 USD/ton-year propylene). These conceptual analysis show the technical, environmental and economic feasibility of the biorefinery concept development.

References

[1] H. Xin, X. Li, Y. Fang, X. Yi, W. Hu, Y. Chu, et al., J. Catal. 312 (2014) 204–215.

[2] G.P. Ortegón, F.M. Arboleda, L. Candela, K. Tamoh, J. Valdes-Abellan, Sci. Total Environ. 539 (2016) 410–9.

[3] O.J. Sánchez, C.A. Cardona, Bioresour. Technol. 99 (2008) 5270–95.

[4] R. Le Van Mao, T.M. Nguyen, G.P. McLaughlin, Appl. Catal. 48 (1989) 265–277.

[5] Ceresana, Market study: Propylene, (2014). http://www.ceresana.com/en (accessed May 1, 2016).

[6] American Association of Cost Engineers, (2015). http://www.aacei.org/ (accessed November 15, 2015).

[7] T. Lehmann, A. Seidel-Morgenstern, Chem. Eng. J. 242 (2014) 422–432.

[8] W. True, Oil Gas J. 110 (2012).

458322 Design and Economical Evaluation of Polygen Process to Co-Produce Synthetic Natural Gas (SNG), Methanol and Ethylene Glycol

Abstract:

Design and Economical Evaluation of POLYGEN process to co-produce Synthetic Natural Gas (SNG), Methanol and Ethylene Glycol

Bor-Yih Yu and I-Lung Chien*

Department of Chemical Engineering, National Taiwan University, Taipei 106, Taiwan

The steady state design and economical evaluation of poly-generation (POLYGEN) process to co-produce synthetic natural gas (SNG), methanol and ethylene glycol (EG) is studied in this work. POLYGEN has been a widely studied topic recently, in which several different chemicals can be produced parallel at the same time. In POLYGEN process, the relative production rates for different chemicals can be adjusted based on different market demands, daily usages, and also changing political strategies. This character enhances the flexibility of plant-wide operation and also the economic performance. Among the products, SNG holds very similar composition and heat value to typical natural gas, and can be used as a replacement in industrial and home usages. Methanol is one of the most important industrial feedstocks, which can be the basis of various kinds of chemicals. Ethylene glycol is also an important organic chemical, which can be used to synthesize polyesters, antifreezes, solvents, and so on.

The items covered by this work are illustrated in the following. In coal-based POLYGEN processes, syngas is first generated from coal gasification in a CO2-rich atmosphere, and the raw syngas is then passes the treating stage. Syngas treating stages contains two section, namely, sour water-gas-shift-reaction (SWGSR) and acid-gas-removal (AGR). In SWGSR, excess amount of steam is added to adjust the H2/CO ratio. After that it is cooled to low temperature to enter AGR. In AGR, SELEXOL method is used to capture 99.7% of H2S and 92% CO2. Part of the captured CO2 is sent back to gasification section in order to reduce the carbon emission, and the other part is compressed and sequestrated. With proper arrangement in configuration, three streams with qualified composition to be the feedstocks for downstream chemical plants are produced.

For downstream chemical plants, the products produced will meet the industrial standard. The SNG has heat value higher than 965 (BTU/scf), the methanol purity is 99.5 (mol%), and EG purity is 99.8 (mol%). EG is traditionally synthesized through petroleum route, a novel route from coal-derived syngas has been widely considered a potential alternative as the oil shortages continuing. Thus in this work, the production of EG from syngas is adopted. Firstly, CO and H2 in syngas are separated, and are used in different reaction stages. In the first stage, CO is reacted to form dimethyl oxalate (DMO), and in second stage, DMO is further reacted with H2 to form ethylene glycol.

After the design, optimization works are studied based on minimized total annual cost (TAC), and the economical evaluation is performed to investigate the overall economic performance. Besides, some waste heat could be recovered by generating steam, and the steam is used to drive turbine to produce electricity for usage inside the plant-wide process. Thus through this work, a better understanding of POLYGEN process, and the design and operation of coal to ethylene glycol route is reached.

458992 Preparation and Properties of Soybean Oil-Based Composites Containing Natural Fillers

Abstract:

Biobased composites have been prepared, which consist of a vegetable-oil based matrix, with ground nutshell or eggshell as natural, rigid, filler materials.

Crosslinked polymers of acrylated epoxidized soybean oil (AESO) and styrene were first successfully produced across the entire composition range via bulk free radical polymerization. Mechanically, these copolymers varied from hard and rigid at high styrene contents, to soft and rubbery at high AESO contents.

The intermediate 50:50 copolymer was chosen as the matrix for composites containing varying amounts of pecan nutshell flour or powdered eggshells. Composites were produced by pre-blending the fillers into the comonomer mixture before reaction and solidification. Through the AESO and filler contributions, these composites contain substantial fractions of biobased starting materials.

Copolymers and their composites exhibited good structural integrity, and were characterized by flexural testing to fracture, dynamic mechanical analysis, SEM imaging, as well as hardness and impact testing.

Addition of the natural fillers to the matrix resulted in substantial increases in hardness and stiffness, with a nearly six-fold increase in modulus observed at 35 volume% eggshell filler. These increases were accompanied by modest decreases in breaking strength and impact energy, and no observed effect on the glass transition temperature of the matrix. Use of coupling agents [1] to increase adhesion and potentially further modify mechanical properties will be discussed.

General comparison of the filled polymer properties to a baseline of neat polystyrene shows improved or similar individual mechanical characteristics, indicating viability of such biobased composites as a replacement material.

[1] Nachtigall, S.M.B., Cerveira G.S., Rosa S.M.L., Polymer Testing, 26(5), 2007.

460812 Solvent Free Sucrose Esters Production in Reactive Systems Containing Emulsifiers

Abstract:

Surfactants are chemicals used in a wide variety of applications, mainly in the manufacture of consumer goods, industrial cleaners, food-feed, and cosmetic products. In the past, efforts have been made to develop biobased surfactants from renewable raw materials in order to reduce the environmental impact of traditional petroleum-based surfactants. Currently, most biobased surfactants available in the market are lipids-based, but still they contain polar moieties generated from fossil resources (e. g. sulfonic acid, ammonium, or poly-ethoxylated groups) (1). In contrast, in completely biobased surfactants (e. g. acylglycerols, alkyl-poly-glucosides, glycosides of fatty acids) both, the polar and non-polar structures are of renewable origin.

Among biobased surfactants, fatty acid sucrose esters stand out as important additives for the food industry (food ingredient number: E473). These are generally produced by transesterification of sucrose (from sugar cane) with fatty acid methyl esters (from vegetable oils). In addition to their renewable character, sucroesters are high value added, biodegradable and biocompatible surfactants, which can also be used in specialized applications such as cosmetics, clean and care products and pharmaceuticals. Recent estimations (2) indicate that the current market of sucroesters is around than 6000 t/year, with prices around 4 to 20 USD/kg depending on the nature of the product (i.e. lipid chain, degree of substitution, etc).

Sucrose is a polyalcohol with eight hydroxyl groups (See Figure 1), so theoretically, sucrose esters could be monoesters, diesters, triesters, and, so on (up to octaesters). As a result, tuning of reaction selectivity to a specific product is a major challenge. Besides, sucroesters properties and application are heavily determined by the degree of esterification. For example, sucrose monoesters are good water-in-oil emulsifiers, while sucrose octaesters are good oil-in-water emulsifiers. In general, the degree of substitution depends of different process conditions such as temperature, reactants molar ratio, catalyst nature and concentration, and the use of other additives (e. g. solvents, emulsifiers).

In addition to controlling the degree of substitution, a major challenge of the process is the incompatibility between reactants; only 0.1 g of sucrose can be dissolved in 100g of FAME at 140°C (3). This reduced solubility limits the mass transfer and reduces the reaction rate and conversion. A processing alternative to overcome this problem is the use of emulsifiers, which enhance contact and solubility of sucrose in FAMEs (4–7). Some reports indicate that fatty acid soaps (mono- and divalent), sucrose esters (the product itself), and mono- and diacylglycerols can be used as emulsifiers for the reaction. However, it is necessary to study the influence of such emulsifiers (nature and concentration) on the transesterification process (i.e. rate of reaction and esterification degree).

In this work, conversion and selectivity of transesterification reaction to produce sucroesters was experimentally evaluated using three different emulsifiers (potassium palmitate, sucrose palmitate and glyceryl monoester) at different concentrations (up to 20%w). Reactions were carried out in a 150 ml glass-jacketed reactor coupled to a methanol total condenser. The system was homogenized with an Ultraturrax® at 15000 rpm prior to reactions, and maintained under agitation at 600 rpm at constant temperature (110-140 °C) until no further evolution of methanol was observed. FAMEs consumption and sucroesters production were tracked by HPLC analysis of samples. Observed conversion were around 40% and mono substitution increased with the concentration of the emulsifying agents. Higher conversions were obtained when potassium palmitate was used as emulsifier, and reaction times were shorter when emulsifier's concentration was increased. Degree of substitution of the final product was higher when a mixture mono- and diacylglycerols was used as emulsifier and lower when sucrose palmitate was used. Results of this study can be used for preliminary process design in a solvent-free production of sucrose esters with different emulsifying properties.

Figure 1. Emulsifiers used to provide contact in sucrose and fatty methyl estermixtures (Molecular structures taken from ChemSpider)

References

1. HAYES, Douglas G. ***Biobased Surfactants : A Useful Biorefinery Product That Can Be Prepared Using Green Manufacturing***. Knoxville, TN, USA, 2012.

2. OTOMO, Naoya. Basic properties of sucrose fatty acid esters and their applications. In : HAYES, Douglas G., KITAMOTO, Dai, SOLAIMAN, Daniel K. Y. and ASHBY, Richard D. (eds.), ***Biobased Surfactants and Detergents***. Urbana, Illinois : AOCS Press, 2009.

3. ZHAO, Rongrong, CHANG, Zhidong, JIN, Qiong, LI, Wenjun, DONG, Bin and MIAO, Xiaowen. Heterogeneous base catalytic transesterification synthesis of sucrose ester and parallel reaction control. ***International Journal of Food Science & Technology***. 2014. Vol. 49, no. 3, p. 854–860. DOI 10.1111/ijfs.12376.

4. PARKER, Kenneth J., JAMES, K. and HURFORD, J. Sucrose Ester Surfactants — A Solventless Process and the Products Thereof. In : HICKSON, John L. (ed.), ***Sucrochemistry***. Washington, DC : American Chemical Society, 1977. p. 97–114.

5. FITREMANN, Juliette, QUENEAU, Yves, MAITRE, Jean-paul and BOUCHU, Alain. Co-melting of solid sucrose and multivalent cation soaps for solvent-free synthesis of sucrose esters. ***Tetrahedron Letters***. 2007. Vol. 48, p. 4111–4114. DOI 10.1016/j.tetlet.2007.04.015.

6. GALLEYMORE, Harry R, JAMES, Kenneth, JONES, Haydn F and BHARDWAJ, Chaman L. Process for the production of a surfactant containing sucrose esters. 4298730. 1981. United States.

7. RIZZI, George P and TAYLOR, Harry M. Synthesis of higher polyol fatty acid polyesters. 3963699. 1976. United States.

461686 Synthesis and Characterization of Biochar-Based Carbon Supported Metal Nanoparticles

Abstract:

Synthesis and characterization of biochar-based carbon supported metal nanoparticles

Sai Teja Neeli and Hema Ramsurn

Russell School of Chemical Engineering, University of Tulsa, Tulsa, OK 74104

Biochar is a carbon-rich solid formed after biomass has been carbonized. It is also a residue from a number of thermochemical conversion processes whereby biomass is converted to bio-oil and/or syngas. Therefore, this material is environmental-friendly and carbon-neutral. Traditionally, biochar has been used as soil amendments for the healthy growth of plants due to its retention capability of nutrients. This investigation aims to use biochar in a different manner: as a support for metal catalysts to promote green chemistry and engineering. The biochar used in this study is obtained from hydrothermal carbonization of model compounds of biomass namely lignin, cellulose and hemicellulose. This means that high pressure and temperature water is used as a benign solvent to depolymerize the model compounds to biochar. The biochar so produced is chemically activated with 0.1 M HNO_3 solution resulting in activated carbon with defined micropore size distribution, high micropore volumes and large specific surface areas. Metal precursors (like iron nitrate) are then added to the biochar and dried in a convection oven at 110 ^{o}C for one week to obtain metal (iron) impregnated biochar. The dried mixture is thermally-treated in a quartz tubular reactor at 1000 ^{o}C for one hour in a nitrogen flow of 500 ml/min. These synthesized biochar based metal catalysts will then be characterized using a number of techniques including scanning electron microscope (SEM), x-ray powder diffraction (XRD), inductively coupled plasma atomic emission spectroscopy (ICP-AES) and Brunauer-Emmett-Teller (BET)

surface area analyzer. Other metal catalysts such as platinum, gold and nickel can also be impregnated on the biochar in the same manner. It is expected to observe distinct XRD patterns corresponding to graphite for the thermally treated metal-impregnated biochar, due to the conversion of the carbonaceous materials to graphite. SEM images will help not only in sizing the metal nanospheres (about 20 -60 nm) but also in determining how the metal is supported: half embedded or completely enclosed in the porous char matrix. Based on the characterization results, these metal-impregnated biochar catalyst could be tested for a number of reactions. These supports can withstand high temperatures and can even be used for gasification reactions and upgrading bio-oil during hydrogenation reactions. There is also the possibility of using bifunctional metal/acid catalysts on the biochar nanoparticles for dehydration and hydrogenation reactions and for aqueous phase processing to produce selectively targeted alkanes from glucose. The ultimate goal is to use these green catalysts to upgrade the oxygenated biofuels by removing the oxygen through various mechanisms and hence increasing the quality and heating value of the desired product.

464335 A Continuous Crystallization Process to Separate Mixtures of Enantiomers Exploiting Two Coupled Fluidized Bed Crystallizers

Abstract:

Crystallization is one of the most effective ways to separate mixtures of enantiomers. So far, primarily batch processes are used. In particular, Preferential Crystallization (PC) has proven to be applicable in batch processes to crystallize selectively pure enantiomers from conglomerate forming racemic mixtures [1]. This contribution presents results of applying a fluidized bed crystallizer configuration for continuous PC of pure enantiomers from a racemic mother liquor [2]. To crystallize both enantiomers simultaneously two fluidized bed crystallizers are coupled via their liquid phases. Saturated racemic mother liquor is supplied from a feed tank with a specific flow rate into the bottom sections of the two double jacketed fluidized bed crystallizers. To initiate the crystallization process in both crystallizers seeds of the opposite enantiomers are added. The preferential growth of the seed crystals reduces the supersaturation in both vessels. In order to allow for continuous production of crystals with a defined size, the lower sections of both crystallizers are conical to offer the opportunity of classified product removal at a certain height. The depleted solutions leaving both crystallizers are returned into the feed tank. In this way always racemic solution is supplied to both crystallizers. Nuclei of the counter enantiomer and other small particles, which have a negative effect on the purity of the product, are taken out with the fluid phases at the tops of the two crystallizers and subsequently dissolved in the heated recycle lines. To assure that the process can be continued continuously, larger particles and formed agglomerates are withdrawn at the bottom, crushed and reinjected to provide permanently new seeds. As a proof of feasibility we will describe the set up developed [3] and present experimental results for the separation of the enantiomers of D/L-asparagine monohydrate [4]. The status of theoretically describing the process using detailed [5] and reduced models [6] will be discussed as well as challenges and opportunities for the separation of other chiral substances. The financial support of German Research Foundation (DFG) within the priority program "Dynamic simulation of interconnected solids processes DYNSIM-FP (SPP 1679)" is gratefully acknowledged.

1. H. Lorenz, A. Seidel-Morgenstern, Angewandte Chemie- International Edition, 53, 2014, 1218.

2. H.-H. Tung, E. L. Paul, M. Midler, J. A. McCauley, Crystallization of Organic Compounds: An Industrial Perspective, Wiley, Hoboken, New Jersey, 2009.

3. D.. Binev, H. Lorenz, A. Seidel-Morgenstern, Chemical Engineering Science 133, 2015, 116.

4. D. Binev, H. Lorenz, A. Seidel-Morgenstern, Crystal Growth & Design, 16, 2016, 1409.

5. K. Kerst, L. Medeiros de Souza, A. Bartz, A. Seidel-Morgenstern, G. Janiga, Computer Aided Chemical Engineering , 2015, 263.

6. M. Mangold, L. Feng, D. Khlopov, S, Palis, P. Benner, D. Binev., A. Seidel-Morgenstern, Journal of Computational and Applied Mathematics 289, 2015, 253.

466054 Tuning the Properties of Crystalline Particles â€“Temperature Cycling in a Continuous Tubular Crystallizer

Abstract:

Tuning the properties of crystalline particles –Temperature Cycling in a Continuous Tubular Crystallizer

P. **Neugebauer**[1], M. **Besenhard**[2], J.G. **Khinast**[1,2]

1 Graz University of Technology, Institute for Process and Particle Engineering, 8010 Graz, Austria

2 Research Center Pharmaceutical Engineering, 8010 Graz, Austria

Correspondence:

e-mail: peter.neugebauer@tugraz.at; tel: +43 (316) 873 - 30424

e-mail: khinast@tugraz.at; tel: +43 (316) 873 – 30400; fax: +43 (316) 873 - 1030400

The utility of tubular reactors for the continuous production of crystalline particles by now has been demonstrated for a wide range of different substrates. Mostly, these systems profit by the ease of supersaturation control by the installation of sections at different temperatures.

Experimental designs employing water baths or jackets have shown to be adequate to benefit from the high surface:volume ratio offered by tubular crystallizers, guaranteeing efficient heat flow from or to the crystallization slurry[1-3].

In the present study we established an easy-to-use platform which allows the modification of crystals from ingredients of pharmaceutical interest under controlled conditions. Our installation facilitates easy changing of process settings such as flow rate, crystal loading and temperature trajectory. Simultaneously, it provides immediate results enabling us to experimentally test a broad range of settings in a short period of time.

Within our setup (Figure 1) a tubular crystallizer with hot and cold zones was shown to be adequate to oscillate dissolution and growth efficiently via temperature cycling. By this, the feasibility of crystal engineering could be shown for pharmaceutical ingredients. By establishing a segmented flow mode interfering effects such as aggregation and breakage could be minimized and by thorough characterization of the temperature cycler unwanted nucleation could be overcome.

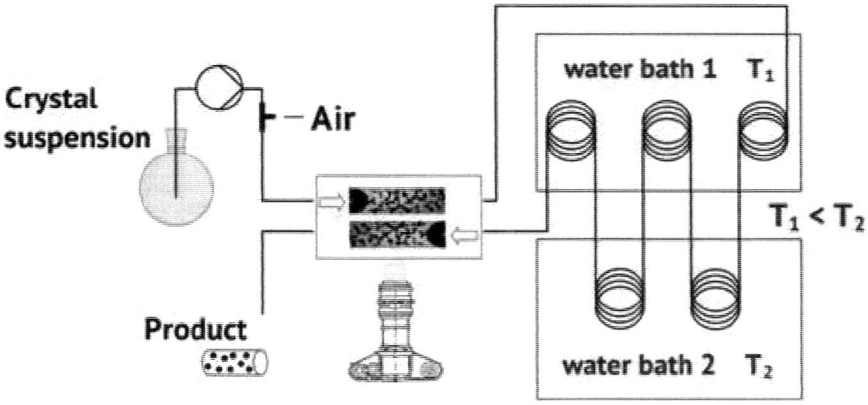

Figure 1: Schematic drawing of the experimental setting

(1) Neugebauer, P.; Khinast, J. G. Continuous Crystallization of Proteins in a Tubular Plug-Flow Crystallizer. *Cryst. Growth Des.* **2015**, *15*, 1089–1095.

(2) Besenhard, M. O.; Neugebauer, P.; Da-Ho, C.; Khinast, J. G. Crystal Size Control in a Continuous Tubular Crystallizer. *Cryst. Growth Des.* **2015**, *15*, 1683–1691.

(3) Jiang, M.; Zhu, Z.; Jimenez, E.; Papageorgiou, C. D.; Waetzig, J.; Hardy, A.; Langston, M.; Braatz, R. D. Continuous-Flow Tubular Crystallization in Slugs Spontaneously Induced by Hydrodynamics. *Cryst. Growth Des.* **2014**, *14*, 851–860.

453235 Continuous Crystallization Using a Sonicated Tubular System for Controlling Particle Size in an API Manufacturing Process

Abstract:

A pH swing crystallization that was very sensitive towards scale-up and required specific equipment due to the non-Newtonian characteristics of the fluid, has been investigated using continuous crystallization (CC) as a means to improve particle size control, scalability and potentially reduce manufacturing cost. A sonicated tubular crystallizer was designed and shown to allow the pH swing crystallization to be operated robustly without blockage, to give a wide range of desired particle size. The results were compared with crystallization in a premixing semi-batch system and a simple addition semi-batch system under similar conditions. Interestingly, it was found that the sonicated tubular system was the only one of the three systems that gave the desired control for small particle size (1~7 Î¼m). Even though the productivity of the sonicated tubular crystallizer still needs to be increased to allow CC and numbering-up to become cost competitive with batch crystallization, we have demonstrated the potential of continuous tubular crystallization as a useful alternative method to conventional batch operation.

463983 Continuous Reactive Crystallisation for Control of Salt Formation Processes

Abstract:

Salt crystallisation is of great importance to the pharmaceutical industry (majority of pharmaceutical products are marketed as salts) but it is often not a well-controlled process. This is because salt formation via a pH shift can be highly sensitive to mixing, and under batch or semi-batch conditions this can lead to severe issues with reproducibility and consistency of quantitative particle attributes, including solid form, crystal size distribution and crystal habit. The salt crystallisation process typically takes the form of a batch titration where the basic drug is dissolved in a solvent before the acid is slowly added. The issue with this procedure is that when the two solutions first come into contact the local supersaturation will build up very quickly and uncontrolled nucleation may occur leading to an undesired salt stoichiometry or crystal size distribution. For this reason, greater control over mixing to achieve greater control over the supersaturation build up is required to improve the consistency of the process and the quality of the particulate product. The aim of this research is to develop a quantitative reactive crystallisation model including a solution speciation model, a mixing model and a population balance model to guide the operation of continuous reactive crystallisation processes. We show the development of this quantitative reactive crystallisation model and how it is used to select operating conditions for the continuous crystallisation of the example 2:1 salt of 3,5-dinitrobenzoic acid and ethylenediamine. By doing this we can control which of the two polymorphic forms (monoclinic & triclinic) is crystallised, the crystal habit and the crystal size distribution.

456813 Process Intensification for Continuous Slug-Flow Crystallization

Abstract:

In the pharmaceutical and chemical industries, the continuous generation of crystals of target size distribution has the potential to improve efficiency for post-crystallization operations [1]. The control of crystallization processes can be challenging when undesirable phenomena such as particle attrition and breakage occur. Slug-flow crystallization has been demonstrated to allow enhanced control of crystal properties such as size and shape [2][3]. The slurry flow is combined with an air flow and fed to a tube to induce a multiphase hydrodynamic instability that spontaneously generates well-mixed slugs where the crystals continue to grow. These slugs are well-mixed without having the mixing blades as in traditional crystallizer designs that induce undesirable particle attrition.

This presentation will elaborate: (1) the process design and strategies for continuous slug-flow crystallization that controls the crystallization phenomena; (2) nucleation systems based on multiphase hydrodynamics from joining two streams in dual impinging jet, coaxial, or radial micromixers, which has been demonstrated to produce seed crystals of uniform crystal size distribution (CSD) [4][5], with a nucleation rate that depends on the liquid flow rate [4][5][6]; and (3) nucleation facilitated by advanced ultrasonication (e.g., focused and indirect) which provides extra degrees of freedom for controlling the CSD by decoupling the nucleation rate from liquid flow rate [3]. Process intensification strategy to improve the crystal growth control is also discussed which makes the whole process more robust. Experimental validation confirms that the proposed crystallizer designs reduce production time and equipment cost by orders of magnitude while suppressing secondary nucleation, attrition, and aggregation—dominant but undesired phenomena that worsen the ability to control the crystal properties.

References

[1] Z. Sun, N. Ya, R. C. Adams, and F. S. Fang, "Particle size specifications for solid oral dosage forms: A regulatory perspective," *Am. Pharm. Rev.*, vol. 13, no. 4, pp. 68–73, 2010.

[2] R. J. P. Eder, S. Schrank, M. O. Besenhard, E. Roblegg, H. Gruber-Woelfler, and J. G. Khinast, "Continuous sonocrystallization of acetylsalicylic acid (ASA): Control of crystal size," *Cryst. Growth Des.*, vol. 12, no. 10, pp. 4733–4738, 2012.

[3] M. Jiang, C. D. Papageorgiou, J. Waetzig, A. Hardy, M. Langston, and R. D. Braatz, "Indirect ultrasonication in continuous slug-flow crystallization," *Cryst. Growth Des.*, vol. 15, no. 5, pp. 2486–2492, 2015.

[4] M. Jiang, Z. Zhu, E. Jimenez, C. D. Papageorgiou, J. Waetzig, A. Hardy, M. Langston, and R. D. Braatz, "Continuous-flow tubular crystallization in slugs spontaneously induced by hydrodynamics," *Cryst. Growth Des.*, vol. 14, no. 2, pp. 851–860, 2014.

[5] M. Jiang, M. H. Wong, Z. Zhu, J. Zhang, L. Zhou, K. Wang, A. N. Ford Versypt, T. Si, L. M. Hasenberg, Y. E. Li, and R. D. Braatz, "Towards achieving a flattop crystal size distribution by continuous seeding and controlled growth," *Chem. Eng. Sci.*, vol. 77, pp. 2–9, 2012.

[6] A. J. Mahajan and D. J. Kirwan, "Micromixing effects in a two-impinging-jets precipitator," *AIChE J.*, vol. 42, no. 7, pp. 1801–1814, 1996.

455203 Continuous Cooling and Membrane-Assisted Crystallization with Airlift Crystallizer

Abstract:

Crystallization plays a substantial role in separation and purification of chemical and pharmaceutical compounds. The common problems in batch crystallisation processes often include large variations in the product quality per batch, inflexibility of the process and complex process controllability. The reason could be difficult handling of wet or semidry crystalline materials and high necessity of operating labour force in batch processes [1]. Furthermore, a common problem in conventional stirred crystallizers is the entanglement of multiple phenomena such as primary nucleation, growth and secondary nucleation.

Initially, batch experiments with airlift crystallizer confirmed that it can suppress secondary nucleation to a higher level than conventional stirred crystallizer [2].

Subsequently, experiments with membrane unit showed its feasibility in assisting crystallization processes by providing sufficient rate and level of supersaturation.

Based on results from membrane unit experiments effective parameters are known as feed temperature and air flow rate. The rate and level of supersaturation can be directly controlled and manipulated by changing these parameters for different process requirements.

In this study the performance of a continuous operated airlift crystallizer is discussed and secondly it is shown that membrane distillation can be applied as an alternative technology for the generation of supersaturation in combination with an airlift crystallizer.

Next, membrane unit and airlift crystallizer were combined. The setup is capable of operating in batch and continuous mode, cooling and membrane-assisted operation. Membrane-assisted Crystallization (MaC), firstly meet the requirements for decoupling the crystallization phenomena toward a more flexible process optimization and secondly it has the potential to operate in continuous mode and to significantly reduce energy consumption and operating labor force.

In this contribution feasibility of performing cooling and membrane-assisted crystallization using airlift crystallizer is presented. It is shown that a continuous membrane-assisted airlift crystallization is believed to be more controllable, flexible and energy efficient.

A model is developed and validated which is capable of predicting the membrane flux and level of supersaturation generation and the crystallization kinetics. The model is used to design and optimize the process.

1. Myerson, A., *Handbook of industrial crystallization*. 2002: Butterworth-Heinemann.

2. Lakerveld, R., J.J. Van Krochten, and H.J. Kramer, *An air-lift crystallizer can suppress secondary nucleation at a higher supersaturation compared to a stirred crystallizer*. Crystal growth & design, 2014. **14**(7): p. 3264-3275.

472727 Continuous Heterogeneous Crystallization of Active Pharmaceutical Ingredients on Excipients Surfaces

Abstract:

Novel continuous heterogeneous crystallization processes were developed, which the active pharmaceutical ingredient is crystallized directly on the surface of an excipient within the crystallizer. The product is then dried, and formed into tablets without the need for complex downstream processing steps such as milling, sieving, granulation, and blending. The APIs and excipients systems selected by study of heteroepitaxial mechanisms, and surface functionality of substrate. The effect of various process parameters, such as temperature, residence time, mode of operation, and excipient loading were studied. Three different process types in two categories of direct crystallization on surface of crystalline excipient and on surface of a carefully selected polymer (with surface modification) were studied. The excipient selection, and process parameters design have significant impact on drug loading, avoiding bulk nucleation and crystallization, control of API crystals shape and size, and control of process. In general, a decrease in temperature as well as the increase in residence time, both increased the drug loading of excipient with API. Within the range of our analysis, proportional drug loading can be easily tailored for a wide range depending on required drug dosage per tablet. The fully continuous processes, and eliminating the downstream steps result in producing crystalline compounds and final form (tablets) in significantly faster, more efficient, with lower footprint, and more economical.

467610 Design of Multi-Actor Distributed Processing Systems:a Game-Theoretical Approach

Abstract:

The manufacturing of a product could be the result of a value chain involving several processing plants distributed across distinct owners who form a network that exchange materials and/or energy among the involved actors. An example of such networks is given by the petrochemical value-chain which consists of refineries, petrochemical crackers, intermediates processing facilities and manufacturers of consumer products. Another example is given by the design of biorefineries, where conversion of biomass into intermediate platform chemicals and upgrading of these into other products will most probably take place at different corporations.

In our previous work, we proposed a novel strategy for the design of distributed manufacturing networks in which, suppliers and consumers linked through different intermediate products (mass or energy) sought maximization of their own profit. Under this assumption, we formulated a multiobjective optimization problem and proved that a 2-level Lagrangian approach could be used to obtain the optimal solution for the overall network (amount of mass/energy to be exchanged as well as transfer prices) through the optimal designs of the individual actors that participate in the network. This approach, assumed the existence of a centralized coordinator that collects the information from all the actors and thus implicitly knows in detail the form and parametric values of their objective functions and internal constraints.

In this presentation we propose and develop a game-theoretical framework and specific methodologies, which, without a centralized coordinator or share of internal information, allows for the optimal design of distributed processing systems. Within this framework the conditions of optimality established through the previously mentioned 2-level Lagrangian approach are satisfied through a negotiation process between suppliers and consumers which leads to a Nash equilibrium point. As this equilibrium point generally leads to an unequal distribution of the benefits between the actors, a Nash bargaining problem is analyzed to reach to an equitable distribution. The presentation will also discuss the use of penalty-term approaches to extend the framework to problems for which the underlying convexity assumptions of the 2-level Lagrangian approach may not be possible to ascertain. A series of case studies that illustrate the application of the proposed ideas to processing networks of various structures will be presented.

This work was funded by the Cooperative Agreement between the Masdar Institute of Science and Technology (Masdar Institute), Abu Dhabi, UAE and the Massachusetts Institute of Technology (MIT), Cambridge, MA, USA - Reference 02/MI/MI/CP/11/07633/GEN/G/00 for work under the Second Five Year Agreement.

458454 Optimization of Reverse Water-Gas Shift Chemical Looping for Continuous Production of Syngas from CO_2

Abstract:

Syngas is widely used in industry as a building block for the production of many bulk chemicals and chemical intermediates. The production method of syngas strongly influences its composition, most importantly the H_2/CO-ratio. Consequently, the syngas composition determines for which application it is suitable. Some of the most important uses of low H_2/CO syngas include the Monsanto process for acetic acid production and hydroformylation processes. High H_2/CO syngas may be used for alcohol synthesis and methanation. Today, syngas is produced almost exclusively from fossil fuels by steam reforming of natural gas or gasification of coal, depending on the desired H_2/CO-ratio. In light of the current and future global problems associated with the rising CO_2 levels in the atmosphere, traditional processes based on fossil fuels must be substituted to minimize the anthropogenic CO_2 production.

One possible way to shift towards a more sustainable future is to use CO_2 as the main carbon source instead of fossil fuels. In the context of sustainable syngas production, the reverse water-gas shift (RWGS) reaction is of high importance, as it converts CO_2 to CO with the help of process heat and hydrogen. The traditional RWGS reaction can be improved by introducing a metal oxide as an oxygen carrier, splitting the original reaction in a reduction and oxidation reaction on the metal oxide. This cyclic two-step process is referred to as reverse water-gas shift chemical looping (RWGS-CL). It has been shown that by using this approach, efficiency improvements can be obtained because the two-step operation yields partially separated gas streams and thus simplifies downstream gas separation [1]. This is especially true for syngas production with low H_2/CO-ratio. Furthermore, unwanted side reactions can be eliminated by separating the reactants spatially or temporally and favorable thermodynamics can be obtained by choosing suitable

oxygen carrier materials. Due to the cyclic switching between two process steps, RWGS-CL is more complex than the traditional RWGS process. A steady state operation in the traditional sense is not possible. However, a cyclic steady state (CSS) is achievable after prolonged cycling. Therefore, simulation and optimization techniques are used to understand the dynamic behavior and to improve the process.

Different reactor designs and configurations for the RWGS-CL process are examined and evaluated. For a fixed bed configuration, a 1D model for the two-step RWGS-CL process is derived based on [2] with kinetic information from [3]. Process simulation is used to compute the cyclic steady state of the process for different flow regimes. On top of the simulation, a multi-objective optimization problem is formulated and solved to maximize the gas conversion and the utilization of the oxygen carrier material. Since both objectives are conflicting they lead to Pareto optimal solutions. The trade-off between gas conversion and the utilization of oxygen carrier material is investigated. The influence of the decision variables and the flow regime on the optimal solution is analyzed and discussed. Constraints on the decision variables are formulated such that continuous production of CO can be achieved with the RWGS-CL process. The simulation and optimization results are compared to those of the traditional RWGS process.

References:

[1] Wenzel, M., Rihko-Struckmann, L. and Sundmacher, K. (2016), Thermodynamic Analysis and Optimization of RWGS Processes for Solar Syngas Production from CO_2. Submitted to AIChE Journal.

[2] Heidebrecht, P., Hertel, C. and Sundmacher, K. (2008), Conceptual Analysis of a Cyclic Water Gas Shift Reactor. International Journal of Chemical Reactor Engineering, 6

[3] Wenzel, M., Dharanipragada, N. V. R. A., Galvita, V. V., Poelman, H., Marin, G. B., Rihko-Struckmann, L. and Sundmacher, K. (2016), CO Synthesis From CO_2 via Reverse Water-Gas Shift Reaction Performed in a Chemical Looping Mode: Kinetics on Modified Iron Oxide. Submitted to Journal of CO_2 Utilization.

463689 Model-Based Comparison of Novel Hollow-Fibre Sorbent Adsorber to Packed Bed Process for Dilute Gas Streams

Abstract:

Adsorption provides a relatively less energy intensive alternative to capture dilute streams from gaseous mixtures. In certain applications such as removal of CO_2 from air, recovery of valuable VOCs from exhaust streams, separating streams with almost identical thermodynamic properties (such as separation of ethylene/ethane); adsorption is often the only route to making the process commercially viable. Process improvement for the recovery of dilute gases has become an important area of research in the recent past. This may be attributed to the unprecedented rise in atmospheric concentration of CO_2 (>400 ppm for the first time in 4 million years[1]) and, the implementation of the Maximum Achievable Control Technology (MACT) provisions of the Clean Air Act, which make the reporting and control of VOC emissions mandatory. Capture of dilute streams also finds applications in upgrading processes for methane, inevitably extracted as a co-stream from natural gas reservoirs.

Traditional processes for adsorption involve use of TSA, PSA or VSA in the packed bed configuration. To improve on the current process, a novel hollow fibre configuration for adsorbents was proposed by Lively et. al. using 13X as a â €œproof of conceptâ€ zeolite[1]. The key advantage of this configuration is the substantial reduction in pressure drop relative to the packed beds. This is of primary importance when dealing with dilute streams where, additional pressure drop may involve significant capital investment. Additionally, the presence of the impermeable lumen layer allows the use of cooling agents without being limited by its adsorption properties, unlike the situation in a packed bed. This is a significant advantage for dilute streams, where TSA is a more economical process alternative compared to PSA. The fast heat transfer in the hollow fibre configuration shortens the cycle time, therefore increasing the throughput and reducing the quantity of adsorbent. One possible drawback of the hollow-fibre configuration maybe the parasitic heat load of the fibre material itself during the desorption step, which can be mitigated by heat integration.

This talk addresses the comparison of an adsorption-desorption cycle (TSA) of a dilute gas stream in both the previously mentioned configurations, packed bed and hollow fibre, using mathematical models. Coupled PDEs are obtained from heat and mass balance. The form of the equation is dictated by the system geometry. Mass transfer models appropriate for each configuration are also used, along with isotherm equations which are configuration independent. The commercial, dynamic process modelling software gPROMS is used to solve the system of equations. The models are used to compare the pressure drop of a dilute gas stream in both configurations, during adsorption. The separation performance is compared for the two configurations taking into account the pressure drop, which is dictated by the gas velocity. Finally, the parasitic heating load of the fibre material is estimated.

[1] 'Modeling of rapid temperature swing adsorption using hollow fiber sorbents', 2014, Rezaei, F, Subramanian, S, Kalyanaraman, J, Lively, R P, Kawajiri, Y, Realff, M J ,*Chemical Engineering Science*, **113**, 62-76

471719 Profitability and Risk in Conceptual Plant Design:Dealing with Key Financial Parameters Rigorously and Simultaneously

Abstract:

Profitability and Risk in Conceptual Plant Design:

Dealing with Key Financial Parameters Rigorously and Simultaneously

Duncan A. Mellichamp

Professor Emeritus

Department of Chemical Engineering

University of California, Santa Barbara, CA USA 93106

e-mail: dmell@engineering.ucsb.edu **Phone**: 805-569-9858 **Fax**: 805-893-4731

Abstract

Many companies impose a minimum *IRR* ("hurdle rate") to justify a "go-forward decision" with a new project, e.g., the *IRR* must be at least 25% to be deemed profitable enough. Presumably, this hurdle rate is

chosen sufficiently higher than the sum of their ongoing *Enterprise Rate, ER*(the average year-over-year after-tax return on investment to the organization from all continuing operations) plus a "risk cushion" to cover uncertainties that statistically arise, causing loss of profitability or even outright failure of some projects.

Most users of discounted cash flow methods understand that the *ER*, when used as *Discount Rate* (*DR*) in net present value calculations imposes an appropriate "opportunity cost" on a potential project —i.e., foregone return that presumably could be obtained from alternative investment in ongoing internal operations. But few users understand just how problematic *IRR* is because the calculated value (i.e., *Discount Rate*) seldom has real physical meaning. Nor can the portion of the *DR* that is greater than *ER* be interpreted as "risk compensation." The key problem is that *IRR* is defined without parameters, so nothing is strictly associated either with "profit" or with "risk." Furthermore, *IRR*is a non-linear function, making any allocation of profit and risk difficult to extrapolate mathematically to different circumstances.In 2013 the author proposed a new index of profitability, $NPV_{\%}$, (*NPV*, evaluated with *DR* = *ER*) normalized by the *Total Capital Investment* and annualized by the *Project Lifetime*).* This metric was shown to be linearly related to ROI_{BT}, thus substantially represents "inverse risk" in the same way that *Pay-Out Time* approximates the length of time a capital investment will be at risk (time to recover the original investment via cash flows). Thus, $NPV_{\%}$ can serve as an inverse measure of business risk while its antecedent, *NPV*, provides a direct estimate of profitability. Further, since $NPV_{\%}$ includes at least one more parameter than *IRR*, it allows for an independent choice of the *Discount Rate*. Because *IRR* and $NPV_{\%}$ are both directly related to the ROI_{BT} of the proposed project, one can fix ROI_{BT}, calculate $NPV_{\%}$ as a function of *ER*, then use a manifold projection (onto the ROI_{BT}, *IRR* axis) to establish equivalence.

Several key outcomes are discussed: (1) Company policy ought to be to adjust the hurdle rate to optimize profitability (*NPV*) and the risk surrogate (inverse of $NPV_{\%}$) while updating *IRR* [as inferred function of $NPV_{\%}(ER)$] *whenever* experience with profits (internal *Enterprise Rate*) changes significantly. (2) A better approach might be to abandon the usual pairing <*NPV* (with *DR* = *ER*), *IRR*> altogether and adopt <*NPV*, $NPV_{\%}$> or a related evaluation metric. Ongoing research shows that such a choice can make evaluating project profitability particularly easy, allowing any reasonable optimization strategy to employ 1, 2, or 3 independent variables--e.g., profitability, risk, and/or capital investment--*simultaneously*.

---

* Mellichamp, D.A., *Computers and Chemical Engineering* **48** (2013) 251– 263.

459874 Characterization and Optimal Site Matching of Wind Turbine: Effects on the Economics of Synthetic Methane Production

Abstract:

Wind can easily provide the energy that mankind needs. However, harvesting energy from wind is challenging. There are only a few regions where it is economically interesting to use it. Furthermore, the variability in the wind velocity and the particular design of the turbine determine the turbine model to use for a particular site location based on its characteristic power curve. Instead of just producing power, with the challenge of its availability, it is possible to store wind energy in the form of chemicals. Davis and Martín (2014a&b) used solar and wind energy to produce methane from renewable energy and CO_2 not only to store both sources of energy in a more handy form, but also to capture and reuse CO_2 as a carbon source. However, the design of such processes must consider seasonal wind variability and uncertainty (Halemane and Grossmann (1983), Grossmann and Sargent (1978)). Thus, the selection of the appropriate turbine type for a specific site location requires accurate modeling of the power curve to address its operation under variable wind velocity conditions. Previous studies on turbine site –matching are based on the capacity factor, using piecewise approximations of the power curve to obtain analytical solutions Jowder (2009) El-Shiny (2010). A more systematic study and power curve modeling is required.

In this work, we evaluate the optimal site matching of wind farms developing a generic single equation model to characterize the power curve of a number of current commercial onshore and offshore wind turbines. Next, mathematical MINLP formulations are developed that allow considering the velocity distribution on a monthly basis, including wind variability using Weibull distribution. We apply the formulation to select the appropriate turbine type and number for the production of synthetic methane. This allows evaluating the effect of the site on the wind turbine selection and the investment and production costs of synthetic methane from hydrolytic hydrogen and CO_2. The chemical facility was developed in a previous work Davis and Martín (2014a&b). In that work it was shown that 94% of the investment cost is due to the wind farm. Thus, we can identify advantages in selecting the proper place and turbine in terms of economic savings for producing an easy to use and store/transport fuel.

The optimal selection of turbine and place result in competitive prices for synthetic methane, below 2€/MMBTU, as long as the appropriate turbine and allocation are selected. By using the formulation to select the allocation of such a facility in Spain, the best place selected corresponds to the one in Davis and Martín (2014a) paper, Cádiz. However, the use of the appropriate turbine design resulted in savings of 42 MM€ compared to the results in Davis and Martín (2014a) and a production cost 25% lower. Therefore, the formulation presented in this work provides a useful tool for the selection of turbines and allocations. Finally, the selection under uncertainty results in the fact that, although in general the cost is expected to be higher, since a more robust solution is expected, sometimes lower production prices are found due to the fact that the more detailed wind distribution fits better with the power curve than just an average velocity.

References

Archer CL., 2004. The Santa Cruz Eddy and U.S. wind power, Ph.D. Thesis, 190 pp., Stanford University, Stanford, 1 April 2004.

Davis, W., Martín, M ., 2014a. Optimal year-round operation for methane production from CO_2 and Water using wind energy. Energy 69, 497-505

Davis, W., Martín, M., 2014b. Optimal year-round operation for methane production from CO_2 and Water using wind and/or Solar energy. J. Cleaner Prod. 80, 252-261.

El-Shimy, M. Optimal site matching of wind turbine generator: Case study of the Gulf of Suez region in Egypt. Renew. Energ. 35 (8), 1870-1878

Grossmann, I. E., Sargent, R.W.H., 1978. Optimum Design of Chemical Plants with Uncertain Parameters, *AIChE J.*, 24 (6), 1021-1028

Halemane, K.P.,Grossmann, I. E. , 1983. Optimal process design under uncertainty *AIChE, J.*, 29 (3), 425-433

470289 Multi-Objective Optimization of Biorefinery and Uncertainty Analysis

Abstract:

Abstract:

The renewable energy industry is growing continuously, and it has several challenges and opportunity for research community. Biomass is a potential source of energy for future as it can be converted to transportation fuels, chemicals and electricity using bio-chemical and thermo-chemical conversion routes. Bio-alcohols production *via* hydrolysis and fermentation can only use cellulosic part, leaving the lignin portion as waste. On the contrary, thermo-chemical processes such as gasification and pyrolysis can use both cellulose and lignin. Gasification produces syngas which can be converted into liquid fuels, chemical and electricity. Fischerâ˜Tropsch (FT) crude, methanol (MeOH) and dimethyl ether (DME) can be synthesised by syngas. Further, solid oxide fuel cell with gas turbine (SOFC-GT) has shown higher thermodynamic performance for converting syngas into electricity and particularly considering the biomass conversion integration. Decision on the best pathways highly depends on bioproducts and biofuels market prices, which are highly uncertain. Further, investment cost and price and availability of raw materials are also uncertain.

The biorefinery superstructure for this study includes following processes: gasification, SOFC-GT, FT, MeOH and DME. Design and operating conditions of this biorefinery superstructure is optimized for profit and annualized capital cost, simultaneously, using stochastic multi-objective optimization method. The final selection of a solution from the obtained Pareto-optimal front depends on its sensitivity to the uncertain parameters, such as feed and product prices, plant life and operating time. Engineers are mainly interested in selecting one or few robust solutions which are less sensitive to the uncertain parameters, and so the uncertainty analysis of the obtained non-dominated solutions may help in identifying robust solutions. In this study, effect of several uncertain operating and market parameters is studied on the performance of selected biorefinery superstructure. The uncertainty analysis is able to identify most promising

biorefinery configurations, based on the profit as main decision crieriton. The results of this study will be presented at 2016 AIChE Annual Meeting.

Keywords:

Multi-objective Optimization, Gasification, Fischer-Tropsch Process, Methanol, Dimethyl Ether, Solid Oxide Fuel Cell, Gas Turbine, Uncertainty Analysis.

References:

1. Caliandro P., Tock L., Ensinas A.V, Maréchal F., Thermo-economic optimization of a solid oxide fuel cell - gas turbine system fuelled with gasified lignocellulosic biomass. Energy Conversion and Management 2014; 85: 764-773.
2. Facchinetti E., Favrat D., Maréchal F., Design and optimization of an innovative solid oxide fuel cell - gas turbine hybrid cycle for small scale distributed generation. Fuel Cells 2014; 14(4): 595-606.
3. Gassner M., Maréchal F., Methodology for the optimal thermo-economic, multi-objective design of thermochemical fuel production from biomass. Computers & Chemical Engineering 2009; 33(3): 769-781.
4. Tock L., Maréchal F., Decision support for ranking Pareto optimal process designs under uncertain market conditions. Computers & Chemical Engineering 2015; 83:165-175.
5. Tock L., Gassner M., Maréchal F., Thermochemical production of liquid fuels from biomass: Thermo-economic modeling, process design and process integration analysis. Biomass and Bioenergy 2010; 34(12): 1838-1854.

467946 Development of a Semantically-Enabled Model-Sharing Biorefinery Platform

Abstract:

The production of chemicals and fuels based on biomass has been steadily increased, as a consequence of the environmental issues that have arisen by fossil fuels in addition to the steady decrease of their reserves. There is a huge potential in capitalizing fragmented work around the world on biorefineries and biorenewables, since this would assist and accelerate the progress in both research and industrial communities. The objective of this work is to integrate biomass pathways, models and tools for biorefineries and create a repository that contains information on these existing biomass pathways and models. The extremely large number of possible pathways as well as the diversity of data consist important challenges that need to be overcome.

An ontology engineering approach will be presented that emulates the types of data as they are available in selected projects, mainly RENESENG[1] but also others. The aim is to develop an intelligent environment able to scale-up to also include and organize resources outside the current project. Ontology engineering has recently attracted a lot of attention in chemical engineering community for managing, searching and enriching knowledge[2,3]. Advantages of ontology engineering include the classification of large amount of information, the classification of complex relations, as well as the generation of new information by the use of reasoners. The services of the repository will include (a) access to all data included in the tool, (b) support for high-throughput search and analysis (c) integration with physical property models and data models, (d) synthesis technologies, and (e) uploading of new data of a variety of scales and types from the user. Having a user-friendly interface, the repository targets users of a wide range; from students to engineers in SMEs or international companies. The state of progress of the development of the repository will be presented and example operations will be demonstrated.

References

1. www.reneseng.com
2. Weisner et al., Computers and Chemical Engineering 35 (2011) 692-708
3. Trokanas, et al., Computers and Chemical Engineering 66 (2014) 259 - 268

459820 Optimal Production of Power in a Combined Cycle from Manure Based Biogas

Abstract:

Most countries have powerful agricultural systems to provide for food. As a result, large volumes of residues are generated. In particular, cattle and pigs farms generate vast amounts of residues (manures). Apart from the difficulty of dealing with such quantities of waste, the composition is also dangerous. Anaerobic Digestion (AD) provides the technology not only to dispose it but also to generate further value. The production of biogas through AD offers significant advantages over other forms of bioenergy production. It has been deemed as one of the most energy-efficient and environmentally beneficial technologies for bioenergy production [1]. Furthermore, biogas generation can drastically reduce greenhouse gases compared to fossil fuels by utilization of locally available resources. Compared to other fossil fuels, methane produces fewer atmospheric pollutants and generates less carbon dioxide per unit energy; as methane is comparatively a clean fuel, the trend is towards its increased use for appliances, vehicles, industrial applications and power generation [2]. Finally, the digestate represents an improved soil conditioner which can substitute mineral fertilizer [3]. However, most of the studies on power production from biogas limit to the use of Stirling engines or basic CHP plant analysis [4-6]

In this work, the production of power using a combined cycle gas turbine/steam turbine for wast heat recovery, which operates with biogas as fuel, is evaluated. The process begins with the production of biogas from pig and /or cattle slurry manure(s) using AD. Afterwards, the gas is cleaned up to remove humidity, hydrogen sulphide, carbon dioxide and ammonia. The cleaned gas (biomethane) is then used in a Brayton cycle (gas turbine) to produce energy. The flue gas that exits the Brayton cycle is typically at high temperature and it is further utilized to produce steam that generates power in a regenerative Rankine cycle (steam turbine). Two alternative steam production schemes are evaluated: either splitting the flue gas to have high temperature gas for the reheating step of the steam or sequential heating up. The model is formulated as a Mixer Integer Nonlinear Programming (MINLP) solved in GAMS® for the optimal production

of power and topology selection.

For a typical production capacity of manure in farms in Spain, 2.6 MW are produced. The hot flue gas is used in sequence for producing superheated steam and next to reheat up the stream after the first expansion in the steam turbine. While typical efficiencies of biogas combustion engines are from 25-40%, the system presented in this work reaches 65% efficiency from the biogas to electricity. The investment for the plant turns out to be 26M€ and the production cost of the electricity is 0.35€/kWh before including the credit from the conditioned digestate, that could be sold as fertilizer. The electricity cost goes down to 0.15€/kWh considering a reasonable credit from the digestate, whose composition depends on the feedstock processed in the facility.

References

[1] Fehrenbach H, Giegerich J, Reinhardt G, Schmitz J, Sayer U, Gretz M, et al. Criteria for a Sustainable Use of Bioenergy on a Global Scale. Germany: Federal Environment Agency; 2008; 245 (prepared by the Institute for Energy and Environmental Research (IFEU), Heidelberg).

[2] Chynoweth DP, Owens JM, Legrand R. Renewable methane from anaerobic digestion of biomass. Renew Energy 2001;22(1):1–8.

[3] Weiland P. Biogas production: current state and perspectives. Appl Microbiol Biotechnol 2010;85(4):849–6

[4] Kang, J. Y., Kang, D.W., Kim, T S., Hur, K.B. (2014) Economic evaluation of biogas and natural gas co-firing in gas turbine combined heat and power systems Appl.Thermal Eng. 2014; 70: 723-731

[5] Japaraju, P., Rintala, J. 17 - Generation of heat and power from biogas for stationary applications: boilers, gas engines and turbines, combined heat and power (CHP) plants and fuel cells 404-427, In The Biogas Hand book. 2013.

[6] Kang JY, Kang DW, Kim TS, Hur KB (2014) Comparative economic analysis of gas turbine-based power generation and combined heat and power systems using biogas fuel Energy 2014; 67: 309-318

464819 Developing Structure-Property Relationships Between Reactant Structures, Ionic Liquids, and Reaction Rate Constant

Abstract:

Recent years have seen a burgeoning interest among researchers in the prediction of reaction rate constants using structure-property relationships. The rate constant has been modeled in terms of reactant structures (Chaudry and Popelier, 2003), solvent structures (Zhou et al., 2014) or a combination of reactant and solvent structures (Datta et al., 2016). The development of models for prediction of rate constants serve two primary purposes. First, such models can be utilized for the computer-aided molecular design (CAMD) of reactants and solvents such that the reactions are accelerated. Secondly, such models can help avoid conducting many cumbersome experiments in order to calculate the rate constant. With regards to CAMD, Zhou et al. (2014) recently designed solvents that enhance the rate of the Diels-Alder reaction. While much work has been carried out to model the rate constant in terms of molecular solvents and molecular descriptors that represent them, comparatively less work has been performed in modeling the rate constant in terms of the structures of ionic liquids. Ionic liquids offer a range of beneficial properties and hence have captured the interest of researchers in recent years. These include, negligible vapor pressure under standard conditions, high thermal stability and structural tunability (Priede et al., 2015). To the best of our knowledge, no previous studies have been published on rate constant modeling with respect to structures of both ionic liquids and reactants. Through our work, we wish to fill this research gap. We have utilized multiple linear regression (MLR), principal component analysis (PCA) and support vector regression in our work to model the reaction rate constant of the Knoevenagel condensation reaction. The rate constant has been modeled in terms of topological indices of reactants and ionic liquids. The model performance has been compared using ten-fold validation.

References:

U. A. Chaudry, P.L.A. Popelier, 2003, J. Chem. Phys. Chem. A, 107, 4578-4582

T. Zhou, K. Mcbride, X. Zhang, Z. Qi, K. Sundmacher, 2014, AIChE Journal, 61, 1, 147-158

468355 Reverse Design of Ionic Liquids for CO_2 Absorption

Abstract:

Chemical engineering design problems have evolved in recent years from large scale bulk processes to targeted small batch approaches. As a result, product design is reaching beyond the traditional solutions to classic chemical engineering problems such as extraction, purification, removal of sulfur from fuel, CO_2 capture and sequestration, energy storage and biomass pre-treatment. Ionic liquids can provide environmentally benign solutions that can be tailored to specific process requirements. Due to the unique structure of ionic liquids, the cation, anion and length of the alkyl chain can be varied to create a molecule with specific physical properties. It has been estimated that over 10^{14} unique molecules exist at room temperature, so experimental trial and error methods will not be effective. More advanced methods must be created determine the ideal combination for a specific set of process requirements.

The purpose of this work is to determine the ideal ionic liquid to capture CO_2, a question which has been the topic of many experimental laboratory studies. However, only a small percentage of the potential ionic liquids have been synthesized and tested for the solubility of CO_2. This innovative approach decouples the solution from time-consuming laboratory study and streamlines the search for the most promising candidates that should be synthesized.

A characterization based group contribution method is combined with density functional theory to determine the ionic liquid that can most effectively absorb CO_2. Infrared spectra data contains descriptor data that can be used to estimate properties of ionic liquids, but does not exist for all ionic liquids. Density functional theory is used to create IR data based on a training set of experimental data. Principal component analysis and partial least square techniques will be employed to reveal important features and patterns in the molecular architecture. A characterization based group contribution method is used to estimate the properties of the new structures. The reverse design of potential ionic liquid molecules will be completed by an exhaustive search of combinations with various cation, anions and lengths of alkyl chains until a candidate molecule is found that provides the highest solubility of CO_2.

461251 Application of Systematic Methodology for Design of Tailor-Made Blended Products: Lubricant Design

Abstract:

Application of systematic methodology for design of tailor-made blended products: Lubricant design.

Marina Fedorova[1], Laurent Grosset[2], Fang Zhou[3], Rafiqul Gani[1]

[1] Department of Chemical and Biochemical Engineering, Technical University of Denmark, Lyngby, Denmark

[2] TOTAL Marketing&Services, Solaize, France

[3] Tsinghua University, Beijung, China

Over past decades, chemical product engineering has received much attention among the chemical engineering community. This is due to the transformation of industries in manufacturing and selling chemical products based on the product performance rather than compositional specifications.

Traditional methods used in new product design/development combine a broad knowledge of existing products with scientific experimentation. The chemical product is developed based on scientific hypothesis, intuition or simple trial-and-error. However, this approach is costly, very time consuming and only a limited number of experiments can be performed. In order to efficiently design chemical products, a systematic methodology followed by focused experimental verification is a better approach. It can be implemented at the initial stage of the design, where it could minimize the required number of experiments, as well as to identify possible solutions that normally would lie outside

the scope of traditional approaches, thus increasing the chances of finding better solutions with less resources.

The systematic methodology for solving mixture/blend design problem consists of four main tasks: 1) problem definition – identification of the final product needs and translations of them into target properties; 2) property models identification; 3) mixture/blend design; 4) experimental or model-based verification. Additional tools and methods needed to perform a specific task in the methodology, which are the property model library, the chemicals database and the mixture/blend design algorithm, also need to be available.

Application of the methodology is highlighted with several case studies related to lubricant design. Lubricants are required in almost all modern machines. Lubricating oil is a substance introduced between two moving surfaces to reduce the friction between them, improving efficiency and reducing wear. The functions of the lubricant depending on application may also include cooling, cleaning and suspending, protection and transfer power.

The application of lubricant is classified into two types – engine lubricants and non-engine lubricants. In this work, the lubricants are designed as engine oils, which is the largest application of lubricants. Typically lubricants are a mixture of base oil (65-98%) and performance additives to achieve required performance and end-user requirements. The target properties for the lubricant blend can be divided into performance properties, long life properties and environmental properties.

Focus of this work is performance properties, as they are central for any lubricant application. These properties include, for example, viscosity, thermal properties, low and high temperature properties, friction coefficient and film thickness that lubricant provides. The first part of the case study is to design the base oil, which is the main component in lubricant. The second part focuses on the development of the specific lubricant additive - its properties are thoroughly studied and the database of the chemicals is created to be used in the mixture design. Currently this database includes 1202 items. Novel blends with besigned functional properties are found in all cases. The method is generic and also be applied for other blended products.

468374 Physical Properties for Lipids Based Process and Product Design

Abstract:

Lipid processing covers several oil and fats technologies such as: edible oil production, biodiesel production, oleochemicals (e.g.: food additives, detergents) and pharmaceutical product manufacturing. New demands regarding design and development of better products and more sustainable processes related to lipids technology, emerge according to consumers demanding improved product manufacturing from sustainable resources and new legislation regarding environmental safety [1]. Physical and thermodynamic property data and models for prediction of pure compound properties and mixtures properties involving lipids represent the basic and most important requirements for process-product design, simulation and optimization. Experimentally measured values of involved compounds are desirable, but in most of the cases these are not available for all the compounds and properties needed. The lack of properties is even larger for mixtures properties. Therefore there is a highly need of predictive properties models.

Oleochemical industry, along with all other lipids processing technologies, presents many challenges in relation to thermodynamic and physical properties as a result of lack of experimental data for pure compounds and their mixtures. Due to these gaps, lipids involved industries are not able to exploit all the advances achieved within computer aided methods and tools, for which property data and models represent the basic requirements to perform process design, simulation and optimization, as well as product design. Another challenge comes from the availability of the data and models within commercial available software [2].

Recent developments within lipid property modelling, as well as new physical and thermodynamic data availability, allowed the development of our in-house SPEED Lipids Database. The database is directly connected with external computer aided tools, such as PROII and ICAS suite, and it is organised within two parts: one is dedicated to the pure compounds and their properties, while the other is dedicated to the mixtures phase equilibria. In product-process design, the chemicals database together with the lipids database is used.

For the development of the database â€" pure compound part, available data for lipids in terms of their molecular description and a range of properties were collected in special attention to the needs for model-based design and analysis within oleochemical industry. Available experimental data was collected for the identified lipids and their corresponding properties. The database contains 330 pure compounds organised within 18 categories (e.g.: fatty acids, triacylglycerides, diacylglicerides, monoacylglycerides, methyl esters, ethyl esters, isopropyl esters, carotenoids, triterpenealcohols, phospholipids, fatty alcohols, sterols, vitamin E, etc.). The experimental data was collected for 7 primary properties (e.g.: critical temperature, pressure and volume, Gibbs free energy of formation, enthalpy of formation, melting point, and normal boiling point), and 6 temperature dependent properties (vapour pressure, liquid heat capacity, liquid density, liquid thermal conductivity, viscosity, and surface tension). To fill in the gaps within the database models for 14 primary properties (e.g.: molecular weight, critical properties, melting point, normal boiling point, enthalpy of formation, enthalpy of fusion, Gibbs fee enthalpy of formation, dipole moment, acentric factor, specific gravity, dipole moment, liquid volume) and 10 temperature dependent properties (e.g.: vapour pressure, liquid enthalpy, liquid viscosity, liquid thermal conductivity, ideal enthalpy, surface tension, latent heat, liquid density, vapour density, vapour thermal conductivity) were implemented. The chemicals database contains 45500 chemicals and more than 27 different properties.

The mixture phase equilibria component of the lipids database contains a collection of over 4500 measured data point corresponding 332 different phase equilibria data sets (92 VLE, 91 LLE, 70 SLE and 79 solubility data) for binary and multi component mixtures. Consistency test were performed for all the VLE and SLE data sets available in the database. From the available VLE data sets, original UNIFAC parameters were regressed, allowing better representation and prediction of the phase equilibria.

The following features are available within SPEED Lipids database: property data consistency test; optimization based data regression to estimate the model parameters; knowledge representation and search engine for database use knowledge representation and search engine for database; computer aided modelling tool to quickly and efficiently develop the necessary property prediction models. Also, a systematic work flow for database maintenance and development is provided. All databases are integrated with process and product simulators. The presentation will highlight the use of the database and property models in case studies involving process-product design.

Reference

[1] Diaz-Tovar C, Mustaffa A., Hukkerikar, A., Quaglia A, Sin G, Kontogeorgis G, Sarup B, Gani R. Lipid Processing Technology: Building a Multilevel Modeling Network. In: Pistikopoulos E. N., Georgiadis M. C., Kokossis A, editors. Proceedings of the 21st European Symposium on Computer Aided Process Engineering 2011, 29:256â€"260.

[2] Sarup B. Advances and Challenges in Modelling of Processing of Lipids. In: Gernaey K. V., Huusom J. K., Gani R., editors. Proceedings of the 25th European Symposium on Computer Aided Process Engineering 2015, 37:117-122.

471395 an Extension of COSMO-Based Methodologies for Computer-Aided Mixture Design

Abstract:

Group contribution methods like UNIFAC have been successful in capturing thermodynamics in mixture design problems while remaining amenable to a mathematical optimization framework. Despite being accurate in many domains, these methods have an inherent limitation: because they are often regressed over large datasets of neutral, ground state structures, they can only consider such structures when applied to predict mixture thermodynamics. This means that UNIFAC-like methods cannot be extended to arbitrary systems as they would fail to capture the electronic complexities of species like transition states, reactive intermediates, and ionic liquids. In this work, we propose the use of quantum chemistry calculations via the COSMO solvation model as a means to incorporate arbitrary systems into mixture design problems. Every component of these systems that is not in the design space can be modelled at an ***ab initio*** level of accuracy and then mapped to a charge density histogram called a sigma profile. In COSMO post-processing methodologies like COSMO-RS[1] and –SAC[2], these sigma profiles represent the main information required to estimate mixture thermodynamics properties. To incorporate these techniques into a mixture design problem, we develop group contribution methods to estimate the sigma profiles of molecules in the design space. That is, we only rely on group contribution estimates for every species that is not fixed in the mixture. We also include several extensions to our existing COSMO-based mixture design methodology[3] to improve the accuracy of the estimates.

We address the optimization of the mixture design problem using derivative-free optimization (DFO) methods. By projecting the problem onto a lower-dimensional space, we exploit the efficiency of DFO techniques at solving problems with few degrees of freedom. We apply these techniques to three case studies: (1) a reaction rates optimization problem, (2) control of a halogen-metal exchange reaction's selectivity, and (3) maximizing the selectivity of an intramolecular nucleophilic aromatic substitution reaction. It is important to note that all of these problems cannot be modeled directly with UNIFAC-like approaches as they either involve species which do not have available UNIFAC groups or exhibit complex electronic properties. We consider all of these problems first from the perspective of unrestricted solvent design, meaning we consider mixtures of solvents which may not exist. Next, we consider all of these problems from a more practical point of view, designing an optimal blend of common laboratory and industrial solvents. Overall, these case studies demonstrate the ability of COSMO-based approaches to incorporate highly accurate quantum chemical information directly into mixture design applications.

[1] F. Eckert and A. Klamt. Fast solvent screening via quantum chemistry: COSMO RS approach.. ***AIChE Journal.*** **2002.** 48(2), 369-385.

[2] S. T. Lin and S. I. Sandler. A priori phase equilibrium prediction from a segment contribution solvation model. ***Industrial & engineering chemistry research.*** **2002.** 41(5), 899-913.

[3] N. D. Austin, N. V. Sahinidis, and D. W. Trahan . A COSMO-based approach to computer-aided mixture design. ***Chemical Engineering Science.*** **2016.** ***submitted***

465531 A Mixed-Integer Linear Programming Approach for the Design of Nanostructured Catalysts

Abstract:

Recent advances in nanotechnology have enabled the fabrication of materials with precision on the micro- or even nano-scale. These precisely fabricated materials have properties that are wildly different from their bulk counterparts, enhancing useful functionalities in a number of application contexts. For metallic nanoparticles, in particular, recent work has highlighted enhanced catalytic activities [1], unique optical properties [2], and the potential for use in drug delivery [3], among other exploitable properties. With a flourishing synthesis capability comes the need to answer the question: "Which design leads to the best catalytic performance?" Unfortunately, the answer to this question is not always obvious due to the combinatorial number of ways in which matter can arrange itself. However, such a material design problem can be formalized as a mathematical optimization problem like those traditionally cast and solved in many process systems engineering contexts. In essence, one seeks to identify a material design (i.e., a specific instantiation of the given class of materials of interest) that maximizes some metric of performance.

In this work, we focus on the design of nanostructured surfaces of transition metal crystals, a material class that has received a lot of attention due to its potential for catalyzing many important reactions [4,5]. We show how correlations that link catalytic activity to site descriptors, such as commonly available volcano plots [6,7], can be used to determine those surface designs that maximize the total reactivity. Volcano plots in practice tend to have steep sides, implying that only the "ideal" sites, i.e., those exhibiting features that lie very close to the peak of the volcano, contribute substantially to the overall reaction rate [7,8,9]. With this observation in mind, the problem of designing the surface of highest reactivity simplifies to that of designing the surface that packs the most ideal sites in the unit area, essentially neglecting the contributions from the remaining, non-ideal sites. Furthermore, when the material designs can be expressed via discrete variables, as is the case with materials conforming to well-defined crystalline lattices, and when appropriate site descriptors can be encoded via linear constraints and disjunctions, as is the case of descriptors based on site coordination numbers, we can cast mixed-integer linear programming (MILP) models and utilize well-developed MILP optimization software in order to determine the optimal surface packings. Besides the tractability advantages it offers, an ancillary benefit of using MILP solvers to address this design problem is that several optimal or near-optimal solutions can be collected at a marginal computational cost via a solution pool approach. In this manner, multiple designs can be identified for comparison against some secondary criteria or further consideration by experts.

The proposed MILP approach depends on a suitably defined graph, which serves as a "design canvas." The nodes of this graph constitute the lattice locations present in a periodic, space-filling tile of a thin layer on the surface of a crystal, while the edges of the graph signify those locations that are considered neighbors for determining a site's coordination number. Such a tile can be enforced to be periodic and replicated ad infinitum along a plane so as to form a full surface. The key decision variables in the model are a set of binary variables which indicate the presence or absence of metal atoms in the crystalline lattice locations. A variety of auxiliary site descriptors can then be encoded, as well as variables to indicate whether a location constitutes an ideal site.

Using our approach, we conducted extensive computational studies involving an array of crystallographic lattices. These design lattices were tested against a set of structure function relationships spanning a range of ideal site descriptors in order to identify optimal surface patterns relating to a range of catalytic systems. Our proposed approach validates certain surface patterns that were previously known to be optimal but also reveals a number of non-intuitive designs. We also demonstrate that the optimal surface patterns depend strongly on the target application, which strengthens the need to keep developing those fabrication capabilities that can yield higher levels of structural control. Additionally, it was found that the periodicity of the design canvas leads to considerable symmetry in our MILP model, which retards the progress of branch and bound search algorithms. To that end, we also investigate several symmetry breaking symmetry strategies and report the ones that appear to work best in practice.

[1] Narayanan R., El-Sayed M. A., "Shape-dependent catalytic activity of platinum nanoparticles in colloidal solution," *Nano Letters*, 4(7):1343–1348, 2004.

[2] Halas N. J., "Playing with Plasmons: Tuning the Optical Resonant Properties of Metallic Nanoshells," *MRS Bulletin*, 30(5):362–367, 2005.

[3] Caldorera-Moore M., Guimard N., Shi L., Roy K., "Designer nanoparticles: incorporating size, shape and triggered release into nanoscale drug carriers," *Expert opinion on drug delivery*, 7(4):479–495, 2010.

[4] Hammer B., Norskov J. K., "Theoretical Surface Science and Catalysis — Calculations and Concepts," *Advances in Catalysis*, 45:71–129, 2000.

[5] Stamenkovic V., Mun B. S., Mayrhofer K. J. J., Ross P. N., Markovic N. M., Rossmeisl J., Greeley J., Nørskov J. K., "Changing the activity of electrocatalysts for oxygen reduction by tuning the surface electronic structure," *Angewandte Chemie - International Edition*, 45(18):2897–2901, 2006.

[6] Nørskov J. K., Bligaard T., Rossmeisl J., Christensen C. H., "Towards the computational design of solid catalysts," *Nature Chemistry*, 1(1):37–46, 2009.

[7] Nørskov J. K., Bligaard T., Logadottir A., Bahn S., Hansen L. B., Bollinger M., Bengaard H., Hammer B., Sljivancanin Z., Mavrikakis M., Xu Y., Dahl S., Jacobsen C. J. H., "Universality in Heterogeneous Catalysis," *Journal of Catalysis*, 209(2):275–278, 2002.

[8] Calle-Vallejo F., Martínez J. I., García-Lastra J. M., Sautet P., Loffreda D., "Fast prediction of adsorption properties for platinum nanocatalysts with generalized coordination numbers," *Angewandte Chemie - International Edition*, 53(32):8316–8319, 2014.

[9] Calle-Vallejo F., Loffreda D., Koper M. T. M., Sautet P., "Introducing structural sensitivity into adsorption-energy scaling relations by means of coordination numbers," *Nature chemistry*, 7(5):403–410, 2015.

466287 First-Principles Rheological Modelling of Smart Drilling Nanofluids

Abstract:

Oil drilling campaigns need rely on reliable, high-performance drilling fluids which can facilitate in-situ control of fluid rheology, making their transport properties tunable. This benefits their application in different downhole environments, where a wide range of cumbersome technical and economic challenges arise. A nanofluid can perform several tasks in the fluid system, and facilitate drilling with a reduction in total solid and/or chemical content of the equivalent mud, whilst also reducing costs. Smart drilling fluids ensure greater filter cake (and thus well) stability in Low Pressure-Low Temperature (LPLT) conditions, thus enabling lower fluid losses, greater well stability and remarkably reduced capital expenditure (Mahmoud et al., 2016).

The development of first-principles models for the rheology of nanofluids is essential because it can effectively guide their tailored preparation: characterising the nanofluid behaviour as a function of shear rate, nanoparticle volume fraction and temperature is critical toward modelling, design and planning of cost-effective drilling campaigns. Upon successful validation, these models have general applicability to all function forms of nanoparticle-enhanced drilling fluids, eliminating the need for limited-value approximations and ad hoc parameterisations of fluid rheology.

First-principle rheological models describing the shear stress and viscosity of novel nanoparticle-based fluids have been derived on the basis of an isoelectric point assumption (Reilly et al., 2016). Model development relies on establishing a force balance between the Stokes drag force exerted upon the (idealised nanoparticle) sphere and the van der Waals interparticle attraction force described by the Hamaker equation. The interparticle distance variation is accounted for by means of an equation which relies on an equilibrium assumption between thickening and thinning rates (Pouyafar & Sadough, 2013). Nanoparticle interactions have been modelled employing an exclusive consideration of particle pairs, with a centre-to-centre separation distance of four radii at zero shear rate. Nevertheless, nanoparticles have been considered adequately dispersed at high shear rates, as a result of the attenuation of local energy gradients because of the interaction of vortices generated by particle rotation. The maximum separation distance can be computed on the basis of a homogeneous particle distribution assumption (Masoumi et al., 2009). The variation of interparticle distances between a minimum and a maximum estimated value has been approximated using a shear-dependant probability distribution, assuming that all particle pairs exist at one of these two states. These dispersion effects have been quantified as additional contributions to base fluid properties.

A univariate equation with a single fitting parameter characterised experimental data from two sources (Barry et al., 2015; Gerogiorgis et al., 2015) with a high coefficient of determination (0.999 in both cases). Coefficients of determination were more rigorously assessed using a logarithmic transformation (0.994 and 0.993, respectively), correcting for a highly asymmetric dataset (a skewness of 2.6 and 2.2 has been computed, respectively). A more elaborate trivariate equation characterised viscosity for variations in shear rate, concentration of nanoparticles, and temperature with high predictive potential (0.986). The coefficient of determination has been rigorously assessed using a logarithmic transformation (calculated to be 0.983). Temperature effects have been explicitly accounted for with an Arrhenius relationship: the activation energy of 11.24 kJ/mol computed is in accordance with values measured in traditional fluids (Khalil et al., 2011). Significant error attenuation improvements in viscosity estimation have been observed when comparing the novel correlation with the standard Herschel-Bulkley model, for a range of shear rates (5 s^{-1} to 1020 s^{-1}); in contrast to the Herschel-Bulkley model which failed to yield a coefficient of determination higher than 0.90, the first-principles model has achieved coefficients over 0.98, for both datasets and univariate as well as trivariate model instances (Reilly et al., 2016).

LITERATURE REFERENCES

1. Barry, M.M. et al., 2015. Fluid filtration and rheological properties of nanoparticle additive and intercalated clay hybrid bentonite drilling fluids. *J. Pet. Sci. Eng.*, **127**: 338–346.

2. Gerogiorgis, D.I., Clark, C., Vryzas, Z., Kelessidis, V.C., 2015. Development and parameter estimation for an enhanced multivariate Herschel-Bulkley rheological model of a nanoparticle-based smart drilling fluid. *Comput.-Aided Chem. Eng.* **37**: 2405-2410.

3. Khalil, M., Jan, B.M. & Raman, A.A.A., 2011. Rheological and statistical evaluation of nontraditional lightweight completion fluid and its dependence on temperature. *J. Pet. Sci. Eng.* **77**(1): 27–33.

4. Mahmoud, O. et al., 2016. Nanoparticle-based drilling fluids for minimizing formation damage in HP/HT applications, SPE Paper 178949, in: *SPE International Conference and Exhibition on Formation Damage Control.*

5. Masoumi, N., Sohrabi, N., Behzadmehr, A., 2009. A new model for calculating the effective viscosity of nanofluids. *Journal of Physics D: Applied Physics*, **42**(5): 055501 (6).

6. Pouyafar, V. & Sadough, S.A., 2013. An enhanced Herschel–Bulkley model for thixotropic flow behavior of semisolid steel alloys. *Metallurgical and Materials Transactions B*, **44**(5): 1304–1310.

7. Reilly, S.I., Vryzas, Z., Kelessidis, V.C., Gerogiorgis, D.I., 2016. First-principles rheological modelling and parameter estimation for nanoparticle-based smart drilling fluids, *Comput.-Aided Chem. Eng.* (in press).

469396 Modelling the Deposition of Actives on Cotton Fabrics during the Washing Process

Abstract:

Many organic textiles such as cotton absorb in the short UV-region of the spectrum extending into the blue range which promotes a yellowing shade of textiles. The extent of this shade depends on several properties of the polymer such as number of conjugated systems, degree of substitution and degradation products [1]. The removal of the yellowish colour is carried out by bleaching of textiles during its processing prior to conduct further finishing steps. However gradual yellowing and whiteness loss of used textile items occur over time due to several causes such as accumulation and aging of unremoved oily soils, being human sebum an important source [2 - 4].

Thus, during the washing process of textiles in a commercial washing machine, whiteness maintenance depends on many different transformations, namely (a) soil removal from fabrics' structure, (b) suspension and anti-re-deposition of removed soil, (c) bleaching of soils remaining in fabrics, (d) deposition of shading dye actives, which are used to shift the yellowness of fabrics to a more preferred blue hue and (e) deposition of Fluorescent Whitening Agents (FWAs), also known as optical brighteners.

Soil Release Polymers (SRPs) play a key role in detergency due to its use for soil release from fabrics and prevention of re-adsorption of soils. FWAs are widely used in the textile and paper industry for improvement and maintenance of whiteness. These molecules absorb light in the UV region (350 nm) and emit on the blue region of the spectrum compensating yellowing of materials [5, 6].

Understanding and modelling the chemical and physical processes driving deposition and retention of the actives driving each of the vectors mentioned above is key for optimising whiteness performance by an optimisation of the detergent formula.

In the present work the adsorption behaviour of two stilbene derivative FWAs typically used in detergent formulas onto flat cotton fabrics has been investigated and modelled as a function of temperature, water hardness, concentration of FWAs in solution, fabric: water ratio and detergent concentration. Furthermore, the impact of SRPs on soil release and soil anti- redeposition on cotton fabrics is investigated.

The adsorption of actives on textiles is tracked by a spectrophotometry based real – time methodology which allows obtaining high time resolution data of the active remaining in solution.

The resulting whiteness of textile garments is assessed by measuring their reflectance under different light conditions by spectrophotometry. Calibration curves that correlate the adsorbed mass of active per area of flat cotton fabrics with the resulting whiteness of textiles have been developed which allow predicting the final whiteness for any adsorbed concentration of actives. This correlation is typically non – linear where the whiteness benefit (delta b* in L*a*b* colorspace) plateaus as we reach high levels of adsorbed active.

Based on experimental data, a mechanistic model for molecular deposition has been developed which considers the following transformations in the system: (I) Dissolution of active in the washing liquor, (II) convective mass transfer flow into the fabrics, (III) Fickian diffusion of molecules to the fabric surface across boundary layer and stagnant layer of water in yarns and (IV) adsorption/desorption on fabrics' structure (adsorption/desorption isotherms).

The time to reach 95% of equilibrium as well as the equilibrium concentration, increase with increasing water to fabric ratio for a fixed active concentration since there is more mass of active per surface area to be adsorbed. Therefore, depending on the type of washing machine (Top Loader vs Front Loader Washing Machine) where the water to fabric ratio varies significantly (typically 0.02 m^3:kg vs 0.004 m^3:kg), washing time plays a key role on the final adsorbed mass of active on cotton fabrics and therefore on the resulting whiteness of textiles requiring almost twice the time for actives in the higher water:fabric ratio machine to reach equilibrium. The importance of these models is that experiments can be run under laboratory conditions with less variable high water to fabric ratios and results scale up to real wash conditions.

A lack of fit is observed for actives depositing on cotton fabrics when Fickian diffusion across a single boundary layer is considered. The hypothesis presented to explain this disagreement between experimental and predicted data is the presence of a complex dual porosity fabric structure [7]. Fabrics have two main porosities, inter-yarn porosity due to the spaces between the yarns of the textiles and intra – yarn porosity due to spaces between the fibres forming the yarns. As a consequence, molecules diffuse across different diffusing distances before they reach the fabric surface. Initially a rapid decay of the bulk concentration is experimentally observed which is thought to correspond to the adsorption of molecules on the outer surface of the fabrics, which progressively decreases until saturation is reached leading to diffusion to deeper areas of fabrics which would promote a slower decrease of the bulk concentration due to larger diffusion distances. The approach followed for modelling the entire process is based on the consideration of two different surfaces with different surface area and diffusion distance where molecules can diffuse and get adsorbed simultaneously in parallel.

References:

[1] Kramer, J.B. (1992), Fluorescent Whitening Agents, The Handbook of Environmental Chemistry, Volume 3, Part F, Ed. O. Hutzinger, Berlin.

[2] Rounds, M.A.J., Purchase, M. E., Smith, B.F. (1978), *Fatty Soil Retention and the Influence of Particulate and Fatty Soils on Yellowing of an Unfinished Dacron/Cotton Blend Fabric,* Textile Research Journal, 517 – 523.

[3] Chi Y-S., Obendorf, S.K., (1998), *Aging of Oily Soils on Textiles Materials: A literature Review.* Journal of Surfactants and Detergents, 1, (3), 407 – 418.

[4] Chi, Y-S., Obendorf, S.K, (2001), *Effect of fibre Substrates on Appearance and Removal of Aged Oily Soils,* Journal of Surfactants and Detergents, 4, (1), 35 – 41.

[5] XianNan, H., et al. (2008). *Study of adsorption kinetics for fluorescent whitening agent on fibre surfaces.* Sci. China Ser. B-Chem., 51 (5), 473 -478.

[6] Iamazaki, E.I. and Atvars, T., (2007). *Sorption of Fluorescent Whitening Agents (Tinopal CBS) onto Modified Cellulose Fibers in the Presence of Surfactants and Salt.* Langmuir, 23 (26), 12886 – 12892.

[7] Warmoeskerken, M.M.C.G, Van der Vlist, P., Moholkar, V., Nierstrasz, V., (2002). Laundry process intensification by ultrasound. Colloids Surf A: Physiochem. Eng. Aspects 210, 277-285

474607 Mass Transfer Scale-up in Melt Polymerization Processes

Abstract:

Many different polymerization processes are of relevance for the production of performance polymers at DSM, from slurry to melt polymerization ones. For all, the understanding of the main limitations to the reaction speed during polymerization is of fundamental importance, especially when new technologies are brought to the market and need to be scaled-up. In this paper, the case of a batch melt-polymerization reactor will be presented. In particular, the interplay of mixing, heating, and mass transfer phenomena on the final batch time will be the focus, for different polymer grades of interest. It is known in literature, that bubbles generation as well as surface renewal of the polymer melt are important factors to determine a complete and fast polymerization in melt processes, by allowing transfer between the melt phase and the gas phase. Both experimental techniques as well as modeling have been used to show the relevance of each of the aforementioned factors, especially when scaling up from laboratory set-up to pilot plant to, finally, commercial scale batch plant. The outcome of the experimental work can be used for the new design of reactors, allowing for an increased productivity of the plant during polymerization.

461198 Development, Control and Modelling of a Scalable Continuous Manufacturing Process for Multiphase Oxidations

Abstract:

In the production of pharmaceuticals and fine chemicals, most reactions are conducted homogeneously in one phase, and typically require additional operations (known as 'work up') to separate the product from byproducts and any remaining starting materials and catalysts at the end of the reaction. Work up/separation procedures can be complicated and time-consuming, and can constitute 40-70% of the costs of chemical processes. They also consume additional resources (energy, materials, solvent, etc.), which is detrimental to the environment [1].

Herein we discuss the design and development of a new type of fully integrated continuous process for multiphase oxidations at scale. Inorganic oxidants are generated electrochemically in an aqueous phase from non-hazardous inorganic salts [2]. The aqueous oxidant solution is subsequently dispensed to mix with reactants in an organic solvent forming a liquid/liquid reactive flow. An unstable emulsion is generated from the reactive flow using either a surface membrane module or a high shear mixer in a recirculation loop. The unstable emulsion is fed forward in to a continuously pulsed flow system as a way of overcoming mass transport limitations by enhancing liquid/liquid interactions [3] and maintaining the emulsion in a kinetically dispersed state. After an appropriate residence time the emulsion is eluted into a coalescence column allowing for a facile separation of the two phases. The aqueous phase is retained and recycled back to the electrochemical cell whereas the organic phase is collected for product analysis, simplifying the workup procedure.

The experimental mini-plant is supported by a predictive multiphase model that has been developed to incorporate different time constants (residence time, droplet formation and coalescence rates as well as intrinsic reaction rates) in order to identify appropriate operating regimes for any given oxidation process.

References

[1] Adler, S., et al., Vision 2020: 2000 Separations Roadmap. **2000**, *American Institute of Chemical Engineers*: New York. 1-99.

[2] Zhu, J., K.K. Hii, and K. Hellgardt, Toward a Green Generation of Oxidant on Demand: Practical Electrosynthesis of Ammonium Persulfate. *ACS Sustainable Chem. Eng.*, **2016**. 4(4), 2027-2036.

[3] Holdich, R.G., et al., Continuous Membrane Emulsification with Pulsed (Oscillatory) Flow. *Ind. Eng. Chem. Res.*, **2013**. 52(1), 507-515.

Acknowledgements

The work is supported by EPSRC grants EP/L012278/1 and EP/L011697/1.

464752 19th Century Meet the 21st Century: Scaling Batch Reactors

Abstract:

The heart of every chemical process is the reactor. The majority of chemical reactions in the chemical processing industry occur in batch reactors. At the laboratory scale, batch reactors are reaction rate limited because they have large heat transfer surface area to reacting volume ratios. The same is generally true for pilot plant batch reactors. However, most commercial-size batch reactors are heat transfer rate limited because their heat transfer surface area to reacting volume ratios are small.

Since the mid-1950â€™s, the cost of computation via digital computers has decreases annually. With the invention of the personal computer and user friendly software since the mid-1980â€™s, the cost of computation has become miniscule. The chemical processing industry has embraced computation as a means for modifying its historic process development procedure. Today, we use reaction data gathered from a laboratory-sized batch reactor to model as large a batch reactor as deemed safe to commission and operate based on that laboratory data.

To model a batch reactor via computation, we first derive the differential equation d[P]/dt = R where [P] is the concentration of product P (moles/m^3) and R is the formation of P by chemical reaction (moles/m^3*s or moles/m^3*min). Our task is to determine R by one of four methods. Those methods are

1. guess the concentration dependence of R_P, solve the above differential equation analytically, plot the resulting dependent variable as a function of time, then calculate k, the rate constant, from the slope of the line;
2. determine the concentration dependence of R_P using the initial concentration experimental method, then calculate k;
3. determine the concentration dependence of R_P using the initial time central difference method, then calculate k;
4. determine the concentration dependence of R_P using the initial time polynomial method, then calculate k.

These procedures were developed between 1890 and 1940. Today's computing power and user-friendly software allow us to develop a different procedure for determining the kinetics of a batch reaction and to scale that reaction into ever-larger batch reactors.

This presentation discusses a new method for determining the kinetics of a reaction occurring in a batch reactor and it discusses how to use this new method to scale the heat transfer requirements of a commercial-sized reactor designed for the reaction in question.

461339 Scale up Methodology and Practical Application of Wet Bead Milling for the Production of Submicron Particles

Abstract:

Particle size reduction via wet bead milling of a temperature sensitive pharmaceutical material to a reproducible submicron endpoint is achievable across the Netzsch Delta Vita mill series. The reduced particle size allows for an improved dissolution rate when compared to unmilled material. Wet bead milling offers the advantage of increased heat dissipation and removal when compared to a dry milling process. A total energy model approach was used to properly scale the process while geometrically scaling equipment and holding key process variables including formulation composition, agitator tip speed, and bead characteristics, among other parameters, constant between scales. An advantage of scaling with constant tip speed is that the power and heat transfer surface area also scale by the diameter of the mill which results in similar cooling across scales thereby avoiding unwanted surprises for temperature sensitive products during scale up. The milling process was smoothly scaled from lab to industrial scale.

451920 Influence of the Energy Criteria for the Development of Industrialization Strategies in Melt Crystallization

Abstract:

Influence of the energy criteria for the development of industrialization strategies in melt crystallization

GUILLOU L., DENAGNON Y., RIGAUT E.

Fives Proabd, 20C rue de Chemnitz – F-68100 Mulhouse

The main purpose of melt crystallization technologies is the purification of organic chemicals. These technologies are in general regarded as being alternatives to other unit operations. Crystallizations from the melt are therefore quite often restricted to limited amount of situations:

- Separation of compounds with close boiling points or forming azeotropic systems
- Purification of heat sensitive or explosive compounds
- Ultrapurifications for 99.95+ grades
- ...

This framework definition implies that crystallization from the melt is eventually at best a complementary technique that is used to overcome issues that would not have been successfully addressed by other approaches. That being said melt crystallization is, within exception of a very limited number of widespread applications, a kind of fifth wheel of the cart in the process designer toolbox.

The melt static crystallization (MSC) is one of the embodiments of the melt technologies. Being by nature a batch process, the MSC is quite often regarded as being, despite its efficiency, applicable only to limited capacities. Quite often a limit set at 10 ktpy is considered. It is associated to applications dealing with ultrapurifications. This combination narrows a little bit more the potential scope of application as it is understood and expressed in most of the reference literature.

As a resulting feature the MSC is regarded as being irrelevant for purifications implying a large composition difference between the feedstock to be treated and the targeted final product. It is indeed considered that these cases will require using fractional crystallization approaches thus multiplying the number of crystallization stages. Such multiplication of recrystallizations would imply additional energetic expenses, lowering of the recovery yield...

Nothing could be anything but wrong. A majority of applications including the ones dealing with ultrapurification requirements or significant concentration of the compound to be purified (purity raised over 50 wt.% or more in concentration difference) are achieved with one or two MSC stages (cf. Fig 1).

Though selected examples, different purification processes organization taken from industrial plants will be presented and rated. The underlying process design strategies will be examined under the scope of energy consumption optimization between processes i) without, ii) including or iii) fully based on MSC. This comparative examination is based on the same performances in terms of capacities achieved and qualities of the feedstock and main products.

We will show how only by optimization of the energy consumption target function the technical solutions built around a MSC core process can match or even overtake the most direct and obvious approaches often favored by process designers.

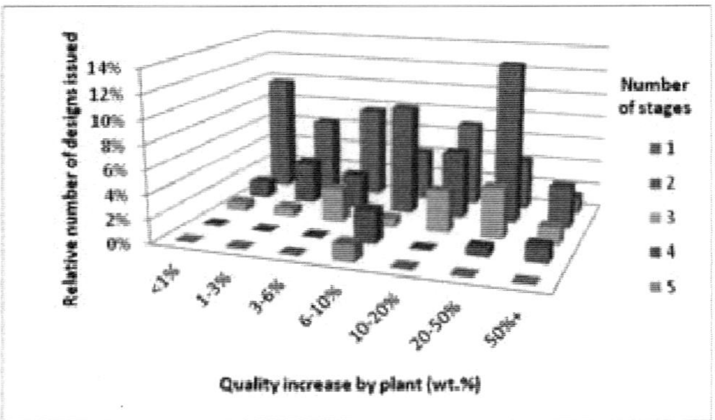

Figure 1: topology of industrial MSC plants (based on Fives Proabd® reference portfolio)

Keywords: Crystallization, purification, energy optimization.

457998 Case Study of a Recrystallization Scale-up with Desirable Psd and Morphology

Abstract:

Scale-up to control particle size of an energetic (in this case nitro-containing compound) is particularly challenging due to its inherent safety hazards. The focus of this work is to scale up a recrystallization process for an energetic compound in order to achieve a block-like morphology with a size in the range of 200 – 300 μm. Due to the energetic nature of the material, strategies such as seeding are undesired. Thus our efforts focused on self-seeding via an optimal anti-solvent dosing strategy.

Efforts began with a solvent screen and construction of solubility curves, using the Avantium Crystal16™ parallel crystallizer. Experiments were conducted at the 100 mL scale using the Mettler Toledo EasyMax™ workstation to optimize dosing and cooling strategies. In situ FBRM and PVM probes were used.

The most promising conditions were then tested at the 2 L scale, using the Mettler Toledo RC1e™. CFD mixing modeling was used to achieve a greater understanding of the effects of mixing regimes on the results of the 2 L recrystallization; this was then used to project the effects of mixing at larger scales up to the target reactor volume of 200 gallons. In addition, FBRM and IR data were used to construct desaturation curves, which aided in reducing the batch cycle time without negatively impacting the crystallization. The 2-L RC-1 resulted in yields greater than 90% and gave isolated solids with improved thermal stability, while meeting the target morphology and size requirements. This process was then used to scale up the crystallization to a 20 L vessel, and finally, to 200 gallons.

451461 Behavior of Crystal Layer Growth in Dynamic Layer Melt Crystallization of Phosphoric Acid

Abstract:

Phosphoric acid (PA) and its derivatives have been widely used in numerous fields such as medical, food and electronic industries with stringent limit of impurity ion contents (e.g., Pb^{2+}, K^{+}, Mg^{2+},). Thus, PA produced by dry or wet process needs further purification to improve the purity by means of separation and purification techniques, such as solvent extraction, chemical precipitation, ion exchange and crystallization. Among these methods, layer melt crystallization is a technically proven method for the preparation of ultrapure substances, forming a crystalline layer on a cooled surface, and commonly further purifying the products by a sweating process to eliminate the impurity inclusions trapped in the crystal layer during its growth. This method is considered as a potential green, environmentally friendly and highly selective separation technology.

It is difficult to analyze the processes of layer melt crystallization quantitatively as the heat and mass transfer occurs simultaneously and changeably during the growth of crystal layer. Fortunately, the necessary thermodynamics properties of PA-H_2O have been measured in previous works, and these required properties can be used as input data in the mass and heat transfer. The growth rate and the effective thermal conductivity of PA hemihydrate during the crystal layer growth process were measured as a function of process conditions to characterize the manufacture efficiency. Besides, the impurity entrapment phenomenon was investigated in order to achieve a desired separation effect of layer melt crystallization for the purification of PA. A reliable model of dynamic layer melt crystallization was developed by the energy balance and some reasonable assumptions. The essence of this model is the description of heat transport during the crystal layer growth in a forced convective melt, showing the implicit relation between crystal layer growth and the forced convection.

During the experiments, the PA melt was mixed by forced convection caused by stirring to eliminate the temperature and concentration difference of the melt, and the variation of the temperatures of the whole process was recorded to have a deeper understanding of the crystallization. An assumed qualitative temperature profile in a cross section of the crystalline system was presented to enable the prediction of profiles of temperature in both the crystal layer and the melt conformed to the experimental data. The initial nucleation and basal crystal layer adhering on the cooled wall were studied as they have a profound and lasting effect on the crystallization process. During the course of crystallization, a recrystallization operation was used after the addition of seeds to guarantee a uniform cylindrical crystal layer around the cooling finger crystallizer. The experimental research on the crystal layer growth was focused on how the operational conditions affect the manufacture efficiency of PA hemihydrate product. Thus, the growth rate of crystal layer was investigated as a function of concentration and the degree of super-cooling. The formed crystalline PA hemihydrate layer is the mixture of the solid crystals and a small amount of mother liquor regarded as a complex, multi-phase, porous material. Then the effective thermal conductivity of PA hemihydrate was calculated by heat transfer coefficient method to characterize the overall heat transfer rate of crystal layer. The effective thermal conductivity of crystal layer of PA is thought to depend strongly on the physical structure affected by temperature and concentration.

451921 Reaction Intensification in Coal Tar Refining

Abstract:

Reaction intensification in coal tar refining

SHASHATY A.[1], GUILLOU L.[2]

1. McGill University, 845 Rue Sherbrooke O, QC H3A 0G4, Montréal, Canada

2. Fives Proabd, 20C rue de Chemnitz – 68100 Mulhouse, France

Processing of coal tar oils has been a way of choice for the recovery of dozens of cyclic, aromatic or polyaromatic compounds: phenol, cresols, indane, coumarone, pyridine bases, thiophenols, carbazole, [1]... The associated coal tar distillation industry has been the flag ship for the development of industrial chemistry up to the 50's. If this industrial area is nowadays considered as being mature the well established associated technology portfolio [2] is ready to evolve. Some opportunities offered by process intensification give access to new options for the tar refiners. Beyond the most obvious opportunities for decreasing the refinery footprints by using some smaller and less energy demanding equipments that are offered by intensified devices, the implementation of these new technologies offers new production alternatives for recovery of chemicals of interest which were given up due to uncompetitive operational costs vs. synthetic routes.

A typical example is considering the so called tar acids. This family of compounds gathers phenol and its homologue molecules (cresols and xylenols). The traditional recovery route is based on a series of reactions with strong bases and strong acids. In a first step, the oil is treated with caustic soda to recover an aqueous phenolate liquor which is then reacidified to reform an organic phenol blend. Such a process is usually either a batch process or a continuous Lurgi phenoraffin process. In the first case, due to the nature of the chemicals involved, a significant investment is required: glass lined reactors, graphite heat exchangers... Serious industrial safety concerns are also raised. This leads to the definition of an economical bottom line which restraint these applications to large coal tar refineries (beyond 400 ktpy whereas most refineries are nowadays within 200-300 ktpy size range). As a matter of fact, shall the phenolate liquor be removed for some process constraints in the distillation train, they are either qualified as a waste or as a low value byproduct which shall be sold to another industry. In such a latter case, these streams are collected throughout many refineries and treated in a single central facility; thanks to economies of scale, these downstream treatment centers can afford moving to continuous processes such as phenoraffin.

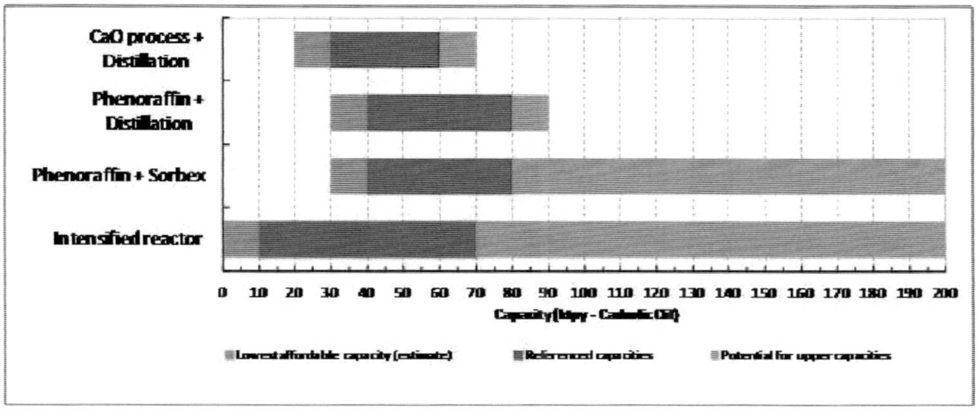

The potential for the recovery of phenols and other tar acids has been investigated at laboratory scale in a microchannel reactor. The process consisted in a two steps:

- A neutralization of a carbolic oil/naphtha fraction with diluted caustic soda
- An acidification with sulfuric acid for forming back phenols for sodium phenolates.

These reactions show some distinctive features such as:

- Large exothermicity both due to the reaction and the dilution of acid or base in the media
- Extremely fast kinetics
- Undesirable formation of bright red coloring agent for uncontrolled temperature rise
- Polyphasic systems

The use of microchannel reactor has been found to be a good alternative to the traditional processing routes. High control of the exothermy has been achieved and phenol recovery has allowed achieving satisfactory yields with decreased water consumption. This technology opens a route for compact reactor systems perfectly compliant with continuous operation in refineries even for small capacities. This new generation of compact reactors opens opportunities for renewal of coal tar refining business model, assisting refiners into a move toward higher value chemicals whatever the size of the distillation train might be.

[1] Franck H.-G., National meeting of American Chemical Society, 13 Jan. 1963, Vol 7:1 Conference 143.

[2] Franck H.-G., Stadelhofer J.W., in "Industrial aromatic chemistry", 1988, Springer-Verlag, Berlin-Heidelberg-N.-Y.

Keywords: reaction intensification, compact reactor, coal tar.

456936 Application of Dynamic Heat Exchanger Design in Natural Gas Liquefaction Process

Abstract:

The paper discusses transient Aspen HYSYS modeling and optimization of two natural gas liquefaction processes and identifies the rate-limiting components during load variations. The optimized model for both processes provides details for comparison. Novel flowrate variations included in this investigation drive transients responses of all units, especially compressors and heat exchanger. Heat exchangers commonly represent the most sensitive components to transients. This sensitivity decreases when using new heat exchanger designs and control methods. Model-predictive control (MPC) responses compare favorably with results using traditional controls. Transient efficiency graphs for both designs illustrate improvements during model predictive control. These new controls and designs optimize natural gas (NG) consumption and liquefied natural gas (LNG) production.

451752 Implementation of Process Intensification (PI) in the Removal of Carbonyl Compounds from the Water/Ethanol Solvent Mixture

Abstract:

Carbonyl compounds are frequently observed ingredients in biochemical, food, chemical and pharmaceutical industries. The high reactivity of such compounds is a great advantage if aldehydes/ketones are important contributors in the synthetic routes towards the final products. However, if the carbonyl compounds act as impurities, they must be removed from the reaction mixture, due to the high tendency of forming genotoxic impurities (GTIs).

Removal of carbonyl compounds could be performed in several ways. Applications of adsorption techniques by using alumina, titanium, or activated carbon, are well-known and even commercially applicable in the petrochemical industry. Implementation of in- or ex-situ removal with pervaporation has also achieved significant advantages with the main application in fermentation processes. Furthermore, combinations of derivatization and separation processes are known, such as reactive extraction with amines or even solid-phase extraction with 2,4-dinitrophenylhidrazine (DNPH). Some new techniques are additionally getting popular, such as applications of extractant impregnated resins (EIRs).

The main aim of this work is to remove benzaldehyde from a solvent mixture consisting of 70 % water and 30 % of ethanol (volumetric ratio). Such a solvent mixture indicates formation of acetals, hemiacetals and hydrates with carbonyl compounds, and could also promote electrophilic and nucleophilic substitution reactions of aldehydes/ketones with other compounds in the reaction mixture. The main approach is focused on finding suitable derivatization agents that would convert carbonyl compounds into new impurities that are harmless. The screening procedure was therefore performed initially with primary/secondary/tertiary amines, inorganic salts, organic acids and their anhydrides. The most desired option was found to be tris(hydroxymethyl)aminomethane (TRIS), which converts carbonyl compounds into oxazolidines. A Process Systems Engineering (PSE) approach including property prediction simulations indicated that such products are more inert compared to carbonyl compounds and additionally more suitable for removal by applying other separation processes.

TRIS was used in a process development study in order to remove ppm levels of benzaldehyde. The initial process design was performed in batch mode with off-line real time process monitoring. A Process Intensification (PI) approach was implemented and the movement from batch to continuous processing was achieved by using a laminar tubular reactor. Real-time process monitoring was also moved from off-line with High Performance Liquid Chromatography (HPLC) towards in-line with UV-VIS spectroscopy in order to satisfy requirements defined within the Process Analytical Technology (PAT) guidance. Both types of processes are compared in order to evaluate advantages of implementing PI, PAT and PSE. Further work will include process control studies and automation, as well as considerations of suitable scale up/out possibilities.

Sensitivity Analysis of the Separation Efficiency of a Vacuum Stripping Microchannel Device

Chafika Adiche, Technische Universität Darmstadt, Thermal Process Engineering, Darmstadt, Germany

Abstract

In this work a two dimensional model describing the transport properties in a vacuum stripping microchannel device was developed. Based on this model, a sensitivity analysis of the separation efficiency of the device to its relevant design and operating parameters was conducted. It is shown that the developed model allows to elucidating the impact of the investigated design and operating parameters on the controlling resistance to mass transfer in the whole device including the liquid-vapor membrane contactor and the liquid channel and thus constitutes a useful tool towards the design optimization of the device.

1 Introduction

The implementation of gas / vapor – liquid separation operations like absorption, stripping, and distillation at microscale constitutes an important milestone for the synthesis of multistep microchemical processes [1]. However, over the last two decades the development of separation microchannel devices for such operations has proven to be a challenging task as compared to the well-established microreactors, micromixers and micro heat exchangers. Indeed the high surface-to-volume-ratio prevailing at microscale results in (i) a poor control of the required thermodynamic vapor-liquid equilibrium [2] and in (ii) a high flow resistance particularly in liquid channel (s). Both phenomena result in the reduction of the separation driving force and thus in the drop of the separation efficiency.

Recent investigations showed that liquid repellent vapor-liquid contactors with hydrophobic surface properties may be successfully used in distillation [3], membrane distillation [4-5] and stripping [6-8] based separation microchannel devices. As compared to wetting vapor-liquid contactors [7-8], the liquid repellent vapor-liquid contactors, may have important potentials to reduce the energy consumption of such microchannel devices [9] as well as the flow resistance in the liquid channels [10]; contributing thereby to a better stabilization of the vapor-liquid interface and thus to the enhancement of the separation efficiency of the device.

In this work, the separation efficiency of a vacuum stripping microchannel device is investigated using a model based approach.

The device is constituted of two micromachined rectangular straight channels separated by a flat sheet microporous hydrophobic membrane (contact area: 80 mm × 6 mm) [Figure 1].

The membrane serves as liquid-vapor contactor, which repels the liquid phase and constitutes therefore a support for the liquid-vapor interface at the entrance of the pores. The liquid feed (aqueous solution of 5 wt. % acetone) is circulated through one of the channels. The other channel called permeate channel is kept under vacuum. Acetone and water evaporate and diffuse through the membrane pores. The separation driving force in the device is created by the partial vapor pressure gradient for each component (acetone, water) across the membrane caused by the vacuum in the permeate channel. The permeated vapor phase is then stripped by vacuum outside the permeate channel.

To avoid liquid intrusion into the membrane pores, the pressure difference across the membrane must be maintained below the liquid breakthrough pressure of the membrane [3-5, 11]. Furthermore, the pressure in the permeate channel must be lower than the vapor pressure of the feed [11].

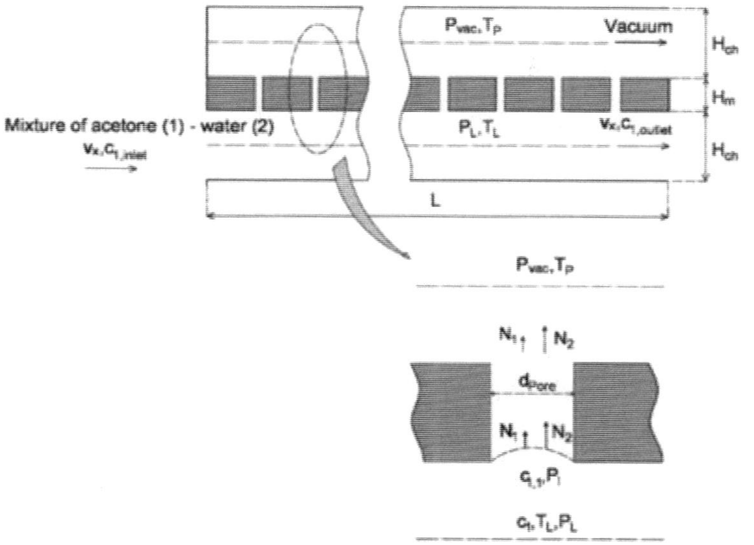

Figure 1. Schematic view of the model of transport properties in a vacuum stripping microchannel device

For the above presented separation device, the following relevant design and operating parameters were identified: (i) the liquid channel depth H_{ch}, (ii) the geometric properties of the considered membrane contactor including porosity ε, mean pore size d_{pore} and thickness H_m, (iii) the feed flow rate or the residence time in the liquid channel τ_{hydr}, (iv) the temperature in the liquid channel T_L and (v) the vacuum pressure P_{vac}.
In this work, two commercial flat sheet microporous hydrophobic membranes provided by Pall (USA) were considered: polyethersulfone polymer cast on a non-woven polyester-support membrane (Supor®-450R) and acryl copolymer membrane on polyester-support (Versapor®-200TR). The contact angles measured for the investigated liquid mixtures on both membrane surfaces were found to be larger than 120 °.

2 Modelling and simulation

Theoretical investigations were carried out for the assessment of the influence of the above mentioned design and operating parameters on the separation efficiency of the vacuum stripping microchannel device at isothermal conditions, and thus to elaborating an optimization approach for the design of the device. For this purpose, a two-dimensional mathematical model describing the transport mechanisms in the device including the liquid channel and the membrane was developed. The model is based on the following assumptions:

a. Since the width of the liquid channel is much larger than the investigated liquid channel depths, the side-wall effects are neglected and the problem may be reduced to two dimensions.
b. For all investigated liquid flow rates, the Reynolds number Re is smaller than 10, so that a hydrodynamically developed laminar flow may be assumed.

c. For all investigated values of vacuum pressure, the Knudson number kn was found to be in the range between 2 and 26, so that the Knudson diffusion may be assumed as the prevailing mass transport mechanism in the membrane pores.
d. Finally, the temperature in the liquid channel is assumed constant and equal to the temperature in the permeate channel i.e. $T_P = T_L$.

Accordingly, the equations of the model are given as follows:

1. The flow velocity profile writes

$$v_x = 1.5\langle v_x\rangle\left(1-\frac{4y^2}{H_{ch}^2}\right). \tag{1}$$

2. The species conservation equation for acetone is given by
3. $$v_x\frac{\partial c_1}{\partial x} = D^L\left(\frac{\partial^2 c_1}{\partial x^2}+\frac{\partial^2 c_1}{\partial y^2}\right), \tag{2}$$

with the diffusion coefficient D^L is calculated using Wilke-Chang correlation.

4. The diffusion flux for acetone at the liquid-membrane interface is formulated as

$$N_1 = \frac{K_m}{\sqrt{MW_1}}\left(P_{1,1} - P_{vac}\, y_{p,1}\right), \tag{3}$$

where K_m is the permeability coefficient of the membrane [11], $P_{1,1}$ the partial vapor pressure of acetone at the liquid-membrane interface and MW_1 the molecular weight of acetone. The permeate composition y_p is computed using the double film theory [11].

5. The liquid-vapor equilibrium relationship at the liquid-membrane interface was determined using Wilson model (*Properties Analysis Tool of Aspen Plus*)

$$P_{1,1} = \frac{1.29c_{1,1}}{c}. \tag{4}$$

6. The boundary conditions for mass transfer are given by

$$c_1\big|_{x=0,y=0} = c_{1,inlet}, \tag{5.a}$$

$$-D^L\frac{\partial c_1}{\partial y}\bigg|_{y=+\frac{H_{ch}}{2}} = N_1, \tag{5.b}$$

$$\frac{\partial c_1}{\partial y}\bigg|_{y=-\frac{H_{ch}}{2}} = 0, \tag{5.c}$$

and $$\frac{\partial c_1}{\partial y}\bigg|_{x=L,\, y=-\frac{H_{ch}}{2};+\frac{H_{ch}}{2}} = 0. \tag{5.d}$$

The previously presented model along with the corresponding boundary conditions has been reformulated using dimensionless variables given by

$$\bar{v}_x = \frac{v_x}{\langle v_x\rangle},\ \bar{y} = \frac{y}{H_{ch}/2},\ \bar{x} = \frac{2x}{H_{ch}Pe},\ \bar{c}_1 = \frac{c_1}{c_{1,inlet}}, \tag{6}$$

with the Peclet number Pe is expressed as

$$Pe = Sc\,Re = \frac{L\,H_{ch}}{2\,D^L\,\tau_{hydr}}. \tag{7}$$

The dimensionless model writes as follows:

- Dimensionless velocity profile:

$$\bar{v}_x = 1.5\left(1-\bar{y}^2\right). \tag{8}$$

- Dimensionless species conservation equation for acetone:

$$\bar{v}_x \frac{\partial \bar{c}_1}{\partial \bar{x}} = \frac{1}{Pe^2}\frac{\partial^2 \bar{c}_1}{\partial \bar{x}^2} + \frac{\partial^2 \bar{c}_1}{\partial \bar{y}^2}. \tag{9}$$

- Dimensionless boundary conditions:

$$\bar{c}_1\Big|_{\bar{x}=0,\bar{y}=0} = 1, \tag{10.a}$$

$$\left.\frac{\partial \bar{c}_1}{\partial \bar{y}}\right|_{\bar{y}=+1} = -\frac{H_{ch}\, N_1}{2\, D^L\, c_{1,inlet}}, \tag{10.b}$$

$$\left.\frac{\partial \bar{c}_1}{\partial \bar{y}}\right|_{\bar{y}=-1} = 0, \tag{10.c}$$

and $$\left.\frac{\partial \bar{c}_1}{\partial \bar{y}}\right|_{\bar{x}=\bar{L},\,\bar{y}=-1:+1} = 0. \tag{10.d}$$

The separation efficiency of the device is measured using the stripping degree. The latter is ranging between 0 and 1, which correspond to 0 and 100 % acetone removal from the liquid phase, respectively.
The degree of stripping σ is determined according to the following equation

$$\sigma = 1 - \frac{\int_{-1}^{+1} \bar{v}_x \bar{c}_1(\bar{L})\, d\bar{y}}{\int_{-1}^{+1} \bar{v}_x\, d\bar{y}}. \tag{11}$$

The established model with all relevant boundary conditions has been implemented using Finite element Method (Comsol Multiphysics), with mesh of triangular elements. A fine mesh was constructed near the membrane surface where significant concentration gradient is expected. To test the precision provided by the selected mesh-size, a mesh-size independence study was conducted for each investigated *Peclet* number and channel depth.

3 Simulation results and discussion

Based on the elaborated model, a sensitivity analysis of the separation efficiency of the device to the previously mentioned design and operating parameters was conducted.

3.1 Influence of the geometric parameters

In this set of simulations, the channel depth was varied between 50 and 500 µm, while all other relevant design and operating parameters were kept constant.
Figure 2 shows that for the membrane Supor®-450R, decreasing the liquid channel depth until 300 µm affects hardly the device performance such that an almost constant and low stripping degree of about 0.18 was determined. For the same range of liquid channel depths, similar behavior was also obtained using Versapor®-200TR with however a slightly lower stripping degree (0.15). Further decrease of the liquid channel depth gives rise to a greater enhancement of the stripping efficiency for each membrane with however a higher extend for Supor®-450R as compared to Versapor®-200TR. For example for the smallest investigated liquid channel depth of 50 µm, the stripping degree obtained for Supor®-450R was 0.77, whereas that determined for Versapor®-200TR was only 0.67.

These results suggest that for liquid channel depths larger than 300 μm, the molecular diffusion in the feed channel is the rate limiting step for mass transfer in the whole device. However, for a channel depth smaller than 300 μm, the resistance to mass transfer in the membrane becomes more important, so that using the membrane allowing a larger acetone permeate flux, here Supor®-450R (with a larger ratio of the product of membrane pore size and porosity to membrane thickness), leads to a better device performance.
According to these results, Supor®-450R membrane contactor will be considered for the next simulation investigations.

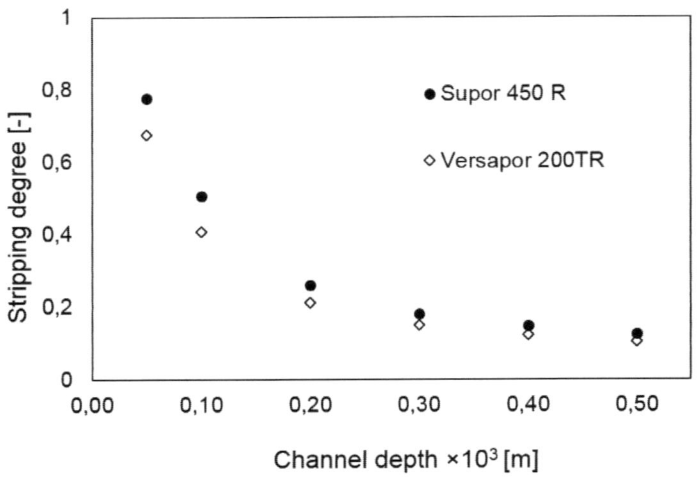

Figure 2. Influence of the channel depth on the stripping degree for different membrane contactors, $x_F = 5$ wt. %, $T_L = 20$ °C, $P_{vac} = 0.01$ bar, $\tau_{hydr} = 5s$.

3.2 Influence of the residence time in the liquid channel

We can see in Figure 3 that for the three investigated channel depths, increasing the residence time in the liquid channel, leads to increasing the stripping degree of the device. Furthermore, the smaller the channel depth, the faster is the increase of the stripping degree until reaching a value of 0.84 for a channel depth of 50 μm and a residence time in the liquid channel of 10 s.

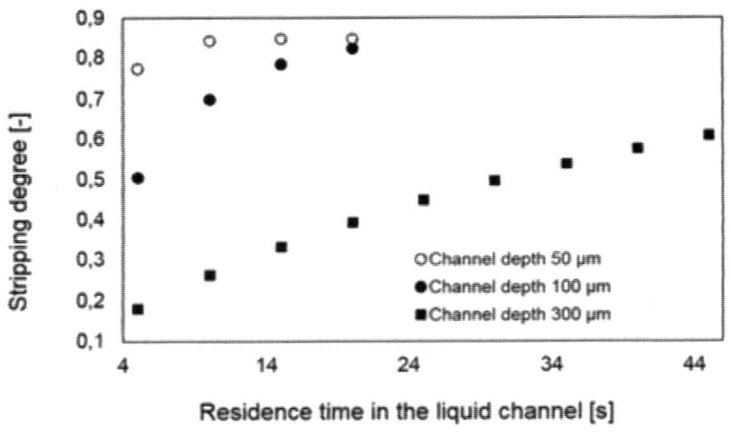

Figure 3. Influence of the residence time in the liquid channel on the stripping degree for different channel depth, $x_F = 5$ wt. %, $T_L = 20$ °C, $P_{vac} = 0.01$ bar, Supor 450R.

This behavior may be attributed to the reduction of the concentration polarization in the liquid channel, resulting therefore in a better mixing of the acetone-water solution (computed diffusion time of acetone in the liquid channel with a depth of 50 µm is about 1.2 s). Consequently, the separation driving force of acetone at the liquid-vapor interface increases, leading therefore to a higher acetone permeate flux, and thus to enhanced stripping efficiency. Similar trend was also reported by Chasani et al. [6] for the stripping of dilute toluene from water using a hydrophobic vapor-liquid contactor into a microchannel device with dry nitrogen as stripping agent.

3.3 Influence of the vacuum pressure

Figure 4 indicates that for each investigated temperature, increasing the vacuum pressure while maintaining all other design and operating parameters constant, results in decreasing the stripping degree. That results in the decrease of the separation driving force, thus in the reduction of the acetone permeate flux and finally in the drop of the stripping efficiency. Furthermore, it is shown that for a given vacuum pressure, increasing the feed temperature improves substantially the stripping efficiency (due mainly to the exponential increase of the vapor pressure of the feed with the temperature), along with broadening the vacuum pressure range.

Similar behavior was also found for the influence of varying the vacuum pressure along with the feed composition, with however the difference that increasing the acetone concentration in the feed for a given vacuum pressure [Figure 5] improves only slightly the stripping efficiency due to the Linear increase of the vapor pressure of the feed with the feed composition.

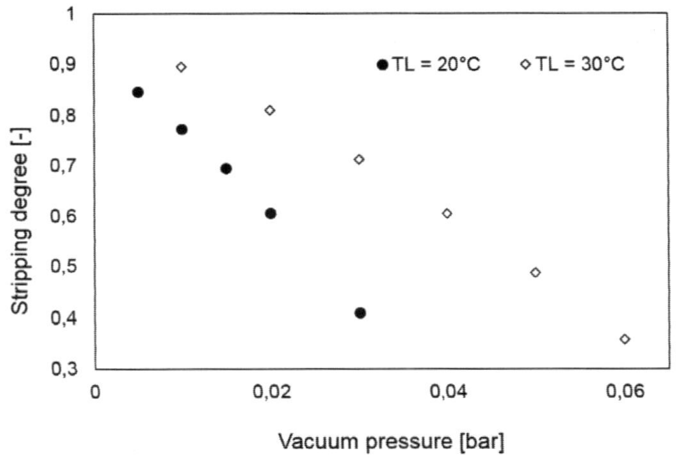

Figure 4. Influence of the vacuum pressure on the stripping degree for different temperatures in the device, $x_F = 5$ wt. %, $\tau_{hydr} = 5s$, $H_{ch} = 50$ µm, Supor 450R.

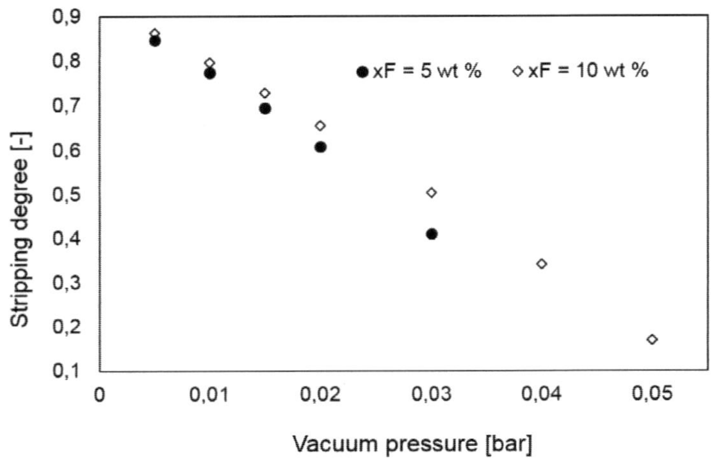

Figure 5. Influence of the vacuum pressure on the stripping degree for different feed compositions, $T_L = 20$ °C, $\tau_{hydr} = 5s$, $H_{ch} = 50$ µm, Supor 450R.

References

1. Pastre, J. C.; Browne, D. L.; Ley, S. V. (2013), '' Flow chemistry syntheses of natural products'', *Chem. Soc. Rev.*, 42, pp. 8849-8869.
2. Hartman, R. L.; Sahoo, H. R.; Yen, B. C.; Jensen, K. F. (2009), ''Distillation in microchemical systems using capillary forces and segmented flow'', *Lab Chip*, 9, pp. 1843-1849.
3. Zhang, Y.; Kato, S.; Anazawa, T. (2009), ''Vacuum membrane distillation on a microfluidic chip'', *Chem Commun.*, pp. 2750-2752.
4. Adiche, C. and Sundmacher, K. (2010), ''Experimental investigation of a membrane distillation based micro-separator'', *Chem. Eng. Proc.: Process Intensification*, 49 (4), pp. 425-434.
5. Adiche, C. and Sundmacher, K. (2010), ''Selection of membrane contactors used in membrane distillation based micro-separator'', *Chem. Eng. Transactions*, 21, pp. 361-366.
6. Chasanis, P.; Kehrmann, K.M.; Kern, J.; Zecirovic, R.; Grünewald, M.; Kenig, E. Y. (2011), '' Investigation of a microstructured high efficiency contactor'', *Chem. Eng. and Process.*, 50, pp. 1244-1251.
7. Constantinou, A.; Ghiotto, F.; Lam, K. F.; Gavriilidis, A. (2014), ''Stripping of acetone from water with microfabricated and membrane gas-liquid contactors'', *Analyst*, 139, pp. 266-272.
8. Cypes, S. H. and Engstrom, J.R. (2004), ''Analysis of a toluene stripping process: a comparison between a microfabricated stripping column and a conventional packed tower'', *Chem. Eng. J.* 101, pp. 49-56.
9. Wootton, R. C. R. and de Mello, A. J. (2004), ''Continuous laminar evaporation: micron-scale distillation'', *Chem. Commun.,* pp. 266-267.
10. Ou, J.; Perot, B.; Rothstein, J. P. (2004), '' Laminar drag reduction in microchannels using ultrahydrophobic surfaces'', *Physics of Fluids*, 16, 12, pp. 4635-4643.
11. Bandini, S.; Saavedra, A.; Sarti, G. C. (1997), ''Vacuum Membrane Distillation: Experiments and Modeling'', *AIChE J.*, Vol. 43, No. 2, pp. 398-408.

466009 Isolation of Biobased Chemicals By Combination of Reactive Distillation and Pervaporation; Process Modeling

Abstract:

Wastewater from pulping contains several acids, mainly acetic acid and formic acid. The ternary mixture acetic acid, formic acid and water forms a binary high boiling azeotrope between formic acid and water and a ternary saddle point azeotrope. The separation step of azeotropic mixtures is energy intensive, especially when water is present as low boiling component with high excess. By changing the substance properties via esterification, reaction products with lower boiling points and low enthalpy of evaporation compared to water as well as the acids are produced.

Reactive distillation is perfectly suited for the separation of acetic acid, formic acid and water by esterification with methanol or ethanol. Therefore low boiling esters are produced which are continuously removed via the distillate. Elevated temperature in the reboiler and simultaneous esterification of both acids enhances the reaction rates. High conversion of the acids is achieved. The distillate contains the esters and excess alcohol. This mixture can further be separated via pervaporation by removing the alcohol with hydrophilic membranes. Chemical conversion of the carboxylic acids simplifies the separation step and produces esters with a higher market value.

Minimization of energy consumption and maximum acid conversion of reactive distillation was optimized by simulation and validated experimentally. Diffusivity of membranes in pervaporation of binary and ternary mixtures was determined experimentally and was modeled with an activity based approach.

467093 Modelling and Optimization of Multistage Separation Process at Oil Offshore Platform

Abstract:

In the era of low oil price, major oil companies strive hard to improve accuracy of the offshore process design and operation efficiency of the platform. For this, this study focuses on the development of optimized offshore topside model, satisfying product specification, maximizing oil production and consequently maximizing profit. First, modeling of topside process that includes three phase separators, gas compressing system and recycle scheme was conducted, and optimal process configuration was selected based on evaluation of multiple case studies. Then, optimization of operating condition (e.g. gas to oil ratio, operating pressure, etc.) was performed by using Mixed Integer Non-Linear Programming (MINLP). Lastly, economic feasibility study was conducted for the design of profitable and eco-friendly offshore platform.

461101 API Particle Formation: Technological Approaches to Forming Unique API Characteristics

Abstract:

The formation of highly differentiated APIs to enable specific downstream filtration, drying and drug product performance, in terms of flowability, blending, content uniformity and in vitro disintegration and dissolution is becoming an ever important feature in pharmaceutical process design.

Within this presentation the application of technological platforms, such as continuous processing, hybrid batch/continuous wet milling enabled system, will be discussed in detail. A series of case studies will show the work flow and design details associated to engineering and population balance models, reactor design, supersaturation monitoring / control to enable the delivery of highly differentiated API particle characteristics. The interface between the drug substance and drug product will be discussed at length to demonstrate the ability of engineering strategies to enable predictable downstream process performance.

466385 Seeding Policy and Optimization of Batch Crystallization Processes

Abstract:

Two important components of a recipe for the operation of a seeded batch crystallization process are the seed properties and the temperature (or supersaturation) profile. When seeds are used in batch crystallization, in most cases the objective is to relieve as much of the generated supersaturation as possible by the growth of the seeds, which corresponds to minimizing the nucleated mass.

Doki et al. (2002) showed that at least for some systems, nucleation could be effectively suppressed by using a sufficient seed mass, called the critical seed loading, and that the required seed mass depends on seed size. Later, Tseng and Ward (2014) showed that the critical seed loading could be determined knowing only crystal nucleation kinetics, which permitted the determination of the critical seed loading ratio for a large number of chemical systems for which nucleation kinetics have been measured. In this contribution we show that similar analysis can be used to determine a feasible range of seed sizes and corresponding seed loadings that will generate product crystals of a desired size (or size distribution) with negligible nucleation.

Furthermore, extending previous work by Hofmann et al. (2010), we show that nearly-analytical solutions for the optimal supersaturation trajectory (and Praeto fronts for multi-objective cases) can be determined for a variety of different objective functions. The nearly-analytical nature of the results allows greater understanding of the effect of the objective function on the shape of the optimal trajectory. This method also requires only knowledge of the nucleation kinetics, and therefore can be applied to determine the optimal supersaturation trajectory for a large number of systems.

Combining the two methods together, we develop a systematic procedure for rapidly determining operating recipes for seeded batch crystallization with a small number of experiments.

References

Doki, N., N. Kubota, M. Yokota and A. Chianese (2002). Determination of critical seed loading ratio for the production of crystals of uni-modal size distribution in batch cooling crystallization of potassium alum. Journal of Chemical Engineering of Japan 35(7): 670-676.

Hofmann, S., and J Raisch (2010). Application of optimal control theory to a batch crystallizer using orbital flatness, 16th Nordic Process Control Workshop, Lund, Sweden, 25â€"27 August 2010; Nordic Working Group on Process Control.

Tseng, Y. T. and J. D. Ward (2014). Critical seed loading from nucleation kinetics. AIChE Journal 60(5): 1645-1653.

466256 Defining a Design Methodology for Conversion of a Batch API Crystallization to a Continuous Crystallization

Abstract:

Crystallization is an essential unit operation for small molecule active pharmaceutical ingredient (API) manufacturing. In recent years there has been an increasing focus on continuous crystallization of such APIs. Advantages of continuous processing include increased control of product attributes, reduced equipment foot print, potential elimination of the requirement for scale-up activities, greater flexibility of equipment infrastructure and reduced expenditure.

This study focuses on creating a defined methodology for converting an existing batch crystallization process to a continuous system. Many previous studies have focused on characterizing continuous crystallization equipment and addressing processing challenges arising from continuous operation such as fouling and attaining steady state operation. The equipment used for this study was a paired CSTR system operated as an MSMPR. The use of this type of system for continuous crystallization has been previously demonstrated in literature [1-4].

The main goal of the design methodology was to transition an existing batch cooling crystallization process to a continuous format, while (at least) maintaining the yield and desired PSD of the existing batch crystallization. The API chosen as the case study for application of the design methodology was produced by a cooling crystallization following a salt formation in solution.

The initial stages of the design methodology consisted of a deeper investigation of the existing batch process, to include solubility screening, solvent selection optimisation, metastable zone width investigation and solid form characterisation. Concurrently, investigations of the existing crystallization process were performed with online monitoring of the process using FBRM and FTIR at 1L scale. From this work an improved understanding of induction times, nucleation temperature and the effect of the cooling rates and hold times on PSD as well as the crystal form was generated.

Temperature selection for continuous cooling crystallization is crucial, and a number of temperature set points were evaluated in batch format to improve confidence for continuous operation. A temperature strategy was selected for first implementation in continuous format, to be sequentially modified as understanding grew.

Following this work a two-step continuous cooling crystallization process was implemented using 2 glass jacketed 1 L reactors. Hot solution was fed to the first crystallizer using a peristaltic pump. Pressure transfer zones were used for intermittent transfer of slurry from the first crystallizer to the second crystallizer and on to a holding vessel for isolation. The effect of start-up strategy on particle size was investigated and was optimized to give the fastest time to steady state for the desired particle size. Various start up strategies were evaluated, including start up with a saturated solution, start up with a seed bed (with various seed loadings), as well as start up with crystal slurry from a previous batch respectively in the first tank. Through systematic trials the initial temperature strategy was sequentially modified until the desired operation and particle size production was achieved. Particle size was measured offline using a Malvern Mastersizer. Crystal form of the product was measured using XRD. The effect of various process parameters such a residence time, agitation and slurry transfer regime on the final PSD were investigated. The product yield was found to compare well with the existing batch process.

This project was made possible by funding from Enterprise Ireland Innovation Partnership Project IP-2014-0356

1. Hou, G., et al., *Development and Characterization of a Single Stage Mixed-Suspension, Mixed-Product-Removal Crystallization Process with a Novel Transfer Unit.* Crystal Growth & Design, 2014. **14**(4): p. 1782-1793.

2. Morris, G., et al., *Estimation of Nucleation and Growth Kinetics of Benzoic Acid by Population Balance Modeling of a Continuous Cooling Mixed Suspension, Mixed Product Removal Crystallizer.* Organic Process Research & Development, 2015. **19**(12): p. 1891-1902.

3. Power, G., et al., *Design and optimization of a multistage continuous cooling mixed suspension, mixed product removal crystallizer.* Chemical Engineering Science, 2015. **133**: p. 125-139.

4. Polster, C.S., et al., *Pilot-Scale Continuous Production of LY2886721: Amide Formation and Reactive Crystallization.* Organic Process Research & Development, 2014. **18**(11): p. 1295-1309.

467662 Tailoring Crystal Size Distribution with Sonocrystallization

Abstract:

Control over the product crystal size distribution (CSD) is very important in industrial crystallization for efficient downstream processing and improved formulation properties[1] Sonocrystallization is a prominent technique to reduce the crystal size and to generate uniform crystal size distributions[2, 3]. Current state of the art includes the sonocrystallization assisting PAT(Process Analytical Technology) to better understand the sonocrystallization process to better control the crystal size distribution to meet the specific requirements of an active pharmaceutical ingredient (API). Various approaches to use ultrasound for size distribution control in a batch cooling crystallization have been investigated. Solute concentration is monitored by In-situ React IC-IR and FBRM used to identify the metastable zone width and crystal chord length distributions. Crystallization of Piracetam in Isopropyl alcohol (IPA) was studied in a batch crystallizer with a working volume of 500ml. A calibration based approach was used for the solute concentration measurement. Crystal size measurement using light microscopy and Image-processing is a promising technique to determine the actual CSD rather than measuring with Laser diffraction technique. Width of the crystal size distribution characterized as Span = (D(90)-D(10))/D(50) determines the final product size distribution. It was observed that, conventional cooling crystallization with operating conditions of agitation rate and cooling rate resulted in bigger and broader crystal size distribution. However, reduced Meta stable zone width (MSZW), lower crystal size and lower Span were observed in case of continuously sonicated crystallization. Application of ultrasound in MSZW at low power of ultrasound and lower cooling rate leads to creation of lower nucleation rate and bigger crystals with narrow distribution. The application of ultrasound prior to nucleation leads to higher nucleation rate (which was observed from FBRM count rate) and smaller crystals with a more narrow distribution. Careful supersaturation control after the nucleation step can lead to the formation of uniform product crystals. The only drawback of application of ultrasound in crystallization process is that the heat liberation due to high power ultrasound, which will be difficult to control by cooling systems. The temperature rise because of liberation of heat of crystallization will also lead to partial dissolution of nuclei which also helps in serve as a feedback control system of the nucleation. Total supersaturation was consumed due to ultrasound induced nucleation and concentration followed the solubility curve, where applied intermittent sonication serve as heating cycles rather than creating more nucleation. For optimization, the influence of time and power of ultrasound, and cooling profile were investigated. Crystal size distribution and span were compared between different control strategies including uninsonated, continuous sonicated, sonicated prior to nucleation and sonicated in MSZW followed by intermittent sonication. Additionally continuous crystallization with ultrasound in a two reactor system of Sono-Nucleator (50ml) and subsequent growth of crystals in a 3liters of reactor was studied.

The results of this research provides better understanding of the sonocrystallization process because of the inclusion of in-situ monitoring systems along with ultrasonic probe in a batch crystallizer. An insight is obtained about how to choose process trajectories to produce smaller crystals with narrow CSD and bigger crystals with narrow CSD, and about implementation in a larger scale batch and continuous crystallization.

References:

1. S. Kim, B. Lotz, M. Lindrud, K. Girard, T. Moore, K. Nagarajan, M. Alvarez, T. Lee, F. Nikfar, M. Davidovich, S. Srivastava and S. Kiang, *Organic Process Research Development*9 (2005), 894–901.

2. Luque de Castro, M. D.; Priego-Capote, F. Ultrasonics sonochemistry 14 (2007), 717–24

3. Jieqiong Li,Ying Bao, Jingkang Wang, Chemical engineering technology 36 (2013), 1341–1346

This work has been supported by Science Foundation Ireland, Grant number: 12/RC/2275.

469553 Process Parameters in the Purification of Curcumin By Cooling Crystallization

Abstract:

Structurally related impurities (often by-products) are of great importance in the pharmaceutical industry. Such impurities are hard to remove completely by crystallization, because of their structural similarity to the solute molecule. Impurities may become incorporated into structures and influence nucleation[1,2] and growth[2], and thus polymorphic outcome, crystal shape and size.[2-4] Often multi-step successive crystallizations are needed to reach a desired purity, significantly lowering the product yield.

Curcumin, a major active component of the Indian spice Turmeric (Curcuma Longa), has a wide range of bioactivities (antimicrobial, anti-inflammatory and anticancerogenic properties). Extract obtained from Turmeric comprises a mixture of curcumin, demethoxycurcumin (DMC) and bisdemethoxycurcumin (BDMC), known as curcuminoids. Commercially available crude curcumin contains significant amounts of the other two curcuminoids, typically about 17% DMC and 3% BDMC. Separation and purification of curcumin is quite challenging, since curcuminoids are structurally very similar, with differences only in the number of methoxy groups.

In this work, purification of crude curcumin by successive cooling crystallizations has been investigated.[5] Statistical experimental design techniques were used to identify the effect of cooling rate, seeding, seed polymorph and agitation on the yield, crystal purity, particle size, shape and structure as well as determining the minimum number of crystallization steps needed to arrive at a particular product specification. Isopropanol was selected as the solvent for the curcumin crystallization based on determined solubilities of the individual curcuminoids. The composition of solid curcumin samples and concentration of individual curcuminoids dissolved in the mother liquor were quantified by HPLC. The polymorphic composition of the crystals was determined by Raman spectroscopy and by X-ray powder diffraction.

Particle size and shape distribution was determined by light microscopy and image-processing analysis, using Feret's diameter and aspect ratio as particle characteristics. Molecular dynamics simulations were performed to calculate curcumin lattice energy as a function of curcuminoid impurity concentration (GAFF 600 molecules, Gromacs).

The results showed that cooling crystallization of curcumin is sensitive to process parameters including seed polymorph, stirring method, cooling rate and the number of crystallization steps, the latter being closely related to impurity concentrations. Manipulation of these process parameters allows for polymorph selection, particle habit and size control, purity and yield control. It was found that removal efficiency of DMC, the major impurity present in crude curcumin, exponentially decreases as a function of the number of successive crystallizations. Crystallization operated in seeding mode and slow cooling rate increases the purity of the obtained curcumin crystals. In particular, using seeds of the metastable Form II improves the purification of curcumin significantly. The computational results showed that incorporated impurity decreases the lattice energy, and that this decrease is more pronounced for the metastable Form II, indicating a weaker affinity of the impurity molecules for the metastable Form II curcumin. At substantially optimal crystallization conditions high purity curcumin (>99.1%) can be obtained by a reduced number of successive crystallization steps. Such a decrease in the total number of successive crystallizations increase the overall curcumin yield from 28% to 50%.

1. Pino-Garcia, O.; Rasmuson, A. Cryst. Growth Des. **2004**, 4, 1025–1037.
2. Shtukenberg, A.; Lee, S.; Kahr, B.; Ward, M. Annu. Rev.Chem. Biomol. Eng. **2014**, 5, 77–96.
3. Cashell, C.; Corcoran, D.; Hodnett, B. K. Cryst. Growth Des. **2005**, 5, 593–597.
4. Liu, J.; Svärd, M.; Hippen, P.; Rasmuson, A. C. J. Pharm. Sci. **2015**, 104, 2183–2189.
5. Ukrainczyk, M.; Hodnett, K. B.; Rasmuson, A. C. submitted to Org. Process Res. Dev. (2016).

This work has been supported by Science Foundation Ireland, Grant number: 12/RC/2275.

460409 Influence of the Composition of Water/Methanol Mixtures and Temperature on the Crystal Form and Solubility of Sodium Dehydroacetate

Abstract:

The slurry transformation experiments for the monohydrate of sodium dehydroacetate with different compositions of water and methanol at varying temperatures were investigated in this study. It was found that the temperature and water content affected not only the solubility but also the crystal form of the precipitated material. The sodium dehydroacetate anhydrate was obtained with a low water content; however, the monohydrate of sodium dehydroacetate was the more stable form with a high water content. At the same time, the solubility of anhydrate and monohydrate of sodium dehydroacetate were measured with different compositions and temperatures. From powder X-ray diffractionï¼ˆPXRDï¼‰patterns, thermal gravimetric analysis (TGA) data and scanning electron microscopy (SEM), a new needle hydrate form of sodium dehydroacetate was shown on the surface of anhydrous phase when the water content was between 2 % and the critical transformation values. In situ Raman spectra and microscope images demonstrated that this needle hydrate was not the metastable intermediate nor was it the result of concomitant crystallization during the transformation. Pure needle hydrate of sodium dehydroacetate was obtained using evaporation crystallization. Moreover, the PXRD patterns and TGA results revealed that this needle hydrate was another new form of sodium dehydroacetate monohydrate. A possible explanation for these observations was found in the relationship between the change of water content, the interaction between solute and solvent and the influence of supersaturation.

463058 Reaching Optimal Productivity in Batch Viedma Ripening Processes Using Minimal Data

Abstract:

In 2004 Viedma reported that nearly racemic mixtures of L-sodium chlorate and D-sodium chlorate undergo a deracemization process in the presence of an attrition source, i.e., the initially racemic mixture reaches chiral purity within several days of simply stirring the suspension[1]. This discovery, subsequently termed Viedma ripening (VR), was subsequently shown to also be applicable to organic molecules that exhibit a conglomerate phase diagram. It was successfully applied to a host of different molecules, among them amino acids[2,3] and pharmaceuticals[4]. VR thus represents a new and promising way to arrive at enantiomerically pure products.

One of the key observations made in early studies on VR was that the rate towards enantiomeric purity proceeds in an autocatalytic fashion, i.e., starting from a low enantiomeric excess there is considerable lag time before the conversion rate speeds up. The rate is finally slowing down again towards the end of the process.

The overall Viedma ripening process was later identified to consist of a complex interplay between racemization in solution, enantioselective agglomeration, attrition/breakage, and size-dependent solubility/Ostwald ripening.[5,6] Being a rather slow process, attempts at optimizing deracemization processes focused on shortening the time needed to reach enantiomeric purity and hence to increase productivity.[7,8] However, no study so far has attempted a thorough optimization of process conditions with the goal of maximizing productivity.

The incentive to optimize a process towards optimal productivity and shortened batch times is evident, but the flexibility to exchange equipment on existing plants is limited. Consequently, the focus of process optimization should be shifted towards the remaining process conditions, such as amounts of seeds (and enantiomeric purity of seeds), seed distribution, and suspension density. Hence, the key objective of this contribution is to introduce a methodology that achieves optimal productivity (maximizes mass of pure enantiomer produced per time) for batch Viedma ripening processes with the use of as little experimental data as possible.

The methodology presented here seeks to optimize a batch-after-batch production of enantiopure crystals. Within such a processing strategy, the current batch can be conveniently run by re-using the mother liquor from the previous batch (minus appropriate purging to avoid impurity build up), adding a defined amount of a racemic mixture of solids and by seeding the process with some enantiomerically pure seeds (part of the product from the previous batch). Due to VR, such a batch will reach enantiopurity after some time. The goal of this methodology is to keep this time as short as possible. It can be shown that such a processing strategy is mathematically equivalent to an ideal continuous plug flow reactor with a recycle stream consisting of part of its effluent stream. Since Viedma ripening exhibits autocatalytic features, this analogy can be used to find an optimal seeding policy for batch Viedma ripening processes. It will be shown that the only data needed for this strategy is a curve of enantiomeric excess versus time, obtained from a single experiment by standard techniques (e.g. HPLC). It will further be shown that an optimal suspension density can be established without additional data.

The strategy is first shown to work based on extensive model results employing population balance equation models and various kinetic expressions (to ensure a wide range of applicability) and then exemplified using an experimental study on Naproxenethylester crystallized from ethanol, which is racemized in solution using a base (sodium ethoxide in this case).

References:

[1] Viedma, "Chiral symmetry breaking during crystallization: complete chiral purity induced by nonlinear autocatalysis and recycling", Phys. Rev. Lett., 2005, 94, 065504

[2] Viedma et al., "Evolution of Solid Phase Homochirality for a Proteinogenic Amino Acid", J. Am. Chem. Soc., 2008, 130, 15274–15275.

[3] Noorduin et al., "Emergence of a Single Solid Chiral State from a Nearly Racemic Amino Acid Derivative", J. Am. Chem. Soc., 2008, 130, 1158–1159.

[4] Noorduin et al., "Fast attrition enhanced deracemization of naproxen by a gradual in situ feed", Angew. Chem. Int. Ed., 2009, 48, 4581–4583.

[5] Iggland and Mazzotti, "A population balance model for chiral resolution via Viedma ripening", Cryst. Growth Des., 2011, 11, 4611–4622.

[6] Gherase et al., "Experimental and Theoretical Study of the Emergence of Single Chirality in Attrition-Enhanced Deracemization", Cryst. Growth Des., 2014, 14, 928–937.

[7] Iggland et al., "Complete solid state deracemization by High Pressure Homogenization", Chem. Eng. Sci., 2014, 111, 106-111.

[8] Noorduin et al., "Scaling Up Attrition-Enhanced Deracemization by Use of an Industrial Bead Mill in a Route to Clopidogrel (Plavix)", Org. Process Res. Devel., 2010, 14, 908-911.

454846 Tailor-Made Solvent and Process Design for the Separation of Azeotropic Mixtures

Abstract:

Tailor-made Solvent and Process Design for the Separation of Azeotropic Mixtures

Marta Gonzalez Garcia[1], Ioulietta Paraskeva[1], Anjan Kumar Tula[1], Emmanouil Papadakis[1], Deenesh Kavi Babi[2], Rafiqul Gani[1]

[1]SPEED, Department of Chemical and Biochemical Engineering, Technical University of Denmark, Soltofts Plads 1, Building 229, DK-2800 Lyngby, Denmark

[2]Utility and Solvents, Department of Insulin Manufacturing 1, Novo Nordisk A/S, Hallas Allé DK-4400 Kalundborg, Denmark

A typical example of a process that does not require the addition of a solvent for achieving the desired separation of an azeotropic mixture is pressure swing distillation. An important condition for its operational feasibility is that the azeotrope must be pressure sensitive over a moderate change in pressure[1]. However, if the azeotrope does not satisfy this condition or the process is energy intensive and/or environmentally unfriendly then a promising alternative is a solvent – based separation process[2]. The solvent – based process configuration for a two phase system (vapor-liquid) consists of an extraction column followed by a solvent recovery column and the process design together with solvent selection is usually obtained through a trial and error approach (heuristics, mathematical programming, and/or hybrid). However, can the design be obtained without any trial and error? It is possible by first determining the process design for a desired (target) solvent and then finding a solvent that matches the desired solvent properties. This decomposition of the design problem is feasible because the process design is based on the properties of the system and does not need the identities of the chemicals involved. On the other hand, the selection problem needs the identities of the chemicals involved but not details of the process. Also, a unique solvent-azeotropic mixture cannot be obtained for all homogeneous azeotropic mixtures. This means that the azeotropic mixtures need to be classified first and for each class, the optimal solvent behavior and the corresponding *apriori* process design determined. Also, can the solvent-based separation problem further improved in terms of energy consumption for the same separation specification?

Assuming a desired solvent-azeotropic mixture behavior and a set of solvent properties, the design of the two column process can be determined, for example, in terms of product purities, number of stages, feed location, and reboiler-condenser duties without knowing the identity of the solvent. This means that having *apriori* designed the separation process, if a solvent that matches the targeted behavior and properties could be found, a simultaneous tailor-made solvent and process design could be very easily achieved. The conceptual basis for this method is that solvent properties have a direct and unique influence on the separation process behavior, for example, separation feasibility is defined by the distillation boundary (corresponding to the ternary system of solvent and the azeotropic mixture); energy consumption is influenced by solvent capacity and solvent pure component properties; waste generation and environmental impact is influenced by solvent selectivity and solvent environmental properties. By using the diving-force concept[3], it is ensured that for the *apriori* design, the extraction and recovery columns operate at the maximum driving force, which translates into the design of a more sustainable energy efficient process.

The following reverse design problem consisting of two sub-problems is formulated and solved: Sub-problem 1: Given ternary phase equilibrium data, driving force diagrams and properties of the chemical system, determine the optimal energy efficient design of the extraction and the solvent recovery columns. Sub-problem 2: Given, a homogenous azeotropic mixture and an *apriori* process design; find the best (optimal) solvent that satisfies the process design in the form of targeted solvent-azeotropic mixture behavior plus solvent properties such as normal boiling point, heat of vaporization, solubility of solute and many more. Note that matching these targets implicitly mean satisfying all the process constraints such as mass and energy balance, separation feasibility and energy demands. Sub-problem 3: Replace the solvent recovery column (SRC) with a liquid-liquid extraction column plus an evaporator (LL+E) by swapping the original solvent with a swap solvent. The swap solvent takes the solute from the original solvent in the liquid-liquid extraction column and the solute is removed by simply heating the solute-swap solvent mixture. The original solvent is recycled back to the original extraction column and the swap solvent is recycled to the liquid-liquid extraction column. The energy needed for the removal of the solute by evaporation is significantly lower than that required by the SRC using the original solvent.

The main features of the method are: classification of the homogeneous azeotropic mixtures, classification of the solvent-azeotropic mixture behavior; classification of the swap solvent mixture behavior, apriori process design based on driving force, solvent-azeotropic mixture behavior and swap solvent-original solvent-solute mixture behavior; finding solvents and swap solvents that match the design targets; and a database containing apriori process design from which designs of other problems can be extrapolated or interpolated. For each class of azeotropic mixture in the database, a list of pre-selected solvents is already available. Therefore, for a separation involving a known azeotrope, the optimal solvent (or swap solvent) and process design are obtained with minimal calculations. For an azeotropic mixture not available in the database, simply identifying the class of the azeotropic mixture, all other necessary details can be extrapolated or interpolated from the available data.

The presentation will highlight the concepts and the method for tailor-made solvent and swap solvent selection and process design, together with tools developed to implement and test this method. The method comprises of 4 –stages. In stage 1, the azeotropic mixture class is identified. For this stage, a database of 43 homogenous azeotropic mixtures that have an azeotropic composition between 30-70% mol is developed using available experimental data. In stage 2, the solvent target behavior is retrieved for the identified azeotropic mixture from a second database. For this stage, a classification of the solvent-azeotropic mixture behavior is developed using a graphical approach based on ternary plots and the position of the distillation boundaries. The distillation boundary of interest is drawn from the azeotropic point (located on the binary vertex of the two compounds that constitute the azeotrope with compound 1 not removed by the solvent) until it touches and overlaps the vertex of compound 2-solvent binary line. The classification is done between compound 2 and solvent, that is, high or low composition (more or less than

50%mol) of compound 2 in the azeotrope and high or low composition (more or less than 50%mol) of solvent in the distillation boundary overlap. For example, if a mixture has an azeotrope composition of 40%mol of compound 2 (low) and a distillation boundary overlap of 60% of solvent composition (high), it will be classified as Low-High. Therefore, the 43 azeotropic mixtures are classified into four groups: High-High, High-Low, Low-High and Low-Low. This behavior also requires that the solvent must not form azeotropes with either of the compounds and that the ternary system must be totally miscible. In stage 3, based on the selection from stages 1 & 2, the corresponding driving force based *apriori* design is retrieved from the database. For this stage, validated process designs by rigorous simulation for all classes of azeotropic mixtures and solvent-azeotropic behavior are determined and stored in a database. Also, additional information that helps to extrapolate from the stored data are provided. For example, plots of solvent flow rate vs composition of solvent in the compound 2-solvent mixture and solvent flow rate vs driving force. In stage 4, the solvent selection problem is solved by database search and/or standard solvent selection-design techniques (3). Final design is verified through rigorous simulation using as initial estimate, the retrieved *apriori* design data. For the swap-solvent selection, a target liquid-liquid phase boundary is defined together with properties such as melting points and boiling points, which must be much greater than the solute.

The method is generic and able to directly determine the best solvent, solvent flow rate, the driving force and the energy required for separation of the classes of homogeneous azeotropic mixtures available in the database. It allows safe extrapolations for azeotropic mixtures not covered in the database but that match the classified behavior. The method has been tested for a wide range of homogeneous azetropes and a collection of these results will be presented. Until the final rigorous simulation verification stage (for which also a good initial estimate is provided), the method hardly needs any calculations and is able to provide more sustainable and near optimal simultaneous selection of solvents and its corresponding optimal process design. The concept and method, can in principle, be applied for many other simultaneous product-process design problems. The savings is not only in energy and material but also time to operate and solve problems. References

1. Seader, J. D., Henley, Ernest J. y Roper, Keith. *Separation Process Principles.* s.l. : John Wiley, 2011.

2. Igor Mitrofanov, Sascha Sansonetti, Jens Abildskov, Gürkan Sin and Rafiqul Gani, 2012, "The Solvent Selection framework: solvents for organic synthesis, separation processes and ionic-liquids solvents", Computer-Aided Chemical Engineering , 30, 762-766

3. *R. Gani*, E. Bek-Pedersen, 2001, "A Simple New Algorithm for Distillation Column Design", *AIChE Journal*, 46 (6), 1271-1274

460397 A Target Upper Bound on Reaction Selectivity Via Feinberg's Cfstr Equivalence Principle

Abstract:

A Target Upper Bound on Reaction Selectivity via Feinberg's CFSTR Equivalence Principle

Mole balances are widely used in the design of chemical plants and processing equipment as well as in conducting techno-economic analyses of potential chemical business ventures [1]. They allow us to calculate unknown flowrates into and out of a system once an appropriate number of specifications are made. On a broader level, mole balances also serve to bound the attainable region of a reaction system. We should not, however, expect that the entire attainable region be accessible for a process on which capacity constraints are imposed. For example, certain areas of the attainable region may be accessible only by permitting extremely low single-pass reactor conversions to achieve high selectivities resulting in impractically large recycle flowrates. It is the underlying kinetics of a chemistry that determine the size and location of these inaccessible regions.

Feinberg and coworkers developed the Continuous Flow Stirred Tank Reaction (CFSTR) Equivalence Principal which allows one to decompose any arbitrary steady-state reactor separator system with total reaction volume "V" into a new system comprising "R+1" CFSTRs (where R is the number of linearly independent reactions) with the same total reaction volume and a perfect separations system [2, 3]. Using this methodology in conjunction with the kinetics of a system of interest, the attainable region given by the mole balances can be refined. Our work aims to further refine the attainable region by introducing flowrate capacity constraints on the CFSTR Equivalence Principle. The constraints on molar flowrates between the "R+1" CFSTRs and the separations system prevent large recycle streams and small reactor conversions. By optimizing this constrained system of CFSTRs, we can determine the maximum possible selectivity of several chemistries completely independent of reactor design given capacity constraints. We have investigated serial reaction networks and more complex reaction networks involving serial and parallel reactions. With the added constraints, the maximum possible selectivity for both types of reaction networks is less than 100%.

We have used this methodology to analyze the maximum possible selectivity of a selection of realistic chemistries and have compared the results to archetypal reactors (e.g., CFSTR, PFR). The results all support the hypothesis that this CFSTR reactor decomposition can be constrained and optimized to obtain a target upper bound on selectivity for chemistries completely independent of reactor-separator design.

References:

[1] Douglas, J. "Conceptual Design of Chemical Processes", McGraw-Hill, New York (1988).

[2] Feinberg, M.; Ellison, P. *Ind. Eng. Chem. Res.* **2001**, *40*, 3181-2194.

[3] Tang, Y.; Feinberg, M. *Ind. Eng. Chem. Res.* **2007**, *46*, 5624-5630.

460985 A Nonsmooth inside-out Algorithm for Robust Flash Calculations

Abstract:

Dependable algorithms for nonideal vapor-liquid equilibrium calculations are essential for effective process design, simulation and optimization. At present, the inside-out formulation[1] for flash calculations serves as the basis for many of the algorithms used by process simulation software due to its robustness with respect to initialization and inexpensive computational cost. However, the original inside-out algorithm will fail if the specified conditions lead to a single-phase result, as the solution is constrained to obey equilibrium relationships which are only valid in the two-phase region. While heuristic modifications of the algorithm have shown effectiveness in allowing process simulators to converge to single-phase solutions, [2] these provide no guarantee of success. Such behavior is particularly undesirable in a process simulation/optimization problem where flash calculations must be performed on streams where the phase regime at the solution is not known ***a-priori.*** To address these issues, we present a modification of the classical inside-out algorithm, which features an additional nonsmooth equation in the inner loop to relax equilibrium conditions when necessary and a modified scheme for updating the outer loop variables. Additional nonsmooth equations are also employed in the outer loop to avoid numerical issues and nonphysical solutions, such as those needed to identify the correct compressibility factor roots from cubic equations of state. These modifications guarantee existence of a solution that is physically correct, regardless of the number of phases present. The nonsmooth subproblems are efficiently solved with a linear-programming based Newton method[3] supplied with automatically calculated elements of generalized derivatives,[4] while the outer loop may be converged either by a similar nonsmooth Newton-type method or simply by successive substitution. Numerical results for simulations involving different flash types and property packages are shown, highlighting the capability of the algorithm to handle both two-phase and single-phase behavior robustly and efficiently. This flash module can also readily be included as a subroutine in more complex problems, since desirable properties of the generalized derivative elements and our nonsmooth framework allow objects which behave in the same way as classical derivatives to be passed around a flowsheet for use in Newton-type methods for simulation and optimization.

---

[1] J. F. Boston and H. I. Britt, "A radically different formulation and solution of the single-stage flash problem," ***Computers & Chemical Engineering***, Vol. 2, pp. 109-122, 1978.

[2] V. S. Parekh and P. M. Mathias, "Efficient flash calculations for chemical process design – extension of the Boston-Britt "Inside-out" flash algorithm to extreme conditions and new flash types," ***Computers & Chemical Engineering***, Vol. 22, No. 10, pp. 1371-1380, 1998.

[3] F. Facchinei, A. Fischer and M. Herrich, "An LP-Newton method: nonsmooth equations, KKT systems, and nonisolated solutions," ***Mathematical Programming***, Vol. 146, pp. 1-36, 2014.

[4] K. A. Khan and P. I. Barton, "A vector forward mode of automatic differentiation for generalized derivative evaluation." ***Optimization Methods & Software,*** Vol. 30, No. 6, pp. 1185-1212, 2015.

461804 Simultaneous Chemical Process Synthesis and Heat Integration with Unknown Cold/Hot Process Streams

Abstract:

In chemical process synthesis, the heat integration between cold and hot process streams reduces the external utility consumptions and leads to economic benefits. Sequential heat integration approaches rely on the assumption that temperatures and flow rates of process streams are known, which means that these methods can only be applied after stream conditions have been determined. On the other hand, simultaneous approaches can handle variable stream temperatures and flow rates, which means that the heat integration subsystem can be optimized along with the process and thus stream conditions. It has been shown that simultaneous approaches lead to better solutions compared to the sequential approaches [1]. While various simultaneous heat integration models have been proposed [1, 2], they require all the process streams to be classified as cold or hot streams (depending on whether the streams require heating/cooling) prior to solving the problem. However, in some cases there exists a subset of process streams that cannot be predetermined as cold or hot streams, especially in the context of superstructure optimization that involves the selection of alternative processing units. For example, the product stream from one of the two alternative reactors (A and B) is sent to one of the two alternative separation units (C and D) for purification. Units A, B, C and D might operate at different temperature ranges. Therefore, we might not be able to determine if the stream that connects the reactor and separator is a cold or hot stream before solving the problem.

Accordingly, in this work, we present a mixed-integer nonlinear programming (MINLP) optimization model for the simultaneous chemical process synthesis and heat integration with general, that is undetermined, process steams. The model extends the transshipment model proposed by Papoulias and Grossmann [3], to account for (1) streams that cannot be classified as hot or cold, and (2) variable stream temperatures and flow rates. The cold/hot stream â €œidentitiesâ€ are represented by classification binary variables which are (de)activated based on the relative stream inlet and outlet temperatures. Variables including stream temperatures and heat loads are disaggregated into two

mutually exclusive parts (cold and hot), and each part is (de)activated by the corresponding classification binary variable. The dynamic temperature intervals constructed from the variable process stream inlet temperatures are implicitly ordered from high to low temperatures and stream inlet/outlet temperatures are assigned to the correct temperature intervals. Stream heat loads at each temperature interval depend on the relative positions of stream inlet and outlet temperatures with respect to the temperatures at the interval boundaries. Preprocessing is used to reduce the size of the problem and speed up the solution process. For example, if a process stream can be predetermined as a cold stream, we can fix the corresponding classification binary variable to one, and deactivate the hot part of the disaggregated variables. It is important to note that the only nonlinearity in this model are the bilinear [flow rates]*[temperature difference] terms. Therefore, in the case that the stream flow rates or temperatures are fixed, the proposed model reduces to a MILP model; and if both of them are fixed, we recover the original transshipment model.

The proposed model is generalized to handle intermediate utilities at arbitrary temperatures. Further, the model is able to deal with both isothermal and non-isothermal phase changes. Should phase changes of pure components are present in a process stream, the original stream can be disaggregated into sub-streams and the heat integration model can be reformulated to include phase detection (i.e. decide at what phase this stream starts and ends).

The heat integration model can be integrated with a process superstructure model. To achieve this integration effectively, we discuss a series of modeling techniques that lead to simplified unit models and connectivity equations. For example, we show that the minimum number of temperature intervals needed for heat integration equals to the minimum number of process streams over all the feasible solutions, which is fewer than the total number of process streams in the original superstructure.

The proposed model is applied on several illustrative examples with variable flows and temperatures, in which some streams have unknown cold/hot classifications. Finally, we present a case study in which the proposed model is integrated with a realistic bioethanol production superstructure and is optimized based on an economic criterion [4].

References

1. Duran, M.A. and I.E. Grossmann, *Simultaneous-Optimization and Heat Integration of Chemical Processes.* Aiche Journal, 1986. **32**(1): p. 123-138

2. Navarro-Amorós, M.A., et al., *An alternative disjunctive optimization model for heat integration with variable temperatures.* Computers & Chemical Engineering, 2013. **56**: p. 12-26

3. Papoulias, S.A. and I.E. Grossmann, *A Structural Optimization Approach in Process Synthesis .2. Heat-Recovery Networks.* Computers & Chemical Engineering, 1983. **7**(6): p. 707-721

4. Kong, L., et al., *A superstructure-based framework for simultaneous process synthesis, heat integration, and utility plant design.* Computers & Chemical Engineering, 2016 DOI: http://dx.doi.org/10.1016/j.compchemeng.2016.02.013.

461915 A Method for Minimization of Total Exergy Loss over the Complete Search Space of Regular Distillation Configurations

Abstract:

Exergy of a stream is its potential to do work when it is brought to the reference conditions via a reversible path. And it is a good metric to account for the temperature level at which heat is used or rejected in a distillation column. The thermodynamic efficiency of a distillation configuration can be derived through exergy analysis. Thus, performing exergy analysis, in conjunction with minimization of total heat duty, will help us identify distillation configurations with low energy requirement and high thermodynamic efficiency. Such configurations are particularly valuable for cryogenic as well as natural gas liquids (NGL) separations.

Here, we present a global optimization algorithm based on Underwood's equations that minimizes the total exergy loss of any regular-column distillation configuration (i.e. distillation sequence separating *n* components using *n*-1 columns) which separates any ideal or near-ideal multicomponent mixture. This objective function tends to favor configurations that use high temperature condensers and low temperature reboilers, relative to other configurations with similar vapor duties. For the first time, a global optimization based rank-list of the complete search space of regular-column configurations based on minimizing total exergy loss is synthesized. The identification of configurations that simultaneously have low vapor duty and high thermodynamic efficiency provides industrial practitioners with strong candidates for heat pump assisted multicomponent distillation.

462457 Shale Gas to Light Olefins: Global Optimization of an Integrated NGL Recovery, Steam Cracking, and Methane Conversion Superstructure

Abstract:

Due to the decreased prices of natural gas in recent years [1], production of "wet" gas that contains natural gas liquids (NGLs), such as ethane, propane, butane, and natural gasoline, is favorable to extract for commercial purposes [2]. In some wet shale plays, the combined ethane and propane composition can exceed 30% [3]. These recent trends motivate industrial efforts to build several new ethane crackers with a combined ethylene production capacity of 12.5 million tonnes/year in the United States [4]. Although the NGLs can be effectively used as petrochemical feedstocks toward light olefins, the optimal processing technology (catalytic and/or thermal), feed composition (mixed feed or a pure feed), and operating conditions are unknown.

Previous work has shown that light olefins including ethylene, propylene, and butene isomers can be produced from natural gas via reforming in a profitable manner out of a superstructure of process alternatives [5]. However, extraction of NGLs prior to methane conversion is an untapped opportunity for these refineries that can simultaneously maximize carbon conversion and avoid high reforming costs. This NGL recovery section is introduced to the existing superstructure of novel/competing process alternatives. A demethanizer column is utilized [6,7] to recover more than 83% of ethane and more than 99% of higher hydrocarbons in the NGL stream. The NGLs can be further separated and sent to steam cracking or catalytic dehydrogenation alternatives. Several steam cracking reactors incorporate a pure or mixture of hydrocarbon feed to produce olefins. The operating conditions and reactor topologies of the steam cracking alternatives are dynamically optimized using the well-established cracking kinetics in the literature [8,9]. Catalytic dehydrogenation process alternatives incorporate commercially available catalysts for propane and/or isobutane dehydrogenation. The methane rich gas from the demethanizer is converted via reforming or direct conversion processes introduced into the process superstructure [5].

The mathematical model of the overall process superstructure is solved using a novel branch and bound global optimization framework. The objective is to maximize the profit of light olefin production from the integrated NGL extraction, cracking, catalytic dehydrogenation, and methane conversion refinery. A technoeconomic analysis will be presented with major topological decisions across several case studies. Different natural gas compositions will be shown to present the topological trade-offs in the NGL recovery section. The net present values can be significantly improved (>10%) through the use of the integrated superstructure.

[1]: Floudas, C. A.; Niziolek, A. M.; Onel, O.; Matthews, L. R. Multi-scale systems engineering for energy and the environment: Challenges and opportunities. AIChE Journal 2016, 62 (3), 602-623.

[2]: Kopalek, M.; High value of liquids drives U.S. producers to target wet natural gas resources, 2014, Energy Information Administration

[3]: Hill, R. J.; Jarvie, D. M.; Zumberge, J.; Henry, M.; Pollastro, R. M.; Oil and gas geochemistry and petroleum systems of the Fort Worth Basin, 2007, AAPG Bulletin, 91(4), 445-473.

[4]: Chang, J.; New projects may raise US ethylene capacity by 52%, PE by 47%, 2014, ICIS Petrochemicals.

[5]: Onel, O.; Niziolek, A. M; Floudas, C. A Optimal Production of Light Olefins from Natural Gas via the Methanol Intermediate. Industrial & Engineering Chemistry Research 2016, 55 (11), 3043-3063.

[6]: Luyben, W. L.; NGL Demethanizer Control, 2013, Industrial and Engineering Chemistry Research, 52, 11626-11638.

[7]: Luyben, W. L.; Effect of Natural Gas Composition on the Design of Natural Gas Liquid Demethanizers, 2013, Industrial and Engineering Chemistry Research, 52, 6513-6516.

[8]: Sundaram, K. M.; Froment, G. F.; Modeling of thermal cracking kinetics – I: Thermal cracking of ethane, propane and their mixtures, 1977, Chemical Engineering Science, 32(6), 601-608.

[9]: Sundaram, K. M.; Froment, G. F.; Modeling of thermal cracking kinetics – II: Cracking of iso-butane, of n-butane, and of mixtures ethane-propane-n-butane, 1977, Chemical Engineering Science, 32(6), 609-617.

467281 An Algorithm for Integrated Molecular and Process Design: Physically-Driven Domain Reduction for Liquid-Liquid Extraction

Abstract:

Computer Aided Molecular and Process design (CAMPD) involves the simultaneous optimisation of processes and processing materials (such as mass separating agents, catalysts) [1]. In CAMPD one considers the minimisation of a process-wide objective subject to process design constraints and molecular feasibility constraints. Molecular-level (material) decisions are represented using discrete variables such as the number of groups of a given kind, whereas process variables can either be continuous or discrete. The resulting design problem is a mixed integer nonlinear problem (MINLP) which is difficult to solve for a number of reasons. The large molecular design space leads to combinatorial explosion [2]; the non-convexity of the domain can lead to poor solutions depending on the starting point; convergence failure can arise for some combinations of molecular and process variables. In particular, numerical difficulties can be encountered in the absence of a good initial guess to the highly nonlinear, and coupled, process and property models. Numerical issues can also occur if a combination of molecular and process inputs violates constraints of phase behaviour implicit in the model. For instance, an evaluation of an absorber model can fail if the stream at its liquid inlet is in the vapour state. Also, the feasible region with respect to these constraints is a function of the solvent: for example, the range of temperatures where a molecule is in the liquid state varies with the choice of solvent.

To address these challenges we have recently a developed a new algorithm for CAMPD of separation systems [3, 4]. The algorithm builds on the concept of embedding screening of molecules within an optimizer [5, 6]. Novel tests were introduced to screen both molecular *and* process variables, in the context of gas-liquid separation. In this approach, we first identify the process domain where the feed is stable. At each major iteration of an outer approximation (OA) algorithm [7], we find a reduced process domain for a solvent. The reduced process domain is an overestimation of the feasible region for that solvent. In a first test, the feasibility of solvent handling and storage is checked. Ina second test, the range of process variables over which the solvent and the feed can form a two-phase mixture is found. The range of process variables over which the required degree of purity of the treated stream may be attained is identified in a third test. If the solvent fails any of the tests it is eliminated. If the tests are feasible, the primal problem is solved using bounds identified by the tests and initial guesses within the reduced process domain. Information from the tests and the primal problem is used to construct the master problem for the OA algorithm.

The extension of this methodology to liquid-liquid extraction systems is presented here. The three tests, initially developed for absorption processes, are recast for the case of liquid-liquid extraction. The approach is illustrated with a case study of the separation of butanol from a fermentation broth. As the broth is dilute in butanol, separation of this mixture by distillation is highly energy intensive. Hence, an optimal solvent for extraction, and the corresponding optimal process variables, are designed using the algorithm proposed here. A group contribution equation of state, the group contribution version of the statistical associating fluid theory with a Mie potential, SAFT-Î³Â Mie[8,9], is used in this work to predict the relevant thermodynamic properties. A systematic study of the performance of the algorithm is undertaken. The algorithm succeeds in avoiding expensive process evaluations for infeasible solvents and enhances convergence to the solution from multiple starting points.

1 Adjiman, C. S., Galindo, A., and Jackson, G. (2014). Proceedings of the 8th International Conference on Foundations of Computer-Aided Process Design FOCAPD 2014, edited by Eden M.R., Siirola J.D., Towler G.P., Computer Aided Chemical Engineering, Elsevier B.V.

2 Joback, K.G. and Stephanopoulos, G. (1995). In Advances in Chemical Engineering, 21: 257-311

3 Gopinath S, Galindo A, Jackson G, Adjiman CS (2016). Proceedings of the 26th European Symposium on Computer Aided Process Engineering, edited by Kravanja Z., Computer Aided Chemical Engineering. Elsevier B.V.

4 Gopinath S, Galindo A, Jackson G, Adjiman CS (2016). Computer-aided molecular and process design for absorption-desorption systems using physically-driven domain reduction (2016). Manuscript submitted

5 Buxton, A., Livingston, A. G., and Pistikopoulos, E. N. (1999). AIChE Journal, 45: 817-843

6 Giovanoglou, A., Barlatier, J., Adjiman, C. S., Pistikopoulos, E. N., and Cordiner, J. L. (2003). AIChE Journal, 49: 3095-3109

7 Duran, M. A. and Grossmann, I. E. (1986). Mathematical Programming, 36: 307-339

8 Papaioannou, V., Lafitte, T., Aveñdano, C., Adjiman, C. S., Jackson, G., Müller, E. A., and Galindo, A. (2014). The Journal of Chemical Physics, 140: 054107

9 Lafitte, T., Apostolakou, A., Avendaño, C., Galindo, A., Adjiman, C. S., Müller, E. A., and Jackson, G. (2013). The Journal of Chemical Physics, 139: 154504

469862 Optimisation-Based Design of a Heat-Integrated Crude Oil Distillation System Using Rigorous Simulation and Surrogate Models

Abstract:

In petroleum refining industry, the crude oil distillation system plays a key role in the overall process. The system comprises a complex heat-integrated distillation column and a heat recovery system where the crude oil feed is partially vaporised. The presence of large number of degrees of freedom (column structure and operating conditions) and the complex interactions existing between individual units makes the design of crude oil distillation unit one of the most challenging tasks within process system engineering. Due to the complex nature of the process, traditional design methods consider designing the complex column and the heat recovery system in separate steps. In this approach, however, interactions between the two subsystems are not taken into account. Some researchers have applied standard optimisation techniques to simultaneously design the crude oil distillation unit and the heat recovery system, but these approaches typically rely on shortcut column models, to avoid the convergence problems associated with rigorous column models. The use of these shortcut models can lead to large errors, as they might fail to predict the behaviour of the complex crude oil distillation unit accurately.

This paper introduce a new framework for the design of crude oil distillation system that integrates a rigorous tray-by-tray crude oil distillation system model implemented in a process simulator (Aspen HYSYS) with an optimiser coded in MatLab. The approach takes advantage of the physical and thermodynamic property models, crude oil characterisation models and column sizing models available in the process simulator.

The proposed approach is implemented in three steps. First, the column superstructure of [1] for optimal feed location, total number of trays and optimal operating conditions is adapted to build the column superstructure using a rigorous tray-by-tray model in Aspen HYSYS. The superstructure consists of conditional and unconditional trays in all column sections, which are represented using Murphree tray efficiencies. In the second step, the problem is formulated as an MINLP with the number of trays in each column section defined as integer variable and the operating conditions as continuous variables. The heat recovery system is represented using Pinch analysis in order to determine the minimum utility requirements that will be used in the calculations. In the final step, an optimisation algorithm (genetic algorithm) is applied to optimise the system, thereby determining the optimal column configuration and operating conditions that minimise a given objective (e.g. total annualised cost).

To expedite the solution procedure, the proposed approach also explored the use of surrogate models based on neural networks that are constructed from a set of samples generated using the original simulation model.

The proposed approach has been applied to the design of a crude oil distillation system that separates crude oil into five products. Two design objectives of interest are considered namely, minimum total annualised cost and minimum energy cost. The results show that significant savings in total annualised cost and energy cost can be achieved using our approach without compromising the solution accuracy.

References

[1]Â Yeomans, H. & Grossmann, I. E. Optimal design of complex distillation columns using rigorous tray-by-tray disjunctive programming models. Ind. Eng. Chem. Res. 2000, 39 (11), 4326â''4335

468420 Aquathermolysis of Waste Triglycerides in a Continuous Flow Reactor for Jet Fuels Production

Abstract:

The unavoidable dwindling fossil fuel and its attendant climate change in our time have given impetus to researchers in the energy sector to delve into different sustainable energy sources. One of such energy sources is the abundant waste triglycerides whose long chain hydrocarbons can be separated from their short oxygen chain via hydrodeoxygenation, decarbonylation, and decarboxylation in the presence of a heterogeneous catalyst and hydrogen. However, the use of gaseous hydrogen is limited by a diffusion process since it is not readily soluble in liquid (polar) components and requires an excess use of hydrogen. To address these challenges, a single-pot â€˜Formic Acid Assisted Autotheromolysis to Jet Fuel (FAT2JF)â€™ process is studied. The use of aqueous phase (formic acid and water mixture) serves as an *in situ* source of H_2 (hydrogen precursor and hydrogen transfer agent) and organic solvent where the reactions in aqueous media (aquathermolysis) helps in reducing mass transfer resistances that results in significant increase in reaction rates in this FAT2JF process. The phase diagram of water, formic acid and some typical triglycerides will be presented.

The present study on aquathermolysis is focused on developing a metal oxide-based low cost catalyst (e.g. Fe-Ni/activated carbon, Fe-Ni-Co/Al_2O_3) system that is stable in aqueous media. The study is conducted in a packed bed continuous flow tubular reactor in the range of 250 to 400âµ'C under hydrothermal conditions. In order to alleviate coke formation and evaluate the stability of the catalyst, the effects of reactants flowrates, feedstock composition, pressure, temperature, and formic acid concentration are studied. GC-MS and elemental analyzer are used to analyze the product compositions. Thermal gravimetric analysis (TGA), energy dispersive X-ray, SEM, BET method, FT-IR, and XRD are used to characterize both fresh and used catalyst. The proposed novel approach has certain advantages over conventional commercial hydrotreating processes including no handling of hydrogen gas, faster reaction times,

and use of renewable formic acid as hydrogen source.

469513 Rheology of Biomass Slurries to Determine Pumpability for Hydrothermal Liquefaction

Abstract:

The pursuit of renewable fuels and chemicals has led to the development of many different biomass conversion processes. One of these processes, hydrothermal liquefaction (HTL), has the advantage of high energy recovery, no drying step, and utilization of a green aqueous reaction media.[1] However, transport of biomass and water slurries during HTL has proven to be a challenge.[2,3] Typical reactors can only process biomass between 10 and 15 wt% solids because higher solids loadings plug reactor systems. To improve the efficiency of HTL, and other processes reliant on pumping biomass slurries, it is essential to increase solids loading. Increasing solids loading reduces the volume of waste streams and increases process throughput. In an effort to increase biomass solids loadings many people have studied the rheological properties of biomass slurries to improve pumpability.[4] Despite these efforts there has yet to be a strong correlation developed between typical rheological properties, like viscosity, and biomass pumpability.

This work takes a unique approach to studying biomass slurry rheology by combining traditional rheometric measurements, such as viscosity and storage/loss moduli, with a novel flow characterization system. Traditional rheometric measurements are made on a rheometer capable of analysis at temperatures up to 300 °C and 2000 psi, which mimics conditions found in some traditional HTL systems. Novel flow studies performed at Idaho National Laboratory are made on an in-house designed syringe attached to an Instron load frame. This system measures the force required to push biomass through a reducing orifice and can study plugging phenomena in biomass slurry transport systems. These studies have shown how plugging can be related to biomass dewatering, followed by bridging across the flow channel. Pumpability studies have been performed on a wide variety of materials from herbaceous to woody crops and residues. In addition to characterizing pumpability in biomass slurries this work has also focused on improving the pumpability of such slurries through preprocessing techniques including: size reduction, blending, and torrefaction. This work will ultimately lead to improvements in feed handling related to hydrothermal liquefaction and other biomass slurry transport systems.

1 Toor, S. S., Rosendahl, L. & Rudolf, A. Hydrothermal liquefaction of biomass: A review of subcritical water technologies. *Energy* **36**, 2328-2342, doi:10.1016/j.energy.2011.03.013 (2011).

2 Elliott, D. C., Biller, P., Ross, A. B., Schmidt, A. J. & Jones, S. B. Hydrothermal liquefaction of biomass: Developments from batch to continuous process. *Bioresource technology* **178C**, 147-156, doi:10.1016/j.biortech.2014.09.132 (2015).

3 Berglin, E. J., Enderlin, C. W. & Schmidt, A. J. *Review and Assessment of Commercial Vendors/Options for Feeding and Pumping Biomass Slurries for Hydrothermal Liquefaction*. (Pacific Northwest National Laboratory, 2012).

4 Stickel, J. J. *et al.* Rheology measurements of a biomass slurry: an inter-laboratory study. *Rheologica Acta* **48**, 1005-1015, doi:10.1007/s00397-009-0382-8 (2009).

469372 The Upgrading of Biomass-Derived Dimethyl Ether to High-Octane Hydrocarbons: The Effect of Process Conditions on Catalyst Performance

Abstract:

Dimethyl ether (DME) and/or methanol can be converted into C_{5+} branched hydrocarbons at low temperatures and pressures over large-pore acidic zeolites. Recently, we have demonstrated that Cu-modification of the existing commercial catalyst (H-BEA) resulted in a 2-fold increase in total hydrocarbon productivity at 200 °C with co-fed H_2.[1] The Cu/H-BEA catalyst incorporated H_2 with DME into the desired branched hydrocarbons, while maintaining the high C_{5+} selectivity of the parent H-BEA. Here, we compare the performance of Cu/H-BEA and the parent H-BEA at 175, 200 and 225 °C, and at weight-hourly space velocities (WHSV) for DME ranging from 0.43-1.4 h^{-1}. The productivity of C_{5+} products increased with temperature over H-BEA, with no significant changes over Cu/H-BEA due to a decrease in C_{5+} selectivity at elevated temperature, which was observed for both catalysts. Operation at 175 °C resulted in a significant increase in C_{5+} selectivity (75-85% for Cu/H-BEA), but lower productivity than at 200 °C due to lower reactant conversion. Lower space velocities increased reactant conversion over both catalysts, but significantly decreased the C_{5+} productivity over Cu/H-BEA. The Cu/H-BEA catalyst was stable for 100 h at 200 °C after an initial 40 h break-in period. In previous research, we demonstrated that C_4-C_8 olefins produced in the DME homologation reaction can be oligomerized to C_{12}-C_{22} branched hydrocarbons with properties that align with a synthetic paraffinic kerosene product.[2] Following a simple vacuum distillation, the fuel properties of the distillate-range product align well with a jet fuel blend when compared with 5 ASTM methods. This DME-to-hydrocarbons reaction represents a promising route to transform renewable biomass sources into commercially viable, high-performance biofuels.[3]

[1]Schaidle, J. A.; Ruddy, D. A.; Habas, S. E.; Pan, M.; Zhang, G.; Miller, J. T.; Hensley, J. E. *ACS Catal.* **2015**, *5*, 1794.

[2]Behl, M.; Schaidle, J. A.; Christensen, E.; Hensley, J. E. *Energy & Fuels* **2015**, *29*, 6078.

[3]Tan, E. C.; Talmadge, M.; Dutta, A.; Hensley, J.; Snowden-Swan, L. J.; Humbird, D.; Schaidle, J.; Biddy, M. *Biofuels, Bioproducts and Biorefining* **2016**, *10*, 17.

462619 A System-Level Analysis on Biomass Thermal Fractionation and Catalytic Upgrading Processes

Abstract:

Biomass presents a promising source for sustainable liquid fuels production. There are three principle routes for converting biomass into liquid fuels, classified by the initial treatment of the biomass: gasification, hydrolysis, or pyrolysis and liquefaction. The pyrolysis route involves thermal decomposition of the biomass at high temperature (500-800°C), and it has demonstrated high yields. Unfortunately, the wide range of components in the resulting pyrolysis vapor (oil) makes it unstable, corrosive, and challenging to catalytically upgrade. Furthermore, the high content of light (<C6) oxygenates in the pyrolysis are unsuitable for liquid fuels and are therefore of limited utility. Multi-stage torrefaction of biomass with catalytic upgrading presents a promising method of overcoming these challenges. Importantly, the decomposition of the biomass over several stages can reduce the number of product species within each fraction, allowing one to mold the catalytic upgrading strategy to target a subset of the chemical functionalities.

Although staged thermal decomposition of biomass coupled with targeted upgrading has a number of advantages over single stage pyrolysis with hydro-treating, there are many unknowns regarding the best way to design such a system. Importantly, the optimization of the number and conditions of each thermal decomposition stage is a complex problem that cannot be solved without knowing the influence on the final yields and process complexity resulting from the subsequent catalytic upgrading. Furthermore, there are a wide variety of chemistries available to perform upgrading. In addition, the reaction sequence must also be configured to optimize the overall process. In order to improve the impact of such an approach, one must also determine if general practices from this strategy may be applicable to different types of biomass feeds.

Accordingly, the ultimate goal of this work is to design a broad roadmap that can be used to identify the key parameters and trade-offs in a bio-refinery process based on multi-stage torrefaction of biomass with catalytic upgrading. In order to investigate the effect of the number of torrefaction stages and the specific chemistries on the yield of fuel-range grade product and the cost, we first derived a series of process alternatives based on experimental torrefaction yields obtained for a 3-stage torrefaction process (270 °C, 360 °C, and 500 °C). In designing the alternatives, we carefully selected the chemistries that are used to upgrade the thermal decomposition fractions, to maximize the yield of fuel-grade product and to minimize the hydrogen consumption, considering the abundance of various chemical functionalities within the biomass fractionation product. We suggested twelve cases where there is 1, 2, or 3 thermal decomposition stages and where each fraction is upgraded 1, 2, 3, or 4 times. Through the analysis, we found that there exists a trade-off between fuel-range carbon yields versus the number of upgrading steps. In general, increasing the complexity of the fraction upgrading systems increases the ultimate yield of C6+ product, thought there are diminishing returns on the increase in product yield versus the complexity of the catalytic upgrading sequences. In contrast, however, the choice of the number of thermal decomposition stages is found to be not simple, requiring careful consideration of the chemistries available to upgrade different components and the relative abundances of these different components. More thermal decomposition stages allow for more fine-tuning of the upgrading processes but lead to more complex and potentially more expensive processes. Thus, the design of the decomposition conditions and number of stages must be done simultaneously with the design of the fraction upgrading system. Finally, we also suggest some hybrid processes by eliminating several upgrading steps that have small marginal effects on the fuel-range carbon yield and discuss about separations that could enhance the process performance via prevention of unwanted side reactions or protection a functionality from conversion.

465151 A Neutron Scattering Study of Ion-Conduction Mechanisms in Nanocomposite Polymer Electrolytes

Abstract:

Polymer Electrolytes are the key to light-weight and energy-dense lithium ion batteries but suffer from low room temperature ion conductivities. Spherical ceramic fillers are known to improve ion conductivity and mechanical properties but are still orders of magnitude away from commercialization. Understanding the ion conduction mechanism enables better design of the electrolyte system. Two important factors that govern ion conduction include electrolyte microstructure and polymer dynamics. While electron microscopy and small angle neutron scattering (SANS) provide structural information, quasi-elastic neutron scattering (QENS) examines polymer dynamics. QENS probes two processes: segmental dynamics of polymer chains and rotation of highly-conductive crystalline channels. While the former process remains unaffected by presence of particles, the latter becomes more restricted. Based on this information, polymer electrolytes were designed with high aspect ratio fillers to enable further stabilization of the highly conducting channels. This results in one-order of magnitude improvement in lithium ion-conduction over spherical fillers.

459477 Mechanical Properties of Graphene–Polymer Nanocomposites

Abstract:

The exceptional mechanical, electronic, and thermal properties of graphene render it a highly promising filler for polymermatrix nanocomposites. While preliminary experiments indeed show significant enhancement in the thermomechanical properties of graphenepolymer nanocomposites compared to those of the polymer matrix even at relatively low loading of filler, our fundamental understanding of structure–property relationships that govern such enhanced material response remains very limited.

Toward this end, in this presentation, we report results of moleculardynamics simulation studies on the mechanical behavior of polymer (high-density polyethylene) nanocomposites reinforced by graphene and fullerenes with the aim of elucidating the underlying mechanisms that govern the mechanical response of these composite materials. Using a united-atom-based model of the glassy polymer matrix, we show systematic trends in the enhancement of the mechanical stiffness of the polymer composite as a function of filler concentration, filler size, and matrix–filler interfacial interaction strength that is governed by dispersive forces. For comparison of different filler reinforcement effects in such polymermatrix nanocomposites, we also present systematic mechanical behavior studies of fullerene–polymer composites and show that, for fullerene fillers, the response is only weakly dependent on filler size with relatively high loadings required for a considerable improvement in the composite's stiffness. In contrast, graphene-polymer nanocomposites, with graphene circular flakes used as fillers, show an appreciable improvement in the stiffness reinforcement with increasing filler size. We explain the differences in the elastic response of (0D) fullerene and (2D) graphene reinforced polymer matrix composites through detailed atomic scale characterization of the nanocomposites in conjunction with the predictions of an analytical continuum-mechanics model for understanding the fillersize dependent response of the polymer nanocomposites. We also derive scaling relations for the dependence of the composite modulus for polymer-graphene nanocomposites as a function of the flake size at given filler concentration. Moreover, we report results of

our continuum mechanical analysis for polymer-matrix reinforcement on the basis of shear-lag modeling and show a parametric dependence of the axial and interfacial shear stress at the graphene-polymer interface on the graphene flake size. Based on our atomistic simulations and this shear-lag analysis, we identify the range of filler length scale over which classical shear-lag theory becomes most effective for describing the polymer-matrix reinforcement.

471794 Probing Multiscale Structure in Aligned Nanotube-Polymer Composites

Abstract:

Composites made from vertically aligned carbon nanotube (CNT) forests are important for a broad range of applications, such as for thermal management, high-flux membranes, protective fabrics, mechanical reinforcement, and energy dissipation. Despite progress, there remains a need for quantitative understanding of how the multiscale structure of CNT forests dictates both the fabrication of advanced composites and ultimately their performance. Drawing such quantitative relationships is challenging because measuring the morphological structure and order in CNT forests is difficult for a) multi-component systems, b) over large range of length scales, and c) without destroying the material.

To nondestructively investigate salient CNT forest characteristics, we built upon a suite of complementary soft and hard X-ray scattering techniques including polarized resonant soft X-ray scattering (R-SoXS â€" micro/meso/nano), small- and wide-angle X-ray scattering (SAXS/WAXS â€" nano/atomic) and X-ray attenuation. R-SoXS is a novel technique at the Advanced Light Source (ALS) ideally suited for composite characterization because it enables selective probing of the CNTs and their polymer matrix *via*scattering with chemical specificity by tuning of incident photon energy. Using our combined synchrotron techniques, we analyzed high-performance nanocomposite membranes fabricated by: 1) synthesizing self-aligned CNT forests by chemical vapor deposition; 2) infiltrating the void space between CNTs with a polymeric material; and 3) then etching open the ends of a large number of CNTs to expose their core volume to fluid transport.

In particular, we mapped structural characteristics in self-aligned CNT forests/composites with *chemical*, *spatial*, and *length-scale* resolution. We developed a unified analytical model to quantitatively describe the structural hierarchy of aligned CNTs and to extract key structural parameters via curve-fitting of our X-ray scattering data. From the bottom

up, our model parameters define the graphitic lattice and wall number (atomic), CNT diameter (nano), CNT bundling and spacing (meso), regular corrugations (micro), and number density (macro). Comparison with electron microscopy reveals good agreement for forests with a wide range of characteristics (e.g., 1-11 walls, 1.5-15 nm diameter, 10^{10}-10^{12} cm^{-2}density). With R-SoXS as our complementary chemical probe, we also analyzed the structure of the polymer matrix around the outside of the CNTs and ions inside the CNT nanochannels.

Finally, we show how the structural characteristics of our CNT forests inform the fabrication of large-area, flexible polymer composites toward their application as ultrabreathable and protective membranes.[1] We leveraged composite structural information to precisely quantify the fluid flow enhancement measured experimentally through the CNT nanochannels in our membranes. This is an exciting advancement because insufficient information regarding the CNT membrane structural parameters is thought to be the primary source of orders-of-magnitude discrepancies in the ultrafast fluid transport inside CNT nanochannels reported in the literature.[2] More precise understanding of the transport in these bulk materials may help elucidate open controversies regarding transport through individual CNT nanochannels.[3] While we focused here on fluidic properties, our methodology is broadly applicable for drawing quantitative, multiscale relationships between structure and other (mechanical, thermal, electrical, etc.) properties in nanocomposite systems.

This work is supported by the Defense Threat Reduction Agency (DTRA) D[MS]2 project under Contract No. BA12PHM123 and was performed under the auspices of U.S. Department of Energy by Lawrence Livermore National Laboratory under Contract DE-AC52-07NA27344.

1. Bui, N.; Meshot, E. R.; Kim, S.; Peña, J.; Gibson, P. W.; Wu, K. J.; Fornasiero, F., Ultrabreathable and Protective Membranes with Sub-5 nm Carbon Nanotube Pores. *Advanced Materials* **2016**.

2. Kim, S.; Fornasiero, F.; Park, H. G.; In, J. B.; Meshot, E.; Giraldo, G.; Stadermann, M.; Fireman, M.; Shan, J.; Grigoropoulos, C. P.; Bakajin, O., Fabrication of Flexible, Aligned Carbon Nanotube/Polymer Composite Membranes by in-Situ Polymerization. *Journal of Membrane Science* **2014,** *460*(0), 91-98.

3. Guo, S.; Meshot, E. R.; Kuykendall, T.; Cabrini, S.; Fornasiero, F., Nanofluidic Transport through Isolated Carbon Nanotube Channels: Advances, Controversies, and Challenges. *Advanced Materials* **2015,** *27* (38), 5726-5737.

460070 Thermal and Mechanical Property Predictions for Conjugated Polymers Using Atomistic Simulations

Abstract:

Paralleling successful application in catalyst design, drug discovery, and protein structure analysis, molecular simulations are emerging as reliable property-prediction tools for synthetic polymers as well. Simulations are not only useful for determining functional properties such as band gaps of π-conjugated polymers, but also for accessing microscopic information that is not easily available from experiments.

In this work, we used atomistic simulations to predict thermophysical and mechanical properties of two p-conjugated polymers, polyacetylene (PA) and poly(*para*-phenylene vinylene) (PPV) [1]. Although the electron transport and optical behaviors of these polymers are well-characterized, information on properties like elastic modulus and glass transition temperature are uncertain. Glass transition temperature, , the temperature above which the polymer softens considerably undergoing a large decrease in stiffness, and the elastic modulus of the polymer, are useful parameters in the design of devices such as flexible LED displays and solar panels that use conjugated polymers.

The literature values of the elastic modulus of PA and PPV at room temperature are scattered over a range spanning two to three orders of magnitude [1]. Similarly, differences in the reported values of the of PPV are greater than 100 °C. Polymer properties are generally quite dependent on their synthesis and processing history. The dispersity in the lengths of the polymer chains, the variations in polymer architecture caused by branching, chemical defects such as oxidation, the presence of plasticizers or other low molar mass impurities, and even the method of measurement can greatly alter the measured property values. These effects seem to be particularly important in the case of the two conjugated polymers investigated. Molecular simulations, on the other hand, allow one to probe a virtual ensemble of perfectly defect-free identical polymer molecules as a basis for comparison with measurements. Architectural variations and chemical defects can be introduced in a controlled manner, if their effects need to be

studied.

We used the variations of specific volume and the potential energy of intermolecular interactions with temperature to identify the glass transition temperature. Properties such as the coefficient of thermal expansion, the torsion angle distributions, and the mean square end-to-end distance, were calculated at different temperatures. π-Conjugation of the sp^2 hybridized carbon atoms in the PPV chain would be expected to impart planarity to the molecule, but a non-planar structure, consistent with previously reported observations in neutron diffraction experiments [2], was predicted.

Conventional mechanical characterization techniques such as tensile testing stretch the polymer specimen to different lengths and measure the force required to produce a given extension of the specimen. However, polymer chains in the sample can align or orient along the stretching direction. We used the technique of dynamic nanoindentation instead to measure polymer stiffness by applying relatively small oscillatory loads, of micronewton magnitude, on the polymer films using a diamond microindenter [3]. The experimental results were in good agreement with the MD predictions employing the static deformation method [4].

This study demonstrates that atomistic simulations can reliably complement existing tools for the prediction of the properties of solid polymers.

References:

1. Venkatanarayanan, Ramaswamy I., Sitaraman Krishnan, Arvind Sreeram, Philip A. Yuya, Nimitt G. Patel, Adama Tandia, and John B. McLaughlin (2016), "Simulated Dilatometry and Static Deformation Prediction of Glass Transition and Mechanical Properties of Polyacetylene and Poly(*para*-phenylene vinylene)," *Macromolecular Theory and Simulations*, 25(3), pp. 315–321.
2. Mao, G., J. E. Fischer, F. E. Karasz, and M. J. Winokur (1993), "Nonplanarity and Ring Torsion in Poly (*p* phenylene vinylene). A Neutron Diffraction Study," *The Journal of Chemical Physics*, 98(1), pp. 712–716.
3. Sreeram, Arvind, Nimitt G. Patel, Ramaswamy I. Venkatanarayanan, John B. McLauglin, Stephan J. DeLuca, Philip A. Yuya, and Sitaraman Krishnan, "Nanomechanical Properties of Poly(*para*-phenylene vinylene) Determined Using Quasi-static and Dynamic Nanoindentation," *Polymer Testing* (2014), 37, pp. 86–93.
4. Theodorou, Doros N. and Ulrich W. Suter (1986), "Atomistic Modeling of Mechanical Properties of Polymeric Glasses," *Macromolecules*, 19(1), pp. 139–154.

467782 Importance of Selecting Right Analytical Tools for Successful Product Development: Case Study of Single and Two Phase Polymerization Reactions

Abstract:

The importance of Chemical Reaction Engineering, which deals with chemical kinetics, transport processes, and subsequent reactor design, for product development cannot be overstated. Understanding the chemical kinetics and underlying transport phenomena is critical for successful reactor scale-up to enable commercial production. The success of a process and hence a product is inherently tied to laboratory reactors and concomitant process analytical techniques which assist in development of intrinsic kinetics devoid of mass and heat transfer limitations.

This talk will focus on two examples which highlight the importance of right analytical tools for process and product development. The first example will discuss capturing â€œintrinsicâ€ kinetics of a potentially hazardous exothermic chemistry. In particular, ethylene and propylene oxide based polyalkylene glycols are widely used in high performance formulated fluids and lubricants. The current alkoxylaton kinetics employed in their reactor design and operation limit optimization at industrial scale due to the need for empirical adjustments. Thus development of intrinsic kinetics, not affected by heat and mass transfer limitations, is necessary for improved process performance and for product optimization. This work addressed the challenge by development of an experimental methodology coupled with an in-situ analytical technique to accurately capture the true kinetics. Kinetic model development, regression of kinetic parameters and model validation for a commercial process will be presented.

The second example will address the role of analytical techniques in development of energy efficient processes and products. Single phase solution polymerization for making polyethylene resins renders itself aptly for efficient processing. In particular, energy intensive solvent recovery after polymerization results in a high energy foot print.

Selection of appropriate solvent to lower polymer solubility *during* polymerization to form two phases i.e. polymer phase and solvent phase provides an energy efficient alternative. The heavier polymer phase can be separated using a decantation step after polymerization to remove up to 50% of the solvent leading to lower solvent separation costs. Evaluation of this new processing route and development of analytical methodology to prevent risk of failure on final implementation will be discussed.

474341 Computer Aided Design of Ionic Liquids through a New Group Contribution Approach Based on Surface Charge Density and COSMO-SAC Predictions

Abstract:

The practical use of computer-aided ionic liquid design (CAILD) approach towards the design of solvents for separation processes, such as CO_2 absorption and liquid-liquid extraction, is greatly hindered by the very limited availability of UNIFAC interaction parameters that cover diverse ionic liquid building blocks (i.e. different cation cores, anions, and side chain groups) and other molecular groups (e.g. CO_2). The non-availability of interaction parameters can be linked to the limited thermodynamic data available for systems involving ionic liquids. COSMO-RS and COSMO-SAC methods provide a fully predictive approach for activity coefficients that rely on surface charge density obtained from quantum chemical calculations. However, these methods are not directly compatible with CAMD or CAILD approach as electronic structure calculations are possible only through *ab initio* quantum chemical programs such as Gaussian or Turbomole and they are specific for each ionic liquid as a whole. Further, these calculations are extremely time consuming and sometimes can take several hours for a single compound. We have recently developed a novel methodology that involves group contribution prediction of ionic liquid surface charge density (sigma profiles) and cavity volumes. This is further integrated with COSMO-SAC methodology to enable prediction of solution thermodynamics (activity coefficient) of systems involving ionic liquids. This approach is fully predictive in nature and does not require fitting group interaction parameters using experimental data or quantum chemical calculations. Further, this method is fully compatible with CAMD approach. This predictive thermodynamic model is integrated with optimization methods to reverse engineer novel ionic liquids. We demonstrate the utility of the model through case studies.

468269 Identification of Magnetic Nanoparticles to Eradicate Cancer Cells Using the Hyperthermia Approach â€“ a Computational Search Approach

Abstract:

In recent years, nanomaterials have received significant attention due to their unique properties. Magnetic nanoparticles (MNPs) in particular are of high importance due to their magnetic properties. These particles are being tested for various applications including cancer cell detection and targeted cell temperature increase.

Hyperthermia treatment of cancer cells has been one of the most promising ideas of last decade. The theory is to design materials which are very efficient in targeting the cancer cells. The designed particles are to have magnetic properties such that at certain magnetic field conditions the particle develops a higher temperature than the surrounding atmosphere/environment.

This project aims to investigate a way to model magnetic nanoparticle behavior in the presence of different magnitudes of magnetic fields. The aim is to model Brownian and Neel relaxation parameters to determine the heat generated by different MNPs to study which is best suited for hyperthermia treatment of cancer cells. For that, approaches were used to determine AC susceptibility of the out-of-phase particles which was then used in a 3-D model containing healthy and cancer cell. The 3-D model is used to determine temperature output for such materials in the presence of an optimal magnetic field.

Selected References:

1. Rudolf Hergt, Silvio Dutz, Robert Muller, Matthias Zeisberger, *Magnetic particle hyperthermia: nanoparticle magnetism and materias development for cancer treatment*, 2006, J. Phys.:Condens. Matter, Vol 18, S2919-S2934.
2. Shihu Wang and Elena E. Dormidontova, *Nanoparticle Design Optimization for Enhanced Targeting: Monte Carlo Simulations*, 2010, Biomacromlecules, Vol 11, 1785-1795
3. Andreas Jordan, Regina Scholz, Peter Wust, Horst Fahling, Ronald Felix, Magnetic fluid hyperthermia: *Cancer treatment with AC magnetic field induced excitation o biocompatible supermagnetic nanoparticles*, 1999, Journal of Magnetism and Magnetic Materials, Vol 201, 413-419
4. Stefan T Bromley, Iberio de P. R, Moreira, Konstantin M. Neyman, Francese Illas, *Approaching nanoscale oxides: models and theoretical methods*, 2009, Chem. Soc. Rev., Vol 38, 2657-2670.

467310 Machine Learning and Natural Language Processing for Pharmaceutical Product Engineering

Abstract:

Pharmaceutical product engineering is a "Big Data" discipline. It requires understanding of details of the drug chemistry during production and within the body, the manufacturing processes and conditions, and the pharmacokinetics of a disease – all data intensive. In fact, a typical New Drug Application (NDA) contains more than 100,000 pages of a variety of information. In this talk, we present a framework, called HOLMES, for the automatic extraction of knowledge from primary sources related to pharmaceutical product engineering. The information extracted is then stored in ontologies. These ontologies are a computer readable semantic knowledge representation used in artificial intelligence. We describe Machine Learning (ML) algorithms and Natural Language Processing (NLP) techniques that are used in HOLMES for Entity and Concept Recognition and Relation Extraction. We will discuss our progress on the creation of an entity-concept-and-relation databank (7968 entities and concepts, 1665 relations); the application of different ML algorithms for the purpose of joint Entity and Concept detection; and the development of a relation clustering algorithm using common feature sets.

ACS APPLIED MATERIALS & INTERFACES

Research Article

www.acsami.org

The Effects of Cross-Linking in a Supramolecular Binder on Cycle Life in Silicon Microparticle Anodes

Jeffrey Lopez,[†] Zheng Chen,[†] Chao Wang,[†,∥] Sean C. Andrews,[†] Yi Cui,[*,‡,§] and Zhenan Bao[*,†]

[†]Department of Chemical Engineering, Stanford University, Stanford, California 94305, United States

[‡]Department of Materials Science and Engineering, Stanford University, Stanford, California 94305, United States

[§]Stanford Institute for Materials and Energy Sciences, SLAC National Accelerator Laboratory, Menlo Park, California 94205, United States

Ⓢ *Supporting Information*

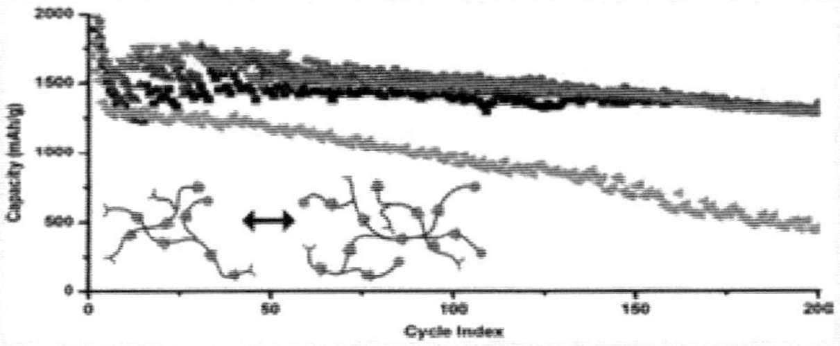

ABSTRACT: Self-healing supramolecular binder was previously found to enhance the cycling stability of micron-sized silicon particles used as the active material in lithium-ion battery anodes. In this study, we systematically control the density of cross-linking junctions in a modified supramolecular polymer binder in order to better understand how viscoelastic materials properties affect cycling stability. We found that binders with relaxation times on the order of 0.1 s gave the best cycling stability with 80% capacity maintained for over 175 cycles using large silicon particles (~0.9 um). We attributed this to an improved balance between the viscoelastic stress relaxation in the binder and the stiffness needed to maintain mechanical integrity of the electrode. The more cross-linked binder showed markedly worse performance confirming the need for liquid-like flow in order for our self-healing polymer electrode concept to be effective.

KEYWORDS: *lithium-ion battery, silicon, microparticles, high capacity anode, self-healing, polymer binder, viscoelasticity*

■ INTRODUCTION

The lithium (Li) ion battery (LIB) is an important energy storage technology. However, the volumetric and gravimetric storage capacity must be increased to meet the needs of electric vehicle (EV) and grid storage applications while also decreasing the cost.[1,2] In order to achieve these performance targets, it is necessary to move to LIB electrode active materials that have increased Li storage capability. Graphite has been the industry standard anode material with a maximum theoretical specific capacity of 372 mAh/g. Silicon (Si) is one of the most promising candidates for high capacity negative electrodes due to its high theoretical specific capacity (3579 mAh/g) and natural abundance.[3,4] Unfortunately, Si electrodes have been difficult to introduce into commercial cells due to the large volume expansion (~300%) that occurs when Si alloys with Li.[4−6] Because Si is a brittle material, the large anisotropic stresses that build up cause the active material to crack, and quickly fail.[5,7,8] This capacity loss results from two mechanisms. First, portions of the Si active material can become electrically disconnected from the rest of the electrode, thus preventing them from further charging and discharging.[9] Additionally, the cracks in the Si provide fresh surfaces for electrolyte decomposition to occur forming new solid electrolyte interphase (SEI).[10]

In order to circumvent this problem, two distinct approaches have been employed. The first involves using structured nanomaterials such as nanowires/tubes,[11−13] core−shell structures,[14−16] nanocomposites,[17,18] or porous particles[19,20] to reduce mechanical stress due to volume expansion and prevent particle cracking. While some of these approaches have been successful in achieving high areal capacity (an important requirement for commercial application) and long cycling life,[13,16] the techniques used to synthesize these materials could be complicated and difficult to produce on a large scale or at a low cost. The other approach focuses on tuning the structure and properties of the polymeric binder employed in the electrode to promote adhesion between the current collector, active material, and conductive additives while also promoting stable SEI formation and aiding in electrode processing. For Si nanoparticle electrodes, a variety of polymers have been examined including polysaccharides,[21−26] vinyl polymers,[27−29] and conjugated conducting polymers.[30−34] Some of the best performing polymers, though in low mass-loading electrodes, are sodium carboxymethylcellulose (CMC),[21] alginate,[22] and poly(acrylic acid) (PAA)[27] due to their strong hydrogen bonding with the oxide surface on the Si active materials,[35] and polyfluorene-conjugated polymers due to their intrinsic electronic conductivity.[30,31] Additionally, covalent cross-linking of these systems has been seen in some cases to improve cycling stability when compared to non-cross-linked sys-

Received: November 23, 2015
Accepted: December 30, 2015
Published: December 30, 2015

DOI: 10.1021/acsami.5b11363
ACS Appl. Mater. Interfaces 2016, 8, 2318−2324

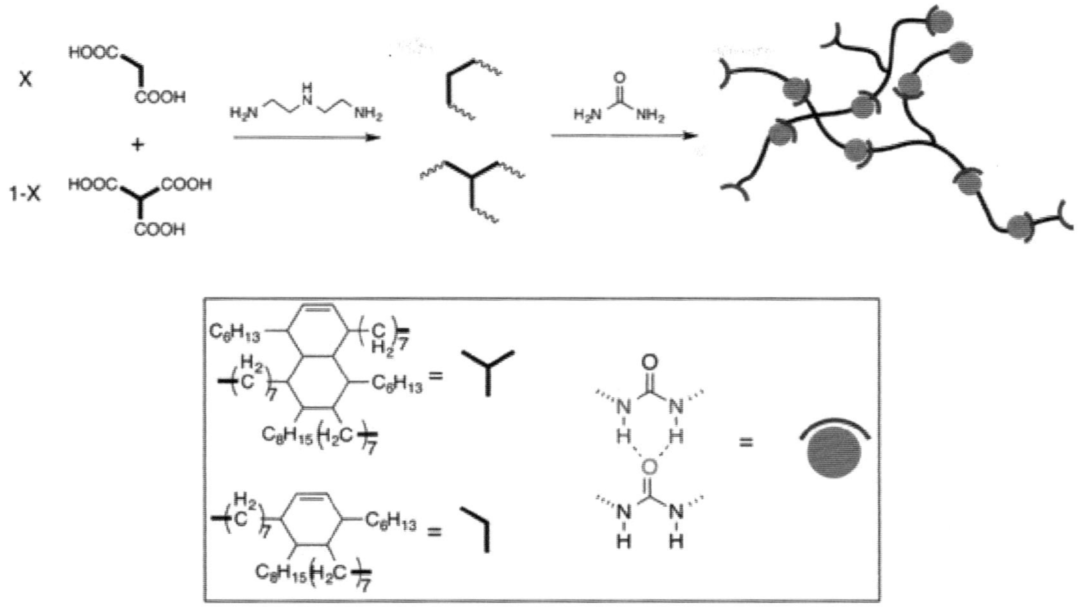

Figure 1. Synthesis of self-healing polymers. A two-step condensation reaction is used to obtain randomly branched oligomers with urea functionalized end groups that associate via hydrogen bonding to form a supramolecular polymer.

tems.[36−38] These works suggest that the properties of the polymer binder can play a critical role in cell performance. Unfortunately, there is still a lack of systematic studies to elucidate the important factors to be considered when choosing or designing a polymer binder.[29,35] Additionally, the above studies have used Si nanoparticles (<150 nm) as the active material. When large Si particles incorporated into electrodes, which is more cost-effective and improves energy density, serious particle fracture and rapid failure of the electrode occurs.[8,39,40]

Our group recently demonstrated the utility of self-healing polymers to enable high areal capacity and long cycle life with large Si particles (~800 nm and larger) by using a hydrogen bonding supramolecular polymer binder.[41,42] To our knowledge, we are still the only group to achieve performance with these large Si particles that compares to the state of the art Si nanoparticle electrodes. These self-healing polymers and other supramolecular binders[26] are highly viscoelastic, exhibiting time dependent strain responses and a combination of solid-like and liquid-like characteristics. Questions arise as to the importance of these viscoelastic properties on battery performance. Here, we evaluate a family of fatty acid derived supramolecular polymers that utilize hydrogen bonding urea end groups to form a polymer network. By varying the concentration of trifunctional cross-linking junctions, a range of viscoelastic properties is achieved in order to elucidate ideal parameters for the design of future generations of polymer binder. We found that increasing the cross-linking results in a more solid-like polymer that allows for improved cycling stability of large particle Si electrodes over a wide range of polymer properties.

■ EXPERIMENTAL SECTION

Synthesis of Self-Healing Polymer (SHP) Binder. Prepolymers were first synthesized from starting materials of either Empol 1016 (donated by Congnis, 80 wt % diacids, 16 wt % triacids, 4 wt % monoacid) or Emery 2045 (donated by BASF, 30 wt % diacids, 70 wt % triacids). The design and synthesis of the self-healing polymer, inspired by Leibler et al.,[43] is a modified approach from previous reports. First, 41.5 g of either Empol 1016 or Emery 2045 was reacted with 17 g of diethylenetriamine at 120 °C for 2 h followed by 24 h at 160 °C with Ar protection to form a randomly branched oligomer terminated with amine groups, as was previously reported. Next, 3 g of the prepolymer or a mixture of the two prepolymers was reacted with 300 mg of urea for 4h at 135 °C to obtain our family of SHPs.

Preparation of Si/SHP Electrode. First, 870 nm Si particles (D50 0.8 μm, Fuzhou Hokin Chemical Technology Co., Figure S1) were dispersed in ethanol solution (200 mg/mL) via sonication and dropcast on a 1 cm^2 copper disc with 10 μm thickness (Fukuda). The electrodes were then dried at room temperature and calendared. The SHP (0.1 g) was dissolved in ethanol (1 mL) and mixed with carbon black (CB, 15 mg) to form a conductive slurry. Five μL of this slurry was dropcast onto the calendared Si electrodes and allowed to dry. The coated electrodes were heated to ~140 °C on a hot plate to melt the SHP/CB composite, and a razor blade was then used to remove excess binder from the surface and help coat Si particles throughout the electrode as was previously reported.[42] The electrodes were then dried overnight in a vacuum oven at 40 °C. The average Si loading for each electrode was 0.878 mg and the average SHP/CB (20:3) composite content was 0.389 mg.

Analysis and Characterization. NMR (1H) spectra were recorded on a Varian Mercury 400 NMR spectrometer at room temperature. Differential scanning calorimetry (DSC) experiments were performed using a DSC-Q2000 (TA Instruments) A temperature range of −70 to 150 °C with a heating and cooling speed of 10 °C/min was used. Rheology experiments were done using a stress-controlled rheometer (TA Instruments Model AR-G2) with a parallel plate geometry for the top plate and a temperature controlled Peltier plate at 25 °C for the bottom plate. The gap height was 1 mm. Adhesion experiments were performed on an Instron 5565 testing station. A 2 cm long by 1 cm wide strip was prepared in the same way as the Si electrodes used in the coin cells. Scotch magic tape (3M) was affixed to the surface and was peeled off at a rate of 20 mm/min as the force was monitored. SEM images were taken on an FEI XL30 Sirion electron microscope with FEG source.

Electrochemical Characterization. Coin cells were assembled by stacking Si working electrode, separator (Celgard 2400), and lithium metal (~1 cm^2 disc) counter electrodes inside an Ar filled glovebox. The mass loading of Si particles in each electrode was controlled

DOI: 10.1021/acsami.5b11363
ACS Appl. Mater. Interfaces 2016, 8, 2318−2324

Table 1. Properties of Synthesized Self-Healing Polymers

trifunctional units (wt %)	Maxwell relaxation time (s)	frequency sweep relaxation time (s)	T_g (°C)	adhesion strength (N/cm)	ionic conductivity ($\times 10^{-4}$ S/cm)	swell ratio (after 24 h)
16	0.04143	0.0858	−6.26	1.442	6.12 (±2.29)	1.480 (±0.005)
30	0.09218	0.1206	−1.34	2.489	5.18 (±1.28)	1.291 (±0.392)
43	0.2027	0.1676	2.74	3.268	4.59 (±1.22)	1.729 (±0.102)
57	0.1820	0.119	0.17	> 3	4.00 (±0.91)	1.914 (±0.234)
70	9.279	2.137	−28.93	> 3	5.55 (±3.99)	2.181 (±0.046)

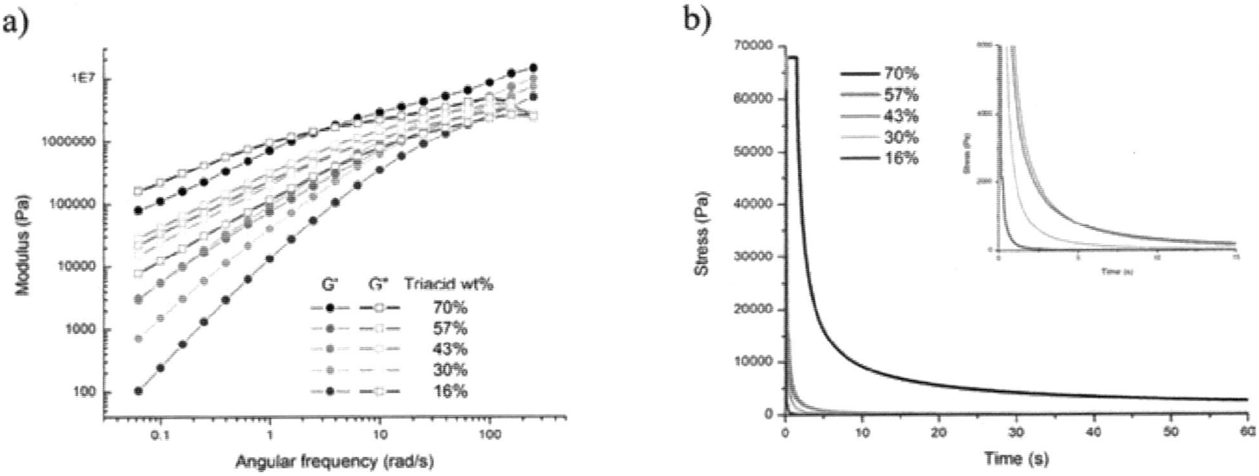

Figure 2. Rheology of self-healing polymer. (a) Frequency sweep at 1% strain of polymers at various cross-linker ratios. Loss modulus dominates at most frequencies and both moduli increase with cross-linking. (b) Stress relaxation of polymers at various cross-linker ratios at 10% strain. Relaxation time increases with cross-linking concentration, but all materials fully relax showing large plastic deformation.

between 0.75 and 1.1 mg_{Si} cm^{-2}. All the electrodes, unless otherwise specified, were cycled at a rate of C/20 (C/n = charge/discharge in n hours) in the first three cycles for activation and at C/10 in the following cycles. The electrolyte for all tests was 1 M $LiPF_6$ in ethylene carbonate/diethylcarbonate/vinyl carbonate/fluoroethylenecarbonate (EC/DEC/VC/FEC, 1:1:0.005:0.05 v/v/v/v). EIS measurements were performed using a VMP3 potentiostat (Bio-Logic) to obtain standard Nyquist plots (1 MHz−100 mHz). The high frequency intercept for the real impedance (Z') of the swollen samples was used as the electrolyte resistance (R). A BT 2000 testing station (Arbin Instruments) was used for galvanostatic charge/discharge cycling. The cycling rates were calculated by assuming theoretical capacity of 3600 mAh g^{-1} for Si. The Coulombic efficiency (CE, %) was calculated as $(C_{discharge}/C_{charge}) \times 100$, where C_{charge} and $C_{discharge}$ were the capacity of the anodes for lithiation and delithiation.

■ RESULTS AND DISCUSSION

By using two different starting materials, Empol 1016 and Emery 2045, with different ratios of mono-, di-, and trifunctional fatty acid molecules, we were able to finely control the density of cross-linking junctions in the final supramolecular polymers and thus vary their viscoelastic properties. Figure 1 illustrates the synthesis of the supramolecular binder. From this synthesis, we obtain a family of low glass transition temperature (T_g) supramolecular self-healing polymers (Table 1 and Figure S5). We first measured the viscoelastic properties of this family of materials at room temperature using frequency sweep and stress relaxation experiments (Figure 2). Creep experiments were also attempted, but the material continuously yielded and the rheometer was unable to maintain a constant load. In each material, the loss modulus dominates for most frequencies and the stress relaxation is complete on very fast time scales. This illustrates significant plastic deformation and the strong liquid-like characteristics of these materials. When observed in bulk at room temperature, the polymers flow slowly over the course of several days. As was expected when comparing the materials, the polymers with a higher density of theoretical cross-linking junctions are stiffer, with a higher resistance to deformation. The more cross-linked materials also have higher storage (G') and loss (G'') moduli. We extracted relaxation times from both the frequency sweep and stress relaxation data. The relaxation time represents a characteristic time scale required for the polymer chains to respond to external stress, and was used to quantify the viscoelasticity of each cross-linking ratio. The crossover point of the storage and loss modulus curves was used to determine the characteristic relaxation time from the frequency sweep data and a four parameter Burger's model was used to fit the stress relaxation data and extract a Maxwell element relaxation time (Table 1 and Figures S6 and S7). We see a good agreement between the frequency sweep and stress relaxation data, and as the percentage of theoretical cross-linking junctions was increased from 16 to 70 wt %, there was an increase in relaxation time of over 2 orders of magnitude. Specifically, the polymer with 16 wt % trifunctional fatty acids was the softest with a relaxation time on the order of 10^{-2} s. The intermediate polymers with 30, 43, and 57 wt % trifunctional fatty acids all had relaxation times on the order of 0.1 s, and the most cross-linked polymer had a relxation time on the order of 1 s.

Because the batteries are flooded with electrolyte, it is important to understand how these polymers behave in the electrolyte environment. Swelling experiments were performed using 1 M $LiPF_6$ with EC:DEC (1:1) as the solvent over the

DOI: 10.1021/acsami.5b11363
ACS Appl. Mater. Interfaces 2016, 8, 2318−2324

course of 24 h (Table 1 and Supporting Information). For each material we saw moderate swelling ratios, defined as $\frac{w_s - w_i}{w_i}\frac{\rho_{solvent}}{\rho_{polymer}} + 1$ where w_s is the weight of the swollen gel, w_i is the initial weight of the dry polymer, and ρ is the density, between 1.3 and 2.2 with an increase in swelling as cross-linking increased. This swelling is important because the uptake of electrolyte is likely the main mechanism of Li^+ ion conduction through the SHP binder. We measured the ionic conductivity of the binder using a three-electrode setup. A measured amount of SHP without CB was coated from ethanol solution onto a 1 cm^2 area of stainless steel and dried to serve as the working electrode. This was placed into a flood cell of EC:DEC (1 M $LiPF_6$) with a Li metal reference and counter electrode. The SHP was coated without the conductive additive so that the Li^+ ions were the only means of current conduction through the film. Impedance spectroscopy was used to measure the high frequency intercept of Z', which was used to calculate the ionic conductivity (Table 1 and Figure S8). Li ion conductivity decreased slightly as cross-linking increased. This could be due to the increased relaxation time, increased tortuosity of the ionic conduction pathways in the more cross-linked materials, or trap sites for the ions that increase in number with cross-linking. The conductivity values of $\sim 5 \times 10^{-4}$ S/cm are similar to values reported for gel-polymer electrolytes and agrees with the hypothesis that electrolyte swelling is important for Li^+ ion conduction through the binder.

The adhesive properties of binders have also been found to be an important factor in their performance in Si negative electrodes.[23,24] The results of the peel test show that the adhesion strength of the SHP binder increased with cross-linking (Figure 3 and Table 1). This likely does not indicate increased adhesion to the Si particles, since the entire family of SHPs has the same urea and amide hydrogen bonding sites, and instead indicates increased cohesion within the SHP coating layer. This is supported by the black SHP composite layer that remains on the electrodes even after the peeling test (Figure S10). The increased cohesion is likely due to the increased stiffness of the more cross-linked polymers and their slower relaxation time causing them to have more solid-like behavior over the time scale of the peeling test. As shown in the photo of the electrodes after the peel test, the adhesion strength of the 57 and 70 wt % trifunctional SHPs exceeded the limits of the tape adhesive and the binder could not be peeled off of the Si electrode (Figures S9 and S10). At these high cross-linking percentages, the high cohesion of the SHP probably excluded the polymer chains of the tape adhesive from penetrating the SHP layer resulting in little to no diffusive adhesion between the tape and SHP binder.

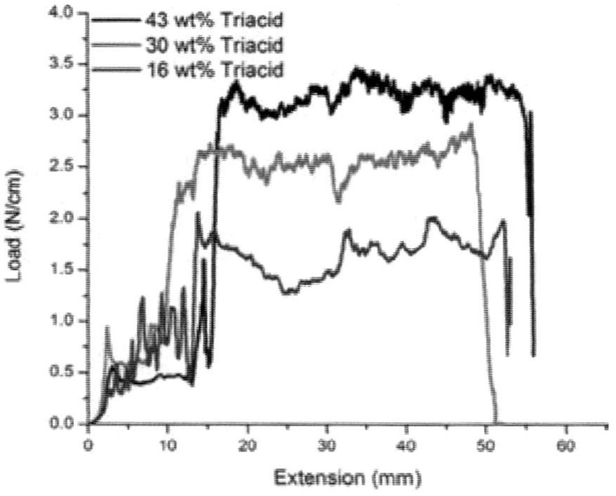

Figure 3. Adhesion of SHP/CB composite to Si electrodes. Adhesion strength increases with cross-linking; 58 and 70 wt % not shown because the samples could not be peeled using the tape available (Figure S8).

Once the materials had been characterized, the electrochemical performance of the Si/SHP electrodes was tested. High mass loading (0.75−1.1 mg cm^{-2}) of 800 nm Si particles (Figure 4a) was used in all cells in an effort to ensure that any positive performance was not just due to low mass loading, which is not relevant for commercial applications. After loading the Si particles, a coating of SHP/CB composite was applied to the electrode (Figure 4b). Voltage profiles in Figure 5b show the characteristic voltage plateau for Si charging around 100 mV vs Li and a mostly sloped discharging profile with a slight plateau around 400 mV that is similar to what we have observed before with the SHP binder. After 3 activation cycles at C/20, the cells were all cycled at C/10 until they reached 80% of their initial capacity. Surprisingly, there was little difference in the performance of the cells with different amounts of cross-linking until the amount of trifunctional groups in the polymer was increased to 70 wt %. Until this point, cells had a specific capacity ~1700 mAh/g after activation (1st C/10 cycle) and reached ~1360 mAh/g (80%) at an average of 133, 146, 178, and 159 cycles, for 16, 30, 43, and 57 wt % trifunctional SHP, respectively (Figure 5a, Figure S11, and Table S1). This relatively low capacity compared to Si nanoparticle electrodes is likely due to poor activation of the large sized Si particles and the increased Ohmic resistance of the thick Si layer. Further improvement to the electrochemical performance of these cells could be achieved by developing techniques for thinner and more conformal SHP coatings, and by mixing in conductive additives with the Si particles to improve the conductivity of the Si layer and improve the activation, but such performance optimization was outside the scope of this study. In contrast to the other electrodes, the cells with the most cross-linked binder only had a specific capacity of ~1300 mAh/g after activation and reached 80% of their initial capacity after only 83 cycles. This result illustrates a further reduced activation of the Si particles. This is due to a reduced ability of the 70 wt % trifunctional SHP to flow over newly exposed Si surfaces after fracture. Additionally, the application of the composite coating was noticeably more difficult for the highly cross-linked binder and resulted in generally thicker binder layers (Table S1). This increased thickness coupled with the decreased ionic conductivity caused fewer of the Si particles to participate in lithiation. Using slower rates (C/40) allowed the capacity to increase to ~2400 mAh/g for the 70 wt % trifunctional SHP and showed that the low capacity is a result of kinetic issues in the electrodes (Figure 5c). The performance of the Si/SHP electrodes is limited at high rates due to the poor ionic conductivity of the binder and the large particle size.

The increased rate of capacity decay is related to the relaxation time. Between the binders with 57 and 70 wt % cross-linking there is an increase in the relaxation time of more than an order of magnitude. As the cross-linking is increased to 70 wt % the bulk material is less able to flow, and its rheological properties are markedly different from those in the rest of the study with a much slower stress relaxation and higher shear

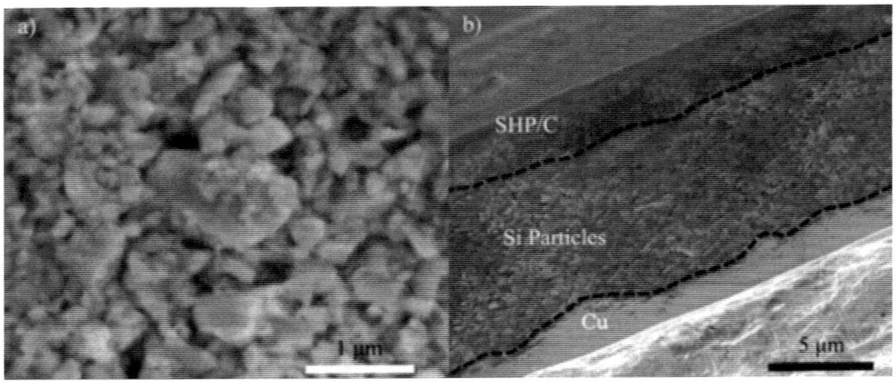

Figure 4. (a) SEM image of 800 nm Si particles. (b) SEM image of the cross section of a Si/SHP electrode before cycling.

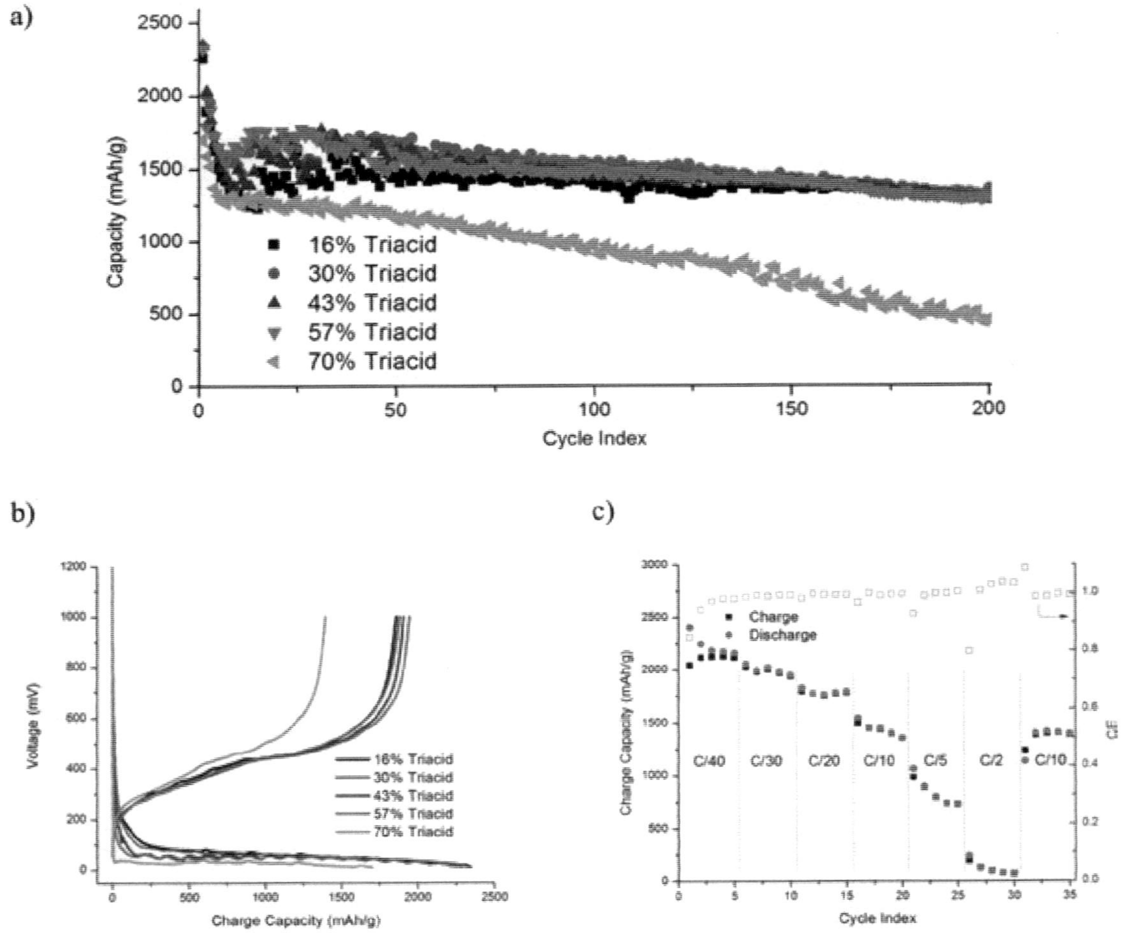

Figure 5. (a) Cycling stability of SHP binder with 800 nm Si particles. (b) Charge/discharge profile of SHP binder showing normal Si lithiation and delithiation profiles. (c) Capacity of 70 wt % Triacid Si/SHP electrode at different C-rates.

modulus. This prevents the polymer from maintaining the electrode's capacity during cycling in the same way that the other polymers do. More specifically, for each of the other binders with lower cross-linking tested in this work the smaller change in relaxation time and other viscoelastic properties only slightly affected cycling performance. The viscoelasticity of our SHP allows the binder to accommodate the stress that builds up at the Si particles expand. The low Deborah numbers (De = τ_{char}/τ_{obs}) of these SHPs also allow it to flow and cover any new Si surface that is exposed as the particles crack. Even the slowest relaxation time measured for the 70 wt % trifunctional SHP was ~10 s, which is 2–3 orders of magnitude faster than the charging or discharging time of a battery. We suspect that this flow property is the main mechanism of improving cycling stability in our self-healing binder and maintains the electrical connectivity of these Si pieces. While the mechanism and performance of the materials with lower cross-linking are similar, the three intermediately cross-linked polymers (30, 43, and 57 wt %) show a slight improvement in cycling stability over the 16 wt % trifunctional material. This suggests that there is a balance between stress dissipation and the maintenance of the mechanical integrity of the electrode. The materials with the slightly slower relaxation time could provide an improvement in the mechanical stability of the electrode during the volume changes that occur with the Li alloying and dealloying processes without sacrificing the flow that allows for long cycling life in the first place.

After the fracture of the large Si particles has been stabilized, the electrodes with different binders all are able to maintain similar capacity decay, except for the binder with the highest amount of cross-linking, due to their identical backbone and end group structure. This similar cycling stability over a wide range of polymer viscosities is an exciting property of our SHP for industry applications, and will make the polymer useful in a variety of processing situations without concern for the final device performance. Each electrode has a similarly high plateau CE of 99.5% (Figure S12), and so the more aggressive capacity loss in the 70 wt % trifunctional SHP is not explained by a lower CE, meaning increased SEI formation. Instead the capacity loss could be due to portions of the electrode becoming electrically isolated after fracture because poor polymer incorporation into the electrode during coating. The slightly higher ionic conductivity of the 16 wt % trifunctional SHP gives the Li access to more of the active material resulting in improved activation of the Si particles and mitigates some of the kinetic issues present. However, this high capacity rapidly drops during the activation cycles so that the more cross-linked binders have a slightly higher capacity during C/10 cycling. The increased adhesion of the more cross-linked binder helps to hold the electrode together during the activation cycles and thus more of the initial capacity is retained.

CONCLUSIONS

In conclusion, this study has shown that the viscoelastic properties of our supramolecular SHP binder can be varied over a wide range with little effect on the binder's performance in Si negative electrodes. This is because the relaxation times measured at all cross-linking ratios, while varied, were still very small when compared to battery cycling times. This ability to vary the viscosity of our polymer without affecting cycling stability makes our binder attractive for industry where it can be modified to meet various processing demands. It was found that a relaxation time on the order of 0.1 s gives the best cycling performance due to a balance of viscous flow and improved mechanical integrity. Large Si particle electrodes could be cycled for over 175 cycles while maintaining above 80% capacity for three of the five binders tested. We also found that, at a high amount of cross-linking, the benefits of the SHP are lost when processing becomes more difficult. Combining this work with previous studies, we can conclude that a good binder for Si negative electrodes should have strong hydrogen bonding and adhesion, good Li ion conductivity, and quick stress relaxation relative to battery cycling. We do see from this study that higher modulus might be important up to a point, as long as other factors are not significantly affected. In future work, we plan to explore modified backbone chemistries to facilitate better rate capability and improve CE while also including covalent cross-linking as a means to further improve electrode stability.

ASSOCIATED CONTENT

Supporting Information

The Supporting Information is available free of charge on the ACS Publications website at DOI: 10.1021/acsami.5b11363.

Additional data includes NMR, DSC, EIS, viscoelastic modeling, adhesion testing, and additional electrochemical characterization. (PDF)

AUTHOR INFORMATION

Corresponding Authors

*E-mail: yicui@stanford.edu.

*E-mail: zbao@stanford.edu.

Present Address

‖(C.W.) Department of Chemistry, University of California Riverside, Riverside, California 92521

Author Contributions

The manuscript was written through contributions of all authors. All authors have given approval to the final version of the manuscript.

Notes

The authors declare no competing financial interest.

ACKNOWLEDGMENTS

This work was supported by the Global Climate and Energy Program at Stanford University. J.L. also acknowledges support by the National Science Foundation Graduate Research Fellowship Program under Grant No. (DGE-114747).

REFERENCES

(1) Armand, M.; Tarascon, J. M. Building Better Batteries. *Nature* **2008**, *451*, 652−657.

(2) Dunn, B.; Kamath, H.; Tarascon, J. M. Electrical Energy Storage for the Grid: a Battery of Choices. *Science* **2011**, *334*, 928−935.

(3) Obrovac, M. N.; Christensen, L. Structural Changes in Silicon Anodes During Lithium Insertion/Extraction. *Electrochem. Solid-State Lett.* **2004**, *7*, A93.

(4) Obrovac, M. N.; Krause, L. J. Reversible Cycling of Crystalline Silicon Powder. *J. Electrochem. Soc.* **2007**, *154*, A103.

(5) Beaulieu, L. Y.; Eberman, K. W.; Turner, R. L.; Krause, L. J.; Dahn, J. R. Colossal Reversible Volume Changes in Lithium Alloys. *Electrochem. Solid-State Lett.* **2001**, *4*, A137.

(6) Liu, X. H.; Wang, J. W.; Huang, S.; Fan, F.; Huang, X.; Liu, Y.; Krylyuk, S.; Yoo, J.; Dayeh, S. A.; Davydov, A. V.; Mao, S. X.; Picraux, S. T.; Zhang, S.; Li, J.; Zhu, T.; Huang, J. Y. In Situ Atomic-Scale Imaging of Electrochemical Lithiation in Silicon. *Nat. Nanotechnol.* **2012**, *7*, 749−756.

(7) Lee, S. W.; McDowell, M. T.; Berla, L. A.; Nix, W. D.; Cui, Y. Fracture of Crystalline Silicon Nanopillars During Electrochemical Lithium Insertion. *Proc. Natl. Acad. Sci. U. S. A.* **2012**, *109*, 4080−4085.

(8) Liu, X. H.; Zhong, L.; Huang, S.; Mao, S. X.; Zhu, T.; Huang, J. Y. Size-Dependent Fracture of Silicon Nanoparticles During Lithiation. *ACS Nano* **2012**, *6*, 1522−1531.

(9) Ryu, J. H.; Kim, J. W.; Sung, Y.-E.; Oh, S. M. Failure Modes of Silicon Powder Negative Electrode in Lithium Secondary Batteries. *Electrochem. Solid-State Lett.* **2004**, *7*, A306.

(10) Saint, J.; Morcrette, M.; Larcher, D.; Laffont, L.; Beattie, S.; Pérès, J. P.; Talaga, D.; Couzi, M.; Tarascon, J. M. Towards a Fundamental Understanding of the Improved Electrochemical Performance of Silicon−Carbon Composites. *Adv. Funct. Mater.* **2007**, *17*, 1765−1774.

(11) Chan, C. K.; Peng, H.; Liu, G.; McIlwrath, K.; Zhang, X. F.; Huggins, R. A.; Cui, Y. High-Performance Lithium Battery Anodes Using Silicon Nanowires. *Nat. Nanotechnol.* **2008**, *3*, 31−35.

(12) Park, M.-H.; Kim, M. G.; Joo, J.; Kim, K.; Kim, J.; Ahn, S.; Cui, Y.; Cho, J. Silicon Nanotube Battery Anodes. *Nano Lett.* **2009**, *9*, 3844–3847.

(13) Wu, H.; Chan, G.; Choi, J. W.; Ryu, I.; Yao, Y.; McDowell, M. T.; Lee, S. W.; Jackson, A.; Yang, Y.; Hu, L.; Cui, Y. Stable Cycling of Double-Walled Silicon Nanotubebattery Anodes Through Solid–Electrolyteinterphase Control. *Nat. Nanotechnol.* **2012**, *7*, 310–315.

(14) Chen, S.; Gordin, M. L.; Yi, R.; Howlett, G.; Sohn, H.; Wang, D. Silicon Core–Hollow Carbon Shell Nanocomposites with Tunable Buffer Voids for High Capacity Anodes of Lithium-Ion Batteries. *Phys. Chem. Chem. Phys.* **2012**, *14*, 12741.

(15) Liu, N.; Wu, H.; McDowell, M. T.; Yao, Y.; Wang, C.; Cui, Y. A Yolk-Shell Design for Stabilized and Scalable Li-Ion Battery Alloy Anodes. *Nano Lett.* **2012**, *12*, 3315–3321.

(16) Liu, N.; Lu, Z.; Zhao, J.; McDowell, M. T.; Lee, H.-W.; Zhao, W.; Cui, Y. A Pomegranate-Inspired Nanoscale Design Forlarge-Volume-Change Lithium Battery Anodes. *Nat. Nanotechnol.* **2014**, *9*, 187–192.

(17) Hu, Y.-S.; Demir-Cakan, R.; Titirici, M.-M.; Müller, J.-O.; Schlögl, R.; Antonietti, M.; Maier, J. Superior Storage Performance of a Si@SiOx/C Nanocomposite as Anode Material for Lithium-Ion Batteries. *Angew. Chem., Int. Ed.* **2008**, *47*, 1645–1649.

(18) Magasinski, A.; Dixon, P.; Hertzberg, B.; Kvit, A.; Ayala, J.; Yushin, G. High-Performance Lithium-Ion Anodes Using a Hierarchical Bottom-Up Approach. *Nat. Mater.* **2010**, *9*, 353–358.

(19) Kim, H.; Han, B.; Choo, J.; Cho, J. Three-Dimensional Porous Silicon Particles for Use in High-Performance Lithium Secondary Batteries. *Angew. Chem., Int. Ed.* **2008**, *47*, 10151–10154.

(20) Yi, R.; Dai, F.; Gordin, M. L.; Chen, S.; Wang, D. Micro-Sized Si-C Composite with Interconnected Nanoscale Building Blocks as High-Performance Anodes for Practical Application in Lithium-Ion Batteries. *Adv. Energy Mater.* **2013**, *3*, 295–300.

(21) Li, J.; Lewis, R. B.; Dahn, J. R. Sodium Carboxymethyl Cellulose: A Potential Binder for Si Negative Electrodes for Li-Ion Batteries. *Electrochem. Solid-State Lett.* **2007**, *10*, A17.

(22) Kovalenko, I.; Zdyrko, B.; Magasinski, A.; Hertzberg, B.; Milicev, Z.; Burtovyy, R.; Luzinov, I.; Yushin, G. A Major Constituent of Brown Algae for Use in High-Capacity Li-Ion Batteries. *Science* **2011**, *334*, 75–79.

(23) Ryou, M.-H.; Kim, J.; Lee, I.; Kim, S.; Jeong, Y. K.; Hong, S.; Ryu, J. H.; Kim, T.-S.; Park, J.-K.; Lee, H.; Choi, J. W. Mussel-Inspired Adhesive Binders for High-Performance Silicon Nanoparticle Anodes in Lithium-Ion Batteries. *Adv. Mater.* **2013**, *25*, 1571–1576.

(24) Jeong, Y. K.; Kwon, T.-W.; Lee, I.; Kim, T.-S.; Coskun, A.; Choi, J. W. Hyperbranched B-Cyclodextrin Polymer as an Effective Multidimensional Binder for Silicon Anodes in Lithium Rechargeable Batteries. *Nano Lett.* **2014**, *14*, 864–870.

(25) Jeong, Y. K.; Kwon, T.-W.; Lee, I.; Kim, T.-S.; Coskun, A.; Choi, J. W. Energy & Environmental Science. *Energy Environ. Sci.* **2015**, *8*, 1224–1230.

(26) Kwon, T.-W.; Jeong, Y. K.; Deniz, E.; AlQaradawi, S. Y.; Choi, J. W.; Coskun, A. Dynamic Cross-Linking of Polymeric Binders Based on Host–Guest Interactions for Silicon Anodes in Lithium Ion Batteries. *ACS Nano* **2015**, *9*, 11317–11324.

(27) Magasinski, A.; Zdyrko, B.; Kovalenko, I.; Hertzberg, B.; Burtovyy, R.; Huebner, C. F.; Fuller, T. F.; Luzinov, I.; Yushin, G. Toward Efficient Binders for Li-Ion Battery Si-Based Anodes: Polyacrylic Acid. *ACS Appl. Mater. Interfaces* **2010**, *2*, 3004–3010.

(28) Park, H.-K.; Kong, B.-S.; Oh, E.-S. Effect of High Adhesive Polyvinyl Alcohol Binder on the Anodes of Lithium Ion Batteries. *Electrochem. Commun.* **2011**, *13*, 1051–1053.

(29) Kwon, T.-W.; Jeong, Y. K.; Lee, I.; Kim, T.-S.; Choi, J. W.; Coskun, A. Systematic Molecular-Level Design of Binders Incorporating Meldrum's Acid for Silicon Anodes in Lithium Rechargeable Batteries. *Adv. Mater.* **2014**, *26*, 7979–7985.

(30) Liu, G.; Xun, S.; Vukmirovic, N.; Song, X.; Olalde-Velasco, P.; Zheng, H.; Battaglia, V. S.; Wang, L.; Yang, W. Polymers with Tailored Electronic Structure for High Capacity Lithium Battery Electrodes. *Adv. Mater.* **2011**, *23*, 4679–4683.

(31) Wu, M.; Xiao, X.; Vukmirovic, N.; Xun, S.; Das, P. K.; Song, X.; Olalde-Velasco, P.; Wang, D.; Weber, A. Z.; Wang, L.-W.; Battaglia, V. S.; Yang, W.; Liu, G. Toward an Ideal Polymer Binder Design for High-Capacity Battery Anodes. *J. Am. Chem. Soc.* **2013**, *135*, 12048–12056.

(32) Wu, H.; Yu, G.; Pan, L.; Liu, N.; McDowell, M. T.; Bao, Z.; Cui, Y. Stable Li-Ion Battery Anodes by in-Situ Polymerization of Conducting Hydrogel to Conformally Coat Silicon Nanoparticles. *Nat. Commun.* **2013**, *4*, 1943–1946.

(33) Kim, S. M.; Kim, M. H.; Choi, S. Y.; Lee, J. G.; Jang, J.; et al. Poly(phenanthrenequinone) as a Conductive Binder for Nano-Sized Silicon Negative Electrodes. *Energy Environ. Sci.* **2015**, *8*, 1538–1543.

(34) Shi, Y.; Peng, L.; Ding, Y.; Zhao, Y.; Yu, G. Nanostructured Conductive Polymers for Advanced Energy Storage. *Chem. Soc. Rev.* **2015**, *44*, 6684–6696.

(35) Erk, C.; Brezesinski, T.; Sommer, H.; Schneider, R.; Janek, J. Toward Silicon Anodes for Next-Generation Lithium Ion Batteries: a Comparative Performance Study of Various Polymer Binders and Silicon Nanopowders. *ACS Appl. Mater. Interfaces* **2013**, *5*, 7299–7307.

(36) Liu, W.-R.; Yang, M.-H.; Wu, H.-C.; Chiao, S. M.; Wu, N.-L. Enhanced Cycle Life of Si Anode for Li-Ion Batteries by Using Modified Elastomeric Binder. *Electrochem. Solid-State Lett.* **2005**, *8*, A100.

(37) Koo, B.; Kim, H.; Cho, Y.; Lee, K. T.; Choi, N.-S.; Cho, J. A Highly Cross-Linked Polymeric Binder for High-Performance Silicon Negative Electrodes in Lithium Ion Batteries. *Angew. Chem., Int. Ed.* **2012**, *51*, 8762–8767.

(38) Song, J.; Zhou, M.; Yi, R.; Xu, T.; Gordin, M. L.; Tang, D.; Yu, Z.; Regula, M.; Wang, D. Interpenetrated Gel Polymer Binder for High-Performance Silicon Anodes in Lithium-Ion Batteries. *Adv. Funct. Mater.* **2014**, *24*, 5904–5910.

(39) Beattie, S. D.; Larcher, D.; Morcrette, M.; Simon, B.; Tarascon, J. M. Si Electrodes for Li-Ion Batteries—a New Way to Look at an Old Problem. *J. Electrochem. Soc.* **2008**, *155*, A158.

(40) Wu, M.; Sabisch, J. E. C.; Song, X.; Minor, A. M.; Battaglia, V. S.; Liu, G. In Situ Formed Si Nanoparticle Network with Micron-Sized Si Particles for Lithium-Ion Battery Anodes. *Nano Lett.* **2013**, *13*, 5397–5402.

(41) Wang, C.; Wu, H.; Chen, Z.; McDowell, M. T.; Cui, Y.; Bao, Z. Self-Healing Chemistry Enables the Stable Operation of Silicon Microparticle Anodes for High-Energy Lithium-Ion Batteries. *Nat. Chem.* **2013**, *5*, 1042–1048.

(42) Chen, Z.; Wang, C.; Lopez, J.; Lu, Z.; Cui, Y.; Bao, Z. High-Areal-Capacity Silicon Electrodes with Low-Cost Silicon Particles Based on Spatial Control of Self-Healing Binder. *Adv. Energy Mater.* **2015**, *5*, 1401826.

(43) Cordier, P.; Tournilhac, F.; Soulié-Ziakovic, C.; Leibler, L. Self-Healing and Thermoreversible Rubber From Supramolecular Assembly. *Nature* **2008**, *451*, 977–980.

468564 Designer Dual Therapy Nanolayered Implant Coatings Eradicate Biofilms and Accelerate Bone Tissue Repair

Abstract:

One of the important problems in the field of orthopedic medicine is the ability to create a stable bone-materials interface with an implant. The most effective treatment requires the delivery of exacting amounts of therapeutics of different types over appropriate timeframes in and around the implant, while maintaining mechanical integrity of the implant materials and allowing for bone integration on their surfaces. This work presents a novel next-generation implant coating for both eradication of an established biofilm within the bone cavity and accelerated bone repair via the controlled delivery of antibiotic and growth factor in sequence from stable nanometer-scale coatings on the implant surface.

Infection is the most common reason for complications, which often lead to complete removal of implants (74%). Infection significantly increases morbidity, and places huge financial burdens on the patient and the healthcare system—projected to exceed $1.6 billion/year by 2020. Because infection is much more common in implant replacement surgeries, these issues greatly impact long-term patient care for a growing population. For revision arthroplasty of an infected prosthesis, a prolonged and expensive two-stage procedure requiring two surgical steps and a 6–8 week period of joint immobilization is today's gold standard. A single-stage revision is preferred as an alternative; however, traditional bulk polymer systems such as bone cement cannot load sufficient amounts of therapeutic to eradicate existing infection, are insufficient or infeasible for the release of sensitive biologic drugs that considerably aid in bone regeneration, and lead to substandard mechanical properties and retarded bone repair.

These issues are addressed by conformal, programmable, and degradable dual therapy coatings (~500 nm thick) in a layer-by-layer fashion using the enabling nanofabrication tool of electrostatic multilayer assembly. The nanolayered construct allows large loadings of each drug, thus enabling ultrathin film coatings to carry sufficient treatment and precise independent control of release kinetics and loading for each therapeutic agent. The coating architecture was designed to enable both early burst and lower sustained release of antibiotics sufficient to eliminate infection, and a long-term sustained release of BMP-2 growth factor to induce more significant and mechanically competent bone formation than a short-term burst release. In rats, the successful growth factor-mediated osteointegration of the multilayered implants with the host tissue improved bone-implant interfacial strength by 15-fold when compared with the bare implant control, and yields a mechanical bond 17-fold higher than that created with the use of clinically available bioactive bone cement.

Here we focused on dual delivery of an antibiotic and a growth factor owing to the urgent need for enhanced infection-reducing and tissue-integrating strategies in orthopedic applications, but the excellent modularity of multilayers for incorporation and release of diverse therapeutics suggests this approach should be also applicable to different implant applications such as vascular graft and artificial heart implants for which the risks of infection are often ignored. Our findings demonstrate the potential of this layered release strategy to introduce a durable implant solution, ultimately an important step forward in the design of biomedical implant release coatings for multiple medical applications.

468861 High Performance Degradable Polymers from Renewable Resources

Abstract:

Synthetic polymers can be used in the manufacture of commercial products as deceptively simple as plastic bottles or as specialized as medical implants and composites for aircraft. Although these versatile materials are essential to our society, their manufacture and disposal can be damaging to the environment. In this talk I will discuss my efforts to develop high performance sustainable polymers for a wide variety of applications. First, I will briefly describe the invention of an efficient and scalable semisynthetic method for the production of a renewable and biodegradable aliphatic polyester, poly(β-methyl-δ- valerolactone) (PMVL). Next, I will explore the use of PMVL as precursor for the preparation of tough plastics, elastomers, pressure sensitive adhesives, and chemically recylcable polyurethane foams. Finally, I will discuss how changes to aliphatic polyester structure relate to several key physical parameters that influence material performance and explain how this knowledge of polymer structure-property relationships can be used to design advanced sustainable polymers tailor-made for specific applications.

468892 Reactive Block Polymer Membranes Tailored for Heavy Metal Ion Separations

Abstract:

The development of novel polymeric materials that enable the fabrication of nextgeneration membranes capable of separating target solutes from solution has been a major focus of recent membrane research. While steric exclusion (*i.e.*, size-selectivity) is the most basic mechanism for membrane separations, new polymeric materials are being developed for chemically-selective membrane adsorbers that allow for efficient, high throughput separations. These adsorptive membranes can be utilized for the removal of pollutants such as heavy metal ions, the concentration of chemical products, or the chromatographic separations of valuable biopharmaceuticals. The novel block polymer materials discussed in this talk are an exciting materials platform for the fabrication of membrane adsorbers because they provide the ability to rationally engineer adsorption capacity through the controlled synthesis of the macromolecular precursor. In this work, the triblock polymer poly(isoprene-*b*-styrene-*b*-*N,N*-dimethyacrylamide) (PI-PS-PDMA) is cast into asymmetric membranes using a technique that can be implemented in roll-to-roll fabrication processes for large scale applications (i.e., the self-assembly and nonsolvent induced phase separation (SNIPS) method). The self-assembled polymer results in a high density of consistently-sized nanoscale pores, which are lined with by poly(*N,N*dimethylacrylamide) (PDMA) moieties. The PDMA groups that lines the pore walls are converted by simple hydrolysis to polyacrylic acid (PAA) groups, which can be easily reacted to other functionalities that are molecularly-designed for target applications. For example, upon reaction with cysteamine at room temperature, the pore-lining groups are converted to N-(2- ethanethiol) polyacrylamide (PASH) moieties that act as high capacity binding sites for heavy metal ions, including lead, gold, cadmium, and silver cations. When placed in a solution containing lead (II) ions, this PASH-functionalized membrane is able to adsorb up to 470 mg Pb^{2+} per gram of membrane, which compares favorably with other state-of-the-art lead adsorbers in the field (500 mg g^{-1}). Alternatively, a distinct membrane functionalization protocol can be executed to affix glutathione, an antioxidant that acts as a ligand for heavy metal ions in the cells of many organisms, to the pore walls. The glutathione-laden membrane is capable of heavy metal adsorption at high capacity even in very low concentration solutions like those that may occur in water treatment systems and natural environments. At concentrations as low as 10 ppm Pb^{2+}, the glutathione-functionalized membrane was able to remove over 95% of the heavy metal content from solution. Furthermore, the adsorption capacity of the membrane could be regenerated upon addition of a mildly acidic solution. Membranes that are based on these novel polymeric materials and are capable of this quality of separation performance provide a scalable and efficient adsorption platform that can be tailored to myriad chemically-selective separations in future applications.

455577 A Hybrid Framework for Process Synthesis-Design: Application to Biorefineries

Abstract:

As energy demand and emissions increase steadily, the need for more sustainable processes is becoming evident. To this end, innovation in synthesis and design of processing routes needs to take into account raw materials availability, existence of appropriate processing technologies and products that have demand and use. When more than one alternative (technology or process interval) exists for one or more processing steps, a processing network results from which the optimal processing route needs to be determined.

The synthesis and design of processing networks for any chemical process in general and biorefineries in particular is an industry-relevant problem. Mathematical programming techniques, where the model equations representing all alternatives present in the network are included as constraints within an optimization problem, are commonly applied to solve this type of synthesis-design problems. The determination of the optimal processing route from a network (superstructure) of alternatives therefore involves decision-making through continuous variables (such as mass flow-rates and separation factors) and discrete variables (such as selection of an alternative in a processing step). The size of the problem depends on the number of alternatives included within the network, while, the complexity of the problem depends on the type of the models and decision variables used to represent the process intervals. Within the context of a biorefinery, considering a network involving a maximum of NS processing steps, NI process intervals (alternative) organized in processing steps, NF different biomass (raw materials), NP different products, and, allowing for processing step bypass and material flow recycle, would result in a mixed integer nonlinear programming problem. The size of the problem depends on the values of NS, NI, NF, NP and NC (number of chemicals). Note also that the models for each process interval also include model parameters that would need to be specified before any problem can be solved.

The objective of a hybrid framework for biorefinery synthesis-design is to provide the means to study the feasibility of establishing biorefineries as a function of biomass availability, location and characteristics, product types (fuels and/or chemicals), available technologies (use of different types of conversion technologies) and many more. The requirements for such a framework are: availability of appropriate numerical solvers for MILP and MINLP problems; a tool to generate the network model equations and to formulate the problem specific mathematical programming problems; and a database of information-knowledge-data, such as, different types and sources of biomass, different types of products, different types of known conversion technologies, processing routes and prices. This hybrid framework is based on a previous work (Quaglia et al. 2012) and has been extended in order to accommodate a wider range of process synthesis-design problems that can be formulated and solved. Key additions to the framework are a data management system that allows the generation of the superstructure (network) based on the biorefinery case study under investigation and retrieval of location-dependent data. A data structure has been designed and implemented in a knowledge base, which contains all necessary information for a range of problem formulations and their solution. The knowledge base currently contains 11 types of biomass, from 10 different locations, 102 intervals representing processing technologies and 9 products. It is a dynamic knowledge base that is ever-growing as more data are being collected and more problems are being solved. Moreover, once data is stored in it, superstructures are generated systematically including only the relevant alternatives for a specific problem definition with respect to a specified location. In addition, a user-friendly interface has been developed covering several of the steps (Bertran et al. 2016).

The main features and application of the hybrid framework are illustrated through the formulation and solution of a multi-feedstock multi-product problem. Six biomass-based raw materials are considered, yielding various products, in eight geographical locations; this leads to a superstructure with 9 processing steps (NS) and 35 processing intervals (NI). The solution is a (set of) raw material(s), a (set of) products, a location and the process configuration. Furthermore, several scenarios, such as minimum waste, maximum profit and different product demands, derived from the main problem have been investigated and the location dependency highlighted. A second case study involving different biomass to produce liquid fuel products is also briefly highlighted, which involves 2 feedstocks (NF), 2 products (NP), 13 processing steps (NS) and 34 processing intervals (NI), which leads to 79118 equations. Note that although the hybrid framework was originally designed for chemical flowsheet synthesis-design and extended for biorefineries, it is generic and has also proven useful for other processes such as carbon dioxide utilization and waste-water treatment networks.

References

Bertran M., Frauzem R, Zhang L, Gani R. A generic methodology for superstructure optimization of different processing networks. In: Kravanja Z, editor. Proc 26th Eur Symp Comput Aided Process Eng â€“ ESCAPE 26. PortoroÅ¾, Slovenia: Elsevier B.V.; 2016.

Quaglia A, Sarup B, Sin G, Gani R. Integrated business and engineering framework for synthesis and design of enterprise-wide processing networks. Computers & Chemical Engineering, 2012, 38:213â€“23.

461880 A New Heat Integration Model for Simultaneous Utility and Total Heat Exchanger Area Targeting

Abstract:

The heat integration between cold and hot process streams is beneficial because it minimizes the utility consumption and increases energy efficiency. There are two approaches to this problem, sequential/decomposition and simultaneous. In the sequential approaches, heat integration is performed after the chemical process has been designed (all the process stream conditions are determined), and it is performed in three sequential steps to determine (1) the minimum utilities for the entire system, (2) minimum number of heat exchangers, and (3) total heat exchanger area and total cost. The major limitations of the sequential approaches are that they neglect (1) the interactions between the chemical process and heat integration subsystems, and (2) the trade-off among utility consumption, number of exchangers, and area. To address the first limitation, simultaneous approaches have been proposed [1, 2], where process synthesis and utility targeting (i.e. minimize total unity consumption under variable stream conditions) are considered simultaneously. However, these approaches do not take into account the tradeoff between energy cost and capital investment (e.g. heat exchanger area) in the heat integration subsystem. Another approach that addresses both limitations is the formulation of rigorous mixed-integer nonlinear programming (MINLP) superstructure heat exchanger network synthesis (HENS) model in which the utility consumption, heat exchanger area, and network configurations are simultaneously optimized [3]. However, this approach can lead to suboptimal solutions and it is computationally expensive when adopted for simultaneous process synthesis and HEN optimization.

To address the aforementioned challenge, we propose a MINLP optimization model for simultaneous utility and area targeting. The model can handle process streams with variable temperatures and flow rates so that it can be integrated with process synthesis models. The model is based on a mathematical representation of the balanced composite curves (BCCs) in the pinch analysis. The cold and hot composite curves (including cold and hot utilities) are each constructed from the cumulative enthalpies from low to

high temperatures using implicit orderings. The BCCs are then constructed by "bringing" the two composite curves together but allowing the heat recovery approach temperature (HRAT) to vary. The BCCs are divided vertically into enthalpy intervals wherever the curves have slope changes, which corresponds to a process stream starting or finishing along the BCCs. Therefore, the total number of intervals needed is twice the amount of process streams. It is important to note that if any one of the two BCCs has a slope change (a "kink" in the graphic representation), we calculate the temperature of the other BCC at the same enthalpy. In each interval, it is assumed that the hot streams and cold streams transfer heat vertically. Using the boundary temperatures and enthalpies at each interval, we can calculate the mean temperature difference and the approximate heat exchanger area needed at each interval. Finally, the objective is to minimize total area cost and utility

The nonlinearities in this model come from either the bilinear [flow rates]*[temperature difference] terms, or the terms in the area cost calculation that appear in the objective function. If either the stream flow rates or temperatures are fixed, the feasible space will be defined strictly by linear constraints, and the only nonlinearities will be in the objective function. In the case that flow rates, temperatures, and HRAT are all fixed, the model reduces to the problem table algorithm in the pinch analysis.

The proposed model can be generalized to consider intermediate utilities. Further, we extend the model so that it handles process streams with unknown cold/hot classifications. In other words, the model can deal with streams that cannot be determined as cold/hot streams prior to solving the optimization problem, which is in particular useful when the proposed model is integrated with a chemical process superstructure model that involves the (de)activation of units.

We apply the proposed model on several examples that involve process streams with variable flow rates and temperatures. Finally, we illustrate its applicability by integrating it with a rigorous biofuel production superstructure model.

References

1. Duran, M.A. and I.E. Grossmann, *Simultaneous-Optimization and Heat Integration of Chemical Processes.* Aiche Journal, 1986. **32**(1): p. 123-138

2. Grossmann, I.E., H. Yeomans, and Z. Kravanja, *A rigorous disjunctive optimization model for simultaneous flowsheet optimization and heat integration.* Computers & chemical engineering, 1998. **22**: p. S157-S164

3. Yee, T.F. and I.E. Grossmann, *Simultaneous optimization models for heat integration—II. Heat exchanger network synthesis.* Computers & Chemical Engineering, 1990. **14**(10): p. 1165-1184

459837 Optimal Composition of Biogas for Methanol Production Via Dry Reforming

Abstract:

Biogas is an interesting fuel that is produced from anaerobic digestion of organic wastes. Its composition is mainly methane, with a certain amount of carbon dioxide and other species in smaller quantities like H_2S or NH_3 [1]. Therefore, it can be considered as a source of methane. Methane or natural gas has been used as energy source or as fuel, but it has also been used as a raw material for the production of syngas. There are a number of alternatives to process methane. We can use steam to reform methane, an endothermic process. Alternatively, oxygen can be used to partially oxidize the methane. Autoreforming combines both, steam reforming and partial oxidation, so that the process is adiabatic. Lately, another technology has been added to the portfolio, dry reforming using carbon dioxide. It is also an endothermic process with a lower yield to hydrogen. The hydrogen to CO ratio varies widely depending on the method used. This ratio is key for the proper production of different species. For instance, the production of methanol requires a H_2:CO ratio of 2, the production of FT-liquids suggests values from 1 to 2 depending on the length of the hydrocarbon chain and the operating conditions at the reactor. The production of ethanol and DME needs a H_2:CO ratio of 1 [2,3].

Biogas already has CO_2 within its composition. However, the syngas produced is expected to show a low ratio H_2:CO. To tune this ratio we can follow different alternatives. Biogas composition is variable and can be adjusted. Furthermore, traditionally dry reforming of methane has been combined with other reforming modes [4-6].

In this paper, we present a mathematical optimization approach for the optimal dry reforming of biogas for the production of methanol. The raw biogas is cleaned up before reforming. We allow for the use of steam and variable biogas composition to increase the hydrogen content of the syngas. Part of the biogas is used to provide energy for the process. Next, the unreacted hydrocarbons and CO_2 are removed. Subsequently, its composition may be adjusted (using either water gas shift reaction or pressure swift adsorption if there is an excess of hydrogen). Finally, methanol is synthesized. The problem is formulated as an NLP with simultaneous heat integration for the optimal biogas composition and methanol production. Two objective functions are considered: a simplified production cost and we proposed an environmental one based on carbon footprint considering the contributions to CO_2 emissions due to the water usage as cooling agent, raw material or steam production and CO_2 consumption and release across the process.

Biogas is expected to have around 50-52% of CH_4 and 45-47% of CO_2, depending on the objective. The production cost of methanol is $1.7/gal, for a plant size that uses 10% of the potential biogas to be produced in Madrid (Spain), with an investment of $46 MM. The production cost is becoming competitive with the current price of methanol ($0.9/gal) with the advantage of its sustainability. The results obtained for both optimizations show a negative carbon footprint, creating a net capture of CO_2 while producing a valuable intermediate for chemicals or fuels.

References.

[1] Japaraju, P., Rintala, J. 17 - Generation of heat and power from biogas for stationary applications: boilers, gas engines and turbines, combined heat and power (CHP) plants and fuel cells 404-427, In The Biogas Hand book 2013

[2] Martín, M., Grossmann, I.E., On the systematic synthesis of sustainable biorefineries Ind. Eng. Chem. Res. **2013,** 52 (9), 3044-3064

[3] Peral, E., Martín, M. Optimal production of DME from switchgrass based syngas via direct synthesis. Ind. Eng. Chem. Res., **2015**, 54, 7464-7475 [4] Lim, Y.; Lee, C.J:; Su Jeong, Y.; Song, I-H.; Lee, C.J.; Han, C. Optimal Design and Decision for Combined Steam Reforming Process with Dry Methane Reforming to Reuse CO_2 as a Raw Material *Ind. Eng. Chem. Res.* **2012**, 51, 4982–4989

[5] Vernon, P. D. F.; Green, M. L. H.; Cheetham, A. K.; Ashcroft, A. T. Partial oxidation of methane to synthesis gas, and carbon dioxide as an oxidising agent for methane conversion. *Catal. Today* **1992**, 13 (2–3), 417–426.

[6] Song, C.; Pan, W. Tri-reforming of methane: a novel concept for catalytic production of industrially useful synthesis gas with desired H2/CO ratios. *Catal. Today.* **2004**, 98 (4), 463–484.

462416 Production of Benzene, Toluene, and Xylenes from Natural Gas Via Methanol: A Process Synthesis and Global Optimization Approach

Abstract:

Efficient utilization of domestic resources, such as natural gas, is crucial toward meeting U.S. energy independence targets and addressing energy security, energy affordability, and the generation of lower-carbon energy. [1] As a result of the developments within the shale gas industry, recent efforts have focused on converting natural gas into more valuable products. [2,3] A subset of these valuable products are aromatics, which represent approximately one-third of the market for commodity petrochemicals. [4] Between 2005 and 2008, the global demand for benzene, para-xylene, ortho-xylene, and meta-xylene was approximately 40, 26, 6, and 0.4 million metric tons per year, respectively. [4] Benzene is a precursor to over 250 different products, including styrene and phenol. Para-xylene, the most valuable xylene isomer, is converted into polyethylene terephthalate (PET) fibers, resins and films. [5] Opportunities exist for widespread penetration of natural gas based chemical refineries, albeit with challenges associated with which types of chemicals to produce and which technologies to invest in. In order to address these challenges, we propose novel methods of producing aromatics using a combined process synthesis and global optimization approach. [6]

We present a deterministic global optimization based process synthesis framework to perform a comprehensive technoeconomic and environmental assessment of a natural gas based aromatics refinery. [6] Several direct and indirect natural gas conversion technologies, including autothermal reforming, steam reforming, and partial oxidation to methanol, are investigated. Numerous novel, commercial, and/or competing technologies are modeled within the framework, including methanol-to-aromatics, toluene alkylation with methanol, selective toluene disproportionation, toluene disproportionation and transalkylation with heavy aromatics, para-xylene separation via adsorptive separation or crystallization, isomerization of xylenes, and dehydrocyclodimerization of liquefied petroleum gas, among others. To address the economic barriers often associated with the utilization of alternative feedstocks, we develop a novel branch-and-bound global optimization algorithm that is capable of determining the optimal technologies to produce aromatics from natural gas at the highest profit. [6,7] Our approach provides an adequate baseline for comparing competing technologies and identifying bottlenecks for natural gas conversion based technologies. Several key aspects of the algorithm are discussed, multiple case studies are presented to investigate the effect of refinery capacity and product output, and the key topological decisions are described.

[1] Floudas, C. A.; Niziolek, A. M.; Onel, O.; Matthews, L. R. Multi-Scale Systems Engineering for Energy and the Environment: Challenges and Opportunities. AIChE Journal 2016, 62, 602-623.

[2] Mokrani, T.; Scurrell, M. Gas conversion to liquid fuels and chemicals: the methanol route-catalysis and processes development. Catalysis Reviews 2009, 51:1-145

[3] Baliban, R. C.; Elia, J. A.; Floudas, C. A. Novel natural gas to liquids (GTL) technologies: Process synthesis and global optimization strategies. AIChE Journal 2013, 59, 505â€“531.

[4] de Klerk, A. Fischer-Tropsch Refining; Wiley-VCH Verlag & Co. KgaA: Weinheim 2011

[5] Niziolek, A. M.; Onel, O.; Elia, J. A.; Baliban R. C.; Floudas, C. A. Coproduction of Liquid Transportation Fuels and C6_C8 Aromatics from Biomass and Natural Gas. AIChE Journal 2015, 61, 831-856.

[6] Niziolek, A. M.; Onel, O.; Floudas, C. A. Production of Benzene, Toluene, and Xylenes from Natural Gas via Methanol: Process Synthesis and Global Optimization. AIChE Journal 2016, 62 (5), 1531 â€“ 1556.

[7] Baliban, R. C.; Elia, J. A.; Misener, R.; Floudas, C. A. Global Optimization of a MINLP Process Synthesis Model for Thermochemical Based Conversion of Hybrid Coal, Biomass, and Natural Gas to Liquid Fuels. Computers and Chemical Engineering 2012, 42, 64-86.

452419 Extensive Sensitivity Analysis and Stochastic Global Optimization for Renewable Energy Businesses Under Operational Level Uncertainties

Abstract:

The current energy supply around the world and the US is highly dependent on fossil fuels. Even though in the past year oil price dropped from its previous attractive peak, cheap oil prices should not be a temptation for stepping away from renewables. The characteristic volatility and future uncertainty of petroleum should encourage governments, investors, shareholders and decision-makers to start preparing for the long farewell mankind has to say to fossil fuels. The modern portfolio of available renewables includes second generation biofuels produced from biomass which appears to be a sustainable solution for mobile energy services giving also an innovative usage of organic waste. However, market volatility, technological uncertainty, and limited experience in renewable energy processes can prevent the development of this burgeoning industry.

By processing biomass in biochemical or thermochemical pathways, renewable energy facilities are able to produce fuels, value-added chemicals, heat and bioelectricity. First and second generation biorefineries are already operating worldwide, and process systems engineering has contributed over the past few years in addressing the complexity of modelling and decision making to optimize and make cost-effective renewable energy projects. Usually, the optimization of renewable energy businesses considers a deterministic design approach in which all the model parameters are assumed to be known from literature. However, during conceptual design there is lack of information and reliability, especially in novel or recently developed processes. When ideality supposed, external factors that might affect the process behavior and the project profitability are mistreated. Underestimating these limitations may lead into non-optimal designs and generate extra expenses or difficulties during startup and operation. Technological risk management should be considered during decision making since it permits to select optimal conditions when uncertainties are taken into account. Therefore, the global optimization problem should be addressed considering risk management at technological level in initial stages. It will allow managers to have a conservative and reliable method

for making and supporting critical decisions. To identify significant uncertainty sources in the process, a sensitivity analysis of the parameters is required.

Our proposed framework is composed of different sections. First, a global sensitivity analysis based on Sobolâ€™s method is performed to identify the significant sources of uncertainty in the kinetic parameters involved in the biological reactions of biofuels and value-added chemicals production. The method is computationally efficient since it averages the results of the evaluated functions and increases the original size of the sample. The first and total sensitivity indices are obtained by calculating the total variance and the individual variances generated when each parameter changes separately. One of the main characteristics of the current work is to explore the efficiency of two different approaches for selecting significant uncertain parameters: 1) evaluating all parameters simultaneously, 2) evaluating parameters independently per product pathway. Then, the identified uncertain parameters are utilized to optimize the profitability of the plant under uncertainty. A stochastic global optimization technique (*ParLMSRBF-R*), similar to a metaheuristic method, is used to maximize the cash flow after tax of a renewable energy business while considering uncertainty and risk management. In this iterative optimization approach the decision variables are selected intelligently with the help of a Radial Basis Function as the response surface or surrogate model. Finally, the optimal points obtained are compared statistically with optimal values from other methods, e.g. Monte-Carlo based optimization. To evaluate the efficacy of this novel decision-making framework, a hypothetical multiproduct lignocellulosic biorefinery is used as a case study.

This optimization framework appears to be more effective than other optimization procedures, e.g. Monte-Carlo based optimization. The algorithm searches for the optimal operating conditions using a fitted and continuously updating Radial Basis Function without losing the clarity of the actual mechanism of the process. The algorithm selects candidate points to be evaluated near the current best solution neighbourhood achieving exploration and exploitation in the solution domain. In terms of computational effort, the optimization algorithm employed requires less expensive function evaluations to reach global optima. In fact, *ParLMSRBF-R* achieves optima with less function evaluations in comparison to Monte Carlo-based optimization approach. In addition, the objective value improves in contrast to the best value achieved with Monte-Carlo based optimization.

464830 Toward a Unified Method for Process Design, Integration and Intensification

Abstract:

Until now, process synthesis and integration methods have focused on classical unit operations, where units such as separators, reactors and storage tanks are integrated but they are not merged to perform multiple tasks *in situ*. Due to the historically standalone development of conversion, separation and storage technologies, excess energy is consumed in isolated units. In this work, we propose an original approach for process design that leverages on the construction of flowsheets using building blocks, where each block performs specific operation (e.g., separation, conversion, storage, etc.) via enabling materials (e.g., membrane, catalyst, adsorbent, etc.). An assembly of blocks of the same type obtains a classical unit, while an assembly of blocks with different types results in a hybrid or intensified unit. Our building block-based flowsheet superstructure not only contains all alternatives typically considered for process synthesis and integration, but it also systematically identifies and incorporates all plausible intensification alternatives at the flowsheet level. To this end, our proposed method can potentially lead to a unified approach for process synthesis, integration and intensification without exhaustive enumeration. The overall problem is formulated as a single mixed-integer nonlinear optimization (MINLP) problem which can be solved using commercial solvers. While the mass and energy flows between the blocks are modeled using continuous variables, the selection of the operations and their enabling materials are discrete decisions. The objective is to synthesize a process with intensified units by minimizing or maximizing a process metric given the feed and product specifications, feed and product prices, material properties and bounds on flow rates. We demonstrate the applicability of the proposed method for the innovative design of several emerging technologies for clean and sustainable energy, such as carbon capture and conversion, natural gas utilization, and multicomponent gas separation. We also demonstrate that the simultaneous synthesis and intensification approach leads to substantially smaller, cleaner, safer, and more energy-efficient designs.

469198 Work and Heat Exchange Networks - from Thermodynamic Insight to Optimization Based Design Procedures

Abstract:

Design of Work and Heat Exchange Networks (WHENs) represents an emerging area of research, and it involves three distinct activities that briefly can be described as: (1) Heat Integration, (2) Work Integration, and (3) Pressure Integration. The first is a classical Process Synthesis problem trying to recover as much heat as possible from hot process streams to heat up cold process streams while selecting an optimal set of hot and cold utilities to provide additional heating and cooling. The second involves the use of work (power) from expansion processes to supply work (power) to compression processes. This work integration can be done on the same shaft (e.g. the use of so-called companders for the extraction of heavier hydrocarbons in natural gas liquefaction plants) or separate shafts where electricity is produced and consumed. The third involves the utilization of additional heating and cooling resulting from changing the pressure of process streams, and this has been the focus of research in our group for the last 10 years [1].

The sparse and relatively recent literature on WHEN design can be grouped into publications on Work Exchange Networks (WENs, e.g. [2]), Work and Heat Exchange Networks (WHENs, e.g. [3]) and Work, Heat and Mass Exchange Networks (e.g. [4]). The extended problem definition for WHEN synthesis that has been used in our research group is:

Given a set of process streams with supply and target states (temperature and pressure), as well as utilities for power, heating and cooling; design a network of heat exchangers, expanders and compressors in such a way that exergy consumption is minimized or exergy production is maximized.

At this stage of methodology development, exergy is used in the objective function since energies of different qualities (thermal and mechanical) are involved. Later, when moving from a thermodynamic approach to an optimization-based approach, a more appropriate objective function involving total annual cost will be used.

The motivation for including compression and expansion processes in the traditional heat recovery problem is that considerable additional energy savings can be obtained. The main issue is to sacrifice modest amounts of mechanical energy to make significant savings in thermal energy, and/or increase work production by efficient utilization of low temperature thermal energy. As such, the research on WHENs has strong relations to heat pumping [5] and refrigeration cycles. It is also closely related to the research on the so-called self-heat recuperation technology [6].

In developing a design methodology for WHENs, the Appropriate Placement concept from Pinch Analysis applied to pressure changing equipment (primarily compressors and expanders) has been a key issue. New challenges compared to traditional Heat Exchanger Network (HEN) design are: (1) The thermodynamic path from supply to target state is unknown (changing both temperature and pressure), (2) Pinch points and utility demand may (often will) change, so that no explicit targeting is possible, and (3) stream identity (hot/cold) may (often will) temporarily change.

A set of 4 theorems have been proposed and proven using thermodynamics and mathematics for expanders above ambient temperature [7], with minor modifications for compressors above ambient as well as expanders and compressors integrated below ambient. Thus, 4 scenarios with a total of 16 theorems are handled with considerable symmetry between the 4 cases. Simultaneous compression and expansion has also been handled with the introduction of a new theorem that deals with the integration sequence for compression and expansion [8]. In line with earlier findings [9], Pinch Expansion and Compression (i.e. expander and compressor inlet temperature should be equal to the Pinch temperature) will often result in significant exergy savings.

An early version of a design procedure for WHENs, based on the Appropriate Placement concept with detailed guidelines from the above mentioned theorems, uses the Grand Composite Curve (GCC) to quantify the maximum amount of heating/cooling that can be accommodated from compression/expansion. The procedure is time consuming even for very small cases and a number of simplifying assumptions. In addition, using the GCC with its modified temperatures ignored the different features for hot and cold streams as well as the relative position of supply and target temperatures to the pinch temperature. As a result, minor (in most cases) errors in estimation of utility consumption were introduced.

To overcome problems related to prohibitive time consumption and inaccuracies in utility estimation due to ignoring the identity of streams (hot/cold), work has been initiated to develop new superstructures with corresponding mathematical models of varying complexity (from LP to MINLP) that can be used in an optimization framework [10]. In fact, early applications of such optimization models have revealed new insight on the WHEN synthesis problem. One example is that Pinch Compression and Expansion should start at the Pinch temperature (hot or cold) that corresponds to the identity (hot or cold) of the stream segment prior to compression or expansion, and not to the original identity of the stream. Two classes of optimization models were developed, i.e. with and without utilizing the insight provided by the theorems. The latter class was used as a final validation of the theorems, while the former class offered advantages in model complexity and thus reduced time to solve the optimization problem.

In conclusion, based on several case applications using the original thermodynamic approach as well as the optimization framework under development, the original hypothesis that Pinch Compression and Expansion represent optimal placement of pressure changing equipment is valid fully or partly in the majority of cases. However, there are cases where expansion/compression at hot/cold utility temperature or ambient temperature provides better solutions. No other temperatures result in WHENs with lower exergy consumption (or higher exergy production).

Preliminary studies on industrial applications such as CO_2 capture processes [11] show that considerable energy savings can be achieved following the new design procedures. An interesting feature of this research is that Exergy Analysis has been moved from post-design evaluation to conceptual stage design.

References

[1] A. Aspelund, D.O. Berstad and T. Gundersen, An Extended Pinch Analysis and Design Procedure utilizing

Pressure based Exergy for Subambient Cooling, *Applied Thermal Engineering*, 27, 2633-2649, 2007.

[2] M.S. Razib, M.M.F. Hasan and I.A. Karimi, Preliminary Synthesis of Work Exchange Networks, *Comput. & Chem. Engng.*, 36, 262-277, 2012.

[3] V.C. Onishi, M.A.S.S. Ravagnani and J.A. Caballero, Simultaneous Synthesis of Heat Exchanger Networks with Pressure Recovery, *AIChE Journal*, 60, 893-908, 2014.

[4] R. Dong, Y. Yu and Z. Zhang, Simultaneous Optimization of integrated Heat, Mass and Pressure Exchange Network using Exergoeconomic Method, *Applied Energy*, 136, 1098-1109, 2014.

[5] C. Fu and T. Gundersen, A Novel Sensible Heat Pump Scheme for Industrial Heat Recovery, *Ind. Eng. Chem. Res.*, 55, 967-977, 2016.

[6] Y. Kansha, N. Tsuru, K. Sato and A. Tsutsumi, A Self-Heat Recuperation Technology for Energy Saving in Chemical Processes, *Ind. Eng. Chem. Res.*, 48, 7682-7686, 2009.

[7] C. Fu and T. Gundersen, Integrating Expanders into Heat Exchanger Networks above Ambient Temperature, *AIChE Journal*, 61, 3404-3422, 2015.

[8] C. Fu and T. Gundersen, Correct Integration of Compressors and Expanders in above Ambient Heat Exchanger Networks, *Energy*, in review, 2016.

[9] T. Gundersen, D.O. Berstad and A. Aspelund, Extending Pinch Analysis and Process Integration into Pressure and Fluid Phase Considerations, *Chem. Engng. Trans.*, 18, 33-38, 2009.

[10] C. Fu, P. Maurstad Uv, B. Nygreen and T. Gundersen, Compression and Expansion at the right Pinch Temperature, *Chem. Engng. Trans.*, 51, 2016.

[11] C. Fu and T. Gundersen, Heat and Work Integration: Fundamental Insights and Applications to Carbon Dioxide Capture Processes, *Energy Conversion and Management*, 2016, DOI: 10.1016/j.enconman.2016.04.108.

470562 Analysis, Synthesis and Optimization of Multiple-Effect Evaporation Systems Using Mathematical Programming

Abstract:

Evaporation processes are used within the process industries in order to produce concentrated products by evaporating part of water from different feeds-diluted water solutions. Concentrated products can represent final products (fruit and vegetable juices) or intermediate products in cases that crystalized (salt, sugar) or dried (milk powder) final products should be produced. Large amounts of steam and cooling water are consumed in these processes. In order to reduce energy and water consumption within evaporation processes different systems can be applied, namely, multiple-effect evaporation, vapor recompression (thermal and mechanical) or their combinations. Additionally, these processes can be integrated with other process subsystems in order to achieve improved energy and water integration.

To address these issues different computer-aided tools have been proposed. However, most studies have focused on analysis and simulation of evaporation processes. Some of the initial studies [1, 2] considered the synthesis of evaporation processes in order to develop tools for computing the minimum utility use for a multiple-effect evaporation system, which was heat-integrated with process hot and cold streams. These studies were based on a modified grand composite curve and heat-path diagram. Also, the focus of the recent works have been on multiple-effect evaporation systems [3] and their energy integration with the background processes in order to minimize the energy consumption within the overall system [4]. These studies have motivated us to further expand research in this direction, by applying mathematical programming approach for the analysis of existing and the design of new evaporation systems as well as their heat integration with other process subsystems or process streams.

The main goal of this paper is to develop models based on mathematical programming that can be applied for the analysis, synthesis and optimization of multiple-effect evaporation systems. The proposed models will be developed in General Algebraic Modeling System (GAMS). The developed models will enable

examination of different scenarios of multiple-effect evaporation in order to address the analysis of existing, retrofit and/or design new evaporation process. Within the proposed framework, a network consisting of a multiple-effect evaporation system and heat exchanger network will be investigated in order to achieve the improved heat integration within the overall system. Two strategies will be considered to achieve this task, namely, sequential and simultaneous. The developed models will be tested on several examples, and also applied to different feed streams. New results are expected to be obtained within this field.

Keywords: multiple-effect evaporation, analysis, synthesis, optimization, mathematical programming.

Acknowledgment

The authors are grateful to the Swiss National Science Foundation (SNSF) and the Swiss Agency for Development and Cooperation (SDC) for providing financial support within the SCOPES 2013 2016 (Scientific Co-operation between Eastern Europe and Switzerland) joint research project (CAPE EWWR: IZ73Z0_152622/1).

References:

[1] Hillenbrand JJB, Westerberg AW. The synthesis of multiple-effect evaporator systems using minimum utility insights—I. A cascaded heat representation. Computers & Chemical Engineering. 1988;12:611.

[2] Westerberg AW, Hillenbrand JJB. The synthesis of multiple-effect evaporator systems using minimum utility insights—II. liquid flowpattern selection. Computers & Chemical Engineering. 1988;12:625.

[3] Khanam S, Mohanty B. Energy reduction schemes for multiple effect evaporator systems. Applied Energy. 2010;87:1102.

[4] Sharan P, Bandyopadhyay S. Energy Integration of Multiple Effect Evaporators with Background Process and Appropriate Temperature Selection. Industrial & Engineering Chemistry Research. 2016;55:1630.

469782 Dynamic Intensification: Periodic Operation of Intensified Process Units

Abstract:

Dynamic Intensification: Periodic Operation of Intensified Process Units

Richard Pattison, Austin Dulaney, Michael Baldea

McKetta Department of Chemical Engineering

The University of Texas at Austin, 1 University Station C0400, Austin, TX 78712

email: mbaldea@che.utexas.edu

One of the core tenets of process intensification is to "do more with less." This includes carrying out more operations in fewer units (integration) and replacing conventional units with equal throughput units with smaller physical dimensions (miniaturization).

In a different vein, periodic operation of process units, particularly chemical reactors, was shown to increase throughput for many reaction schemes in comparison to steady state operation of a plant of the same capacity [1]. However, the potential benefits of periodic operation are often limited by material and energy transport constraints. Intensified units with small physical dimensions that combine both reaction and separation can eliminate some of these physical constraints by increasing heat and mass transfer rates through high surface area to volume ratios.

In this paper, we focus on the broad class of intensified processes that feature reaction and separation, such as reactive distillation columns and membrane reactors. We develop a prototype system representation under the assumption on unconstrained inlet and outlet material fluxes [2] (i.e., the reacting system is not limited by transport phenomena, which provides a design target for intensified unit operations). On this basis, we utilize the Pi criterion to define conditions under which imposing an oscillatory behavior (e.g., by periodically varying the manipulated variables) results in superior productivity compared to steady state operation. Our results are based on analytical solutions for several common reaction schemes, including reversible, consecutive and higher-order reactions.

We illustrate our theoretical framework with several detailed case studies. Moreover, we use numerical analysis to tackle the more complex cases whereby the material fluxes to the intensified unit are constrained. Our simulation results show that intensified reaction-separation systems can benefit from periodic operation for most reaction schemes, proving that dynamic intensification is a highly promising avenue for advancing the state of the art in the process intensification realm.

We will present the framework for a general reaction scheme and provide analytical conclusions for specific cases, then present results of detailed and optimized case studies, where we no longer assume that the material flux is unconstrained.

[1] Sterman, L.E.; Ydstie, B.E. Periodic forcing of the CSTR: An Application of the generalized Π criterion. *AIChE J.,* **1991**, 37, 986-996.

[2] Peschel, A.; Freund, H.; Sundmacher, K. Methodology for the design of optimal chemical reactors based on the concept of elementary process functions. *Ind. Eng. Chem. Res.*, **2010**, *49*, 10535-10548.

460244 Process Intensification and Distillation: How to Control a Fixed Design Dividing Wall Distillation Column

Abstract:

An example of one of the emerging technologies in the field of process intensification is dividing wall distillation columns or DWCs. As part of an ongoing experimental study of dividing wall columns at The University of Texas at Austin, the design and selection of control strategies for a fixed design DWC will be presented. A dynamic process simulator was used to test various feed compositions on a dividing wall column to understand the operational flexibility of a fixed design DWC. A fourth trace component was added to the feed and the selection of and changes in manipulated variables necessary to move the trace component between product tanks was investigated. Conventional distillation temperature control techniques were tested on the dividing wall column as well as more advanced control strategies, such as model predictive control, thought to be better suited to capture the inherent interactions of the system caused by the intensified nature of the process. The ability of these control systems to handle setpoint changes and feed disturbances was compared. To verify the results from the simulation, the control systems were implemented on a pilot scale dividing wall distillation column from which dynamic responses were obtained.

460843 Techno-Economic Analysis of a Single-Pass Methanol Synthesis Process

Abstract:

< style="padding-left: 40pt; text-align: center;">Techno-Economic Analysis of a Proposed Hybrid Methanol Synthesis Process

K.M. Bentaher, R.J. Machiela, T.N. Rogers(*)

Michigan Technological University, Department of Chemical Engineering 1400 Townsend Drive, Houghton, MI 49931-1295

(*)Corresponding Author

< style="padding-left: 5pt; text-align: left;">Abstract

Recent advances in drilling technologies, such as horizontal drilling and hydraulic fracturing, have led to shale gas revolution and considerably boosted U.S. natural gas production. Between 2005 and 2011, U.S. dry gas production increased by about 27 percent. This significant increase in the supply of natural gas has driven down its spot market price by more than 50 percent in the past 7 years. Since methanol is a key domestic transportation fuel, gasoline additive, and intermediate for many other important industrial chemicals, methanol production methods that employ a natural gas feedstock are re-evaluated in this paper to take advantage of the sharp drop in natural gas prices.

In this study, the starting point is the conventional methanol synthesis process which has a recycled syngas stream merging with the methanol synthesis reactor's feed stream. This recycle loop requires a compressor that represents a significant operating expense. To eliminate the syngas recycle loop, a new hybrid process is proposed (Figure 1) that consists of a conventional methanol synthesis reactor (R-1) followed by a reactor (R-2) that condenses methanol vapor as it forms to achieve nearly complete one-pass conversion of syngas to methanol. By condensing methanol vapor on a cooling surface that spirals down the reactor's interior, the equilibrium limitation on methanol conversion is removed. The small amount of syngas that is not converted, rather than being recycled, is used as fuel in the reformer. The upstream conventional methanol synthesis reactor R-1 is included to decrease the R-2 reactor's feed rate, thereby reducing its size and cutting its high capital cost. The novelty of the new hybrid process is in the arrangement of conventional, proven equipment and technologies.

After-tax discounted cash flow analysis is used in this study to incrementally compare the new hybrid process design to two alternative approaches: conventional methanol synthesis with recycle and a process employing a single pass condensing reactor only. Cash flow analysis is also used to establish the optimum ratio of the sizes of reactors R-1 and R-2. Several standard metrics for profitability, such as NPV, ROI, IRR, and payback period, are reported for each candidate process over a 10-year analysis window.

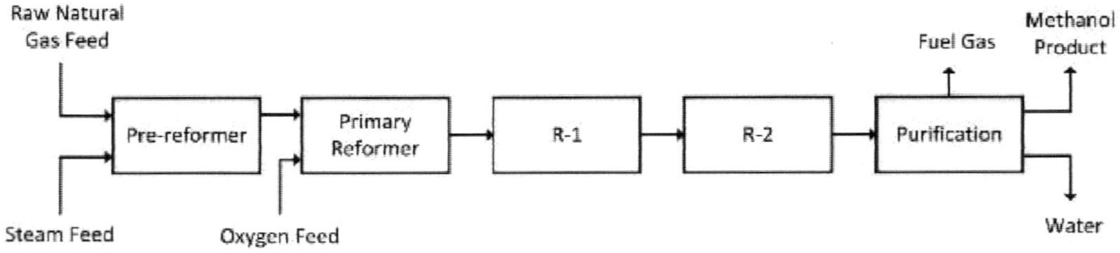

Figure 1. Proposed Hybrid Process for Converting Natural Gas to Methanol

461895 Comparison of Light Gas Separation By Membranes and Conventional Technology: A Case Study

Abstract:

In this presentation, an energy-intensive separation of light gases is selected as a case study, to compare the advantages and weakness of membrane-based separation with conventional technology. The study will determine the target membrane property (e.g. permeance, selectivity) for a variety of process configurations, and provide guidelines for membrane material development.

Continuous Vacuum Filtration Technologies for Sulfuric and Hydrochloric Acid Slurries

Garrett Bergquist, BHS-Sonthofen Inc., Charlotte, NC

Introduction

In many chemical manufacturing processes there exist slurry streams where the separation of solids from liquids is required. The range of equipment available for these streams is vast and the decision between equipment choices is not always easy. For clarification of low solids streams there exists technology such as candle filters, plate filters, and cartridge filters. These are typically a batch operation but can be made continuous by using multiple units in parallel. While one unit is actively filtering, the second unit is on side time operation such as filling, draining, or discharge. For higher solids applications continuous filters can be employed, including centrifuges, rotary vacuum filters, and horizontal belt filters.

Regardless of the specified filter type, when designing the filter, consideration must be made in the selection of appropriate materials. Materials of construction should be selected to withstand the temperature and chemical corrosion properties of the surrounding environment and liquids that come in contact with the filter.

This article discusses the use of alternate materials available in the construction of vacuum belt filters to address chemical compatibility concerns in applications using sulfuric acid and hydrochloric acid as the liquid constituent in a slurry. This article continues with a case study where the challenge is to process 2,200 kg/h dry solids that have been precipitated from a sulfuric acid solution.

Sulfuric Acid

Sulfuric acid is commonly and widely used in the production of fertilizer. Other manufacturing processes employ sulfuric acid to produce automotive batteries, pigments, iron, paper pulp, etc. Sulfuric acid can be a tricky liquid when it comes to designing robust systems for chemical compatibility. At high concentrations, sulfuric acid acts as a strong oxidizing agent. At low concentrations sulfuric acid acts as a reducing agent. For that reason, a material that is suitable for sulfuric acid at a high concentration may not be suitable for sulfuric acid at a low concentration and vice versa. Even at temperatures below 140°F (60°C), sulfuric acid at concentrations between 30wt% and 75wt% will corrode 316 stainless steel at rates greater than 200 mils per year (Figure 1).

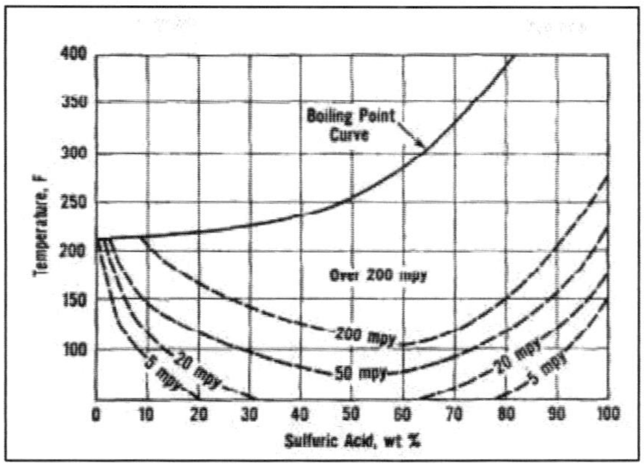

Figure 1. Corrosion of 316 Stainless Steel by sulfuric acid as a function of temperature. (Fontana, 1952)

When it comes to safety, sulfuric acid is an irritant and is considered very hazardous. In cases of skin contact and eye contact, burns may occur. Inhalation can result in severe irritation of the respiratory tract. Engineering controls are recommended to prevent physical interaction and to keep airborne concentrations of vapors below 1 mg/m^3 (NIOSH, 1997).

Hydrochloric Acid

Hydrochloric acid, also known as muriatic acid, is widely used in steel pickling, food production, calcium chloride production, gold mining, catalyst regeneration, dyes and pigment production, etc. Like sulfuric acid, hydrochloric acid is extremely corrosive to metals, including carbon steel, stainless steel, nickel and aluminum. In processes using hydrochloric acid, piping is often lined in thermoplastic and storage tanks rubber lined or made from FRP.

Hydrochloric acid is a highly hazardous substance and corrosive to mucus membranes. Exposure, even to vapors, can cause severe burns to exposed skin. The American Conference of Governmental Industrial Hygienists (ACGIH) recommends that the threshold limit value (TLV) for hydrochloric acid vapors to be 2 parts per million (ACGIH, 2003).

Vacuum Filtration Technology

In all vacuum filtration technologies, the concept and mechanism of filtration is the same. Vacuum is pulled through a filter cloth backed by a porous supporting structure while slurry is fed to the cloth. The liquid filtrate is pulled through the cloth by the vacuum pressure differential while the solids are retained on the surface of the cloth. The gas and liquid that pass through the cloth are then piped together to a gas/liquid separator to allow the liquid to fall via gravity which is then pumped from the bottom of the receiver. The separated gas is piped from the top of the separator to the vacuum pump and subsequently exhausted (Figure 2). The configuration in which this process is accomplished is what sets apart the different vacuum filtration technologies.

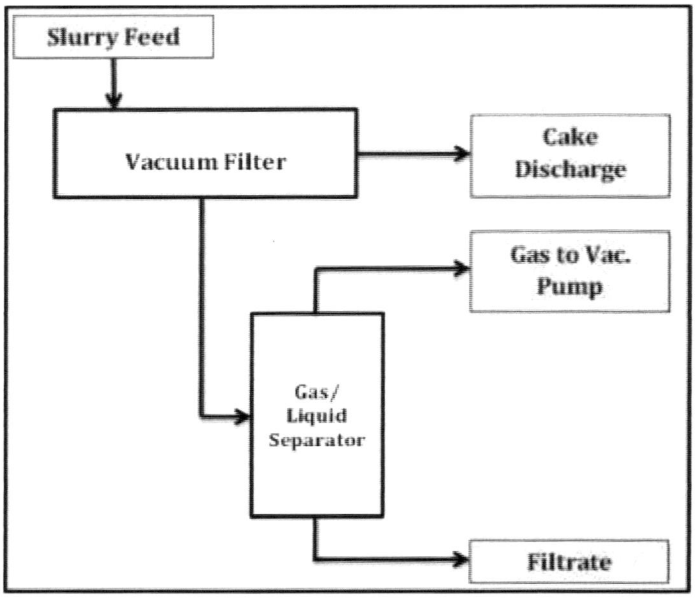

Figure 2. Process flow diagram for basic vacuum filtration

Rotary Vacuum Drum Filter

In a rotary vacuum drum filter, filter cloth surrounds a cylindrical segmented drum. The drum itself is constructed from an alloy material and acts as the porous supporting structure for the filter cloth. The bottom part of the drum is submerged in an open slurry trough. With vacuum being pulled through the cloth, filtration begins to occur as the drum rotates into the slurry trough. Cake heights are determined by the residence time spent submerged in slurry. As the drum rotates the cake can undergo washing via spray nozzles and dewatering by pulling residual moisture from the cake via gas flow from the vacuum pressure differential. The cake is then discharged from the media by a scraper knife or discharge roller.

Benefits in using a vacuum drum filters include having a large amount of filter area in a small footprint relative to horizontal belt filters. Additionally with the discharge design, a cake heel can be left on the drum which allows for slight depth filtration at the beginning of the following filtration cycle. When it comes to operation using hazardous chemicals, specifically sulfuric acid and hydrochloric acid, the vacuum drum filter has some major shortcomings. For one, the slurry trough and acid rich filter cake are open to the surrounding environment. If the acid is at all heated above ambient temperature, then a significant quantity of acid vapor will fill the surrounding environment. Given the height of the filter, providing a ventilated enclosure to keep acid fumes below the respective exposure limits is both difficult and expensive.

A second drawback comes when considering compatible materials for the vacuum belt filter piping and drum. Stainless steel alloys such as 304 and 316 are not suitable for use in sulfuric and hydrochloric acid solutions. As previously stated, with relatively low temperature a broad range of sulfuric acid concentrations can corrode stainless steel over 200 mils per year. This corrosion consideration leads to the material of construction specification of higher grade alloys such as Alloy 20 or Hastelloy. By using more exotic alloys the cost of the filter is increased significantly. When comparing the cost of Alloy 20 and Hastelloy C-276, the relative cost is over 3 and 7 times that of 316 Stainless Steel respectively (Grocki, 2012).

Continuous Horizontal Rubber Belt Filter

In the continuous horizontal rubber belt filter, an endless filter cloth is backed by an endless, perforated rubber drainage belt which in turn is supported by a deck attached to the filter frame. The supporting deck is made of rollers or the rubber belt is floated by blowing air or slide water to reduce friction and wear as the rubber belt travels across it. The surface of the rubber belt is grooved to allow filtrate flow towards the belt perforations which are sealed to a vacuum box underneath the rubber belt. Motion of both the rubber belt and filter cloth are continuous at a constant speed.

The horizontal rubber belt filter is well utilized in fast settling slurries as the solids are fed onto the top of the belt vs the bottom feed mechanism of the vacuum drum filter. Additionally if intensive cake washing or longer drying times is required, the length of the belt filter can be designed to accommodate these process requirements.

When considering the use of the horizontal rubber belt filter for slurries with sulfuric and hydrochloric acid, some of the concerns regarding environmental exposure and material of construction can be addressed. A fume hood over the filter can be constructed more easily given the horizontal surface. However since the rubber belt must periodically be replaced via the side of the machine, the hood can only be suspended over the filter and not attached to the frame. Curtains can be hung from the edges of the fume hood to cover the remaining space but do not eliminate the risk of someone reaching in to the filter or overflow at the slurry feed zone in an upset condition. Additionally, to ensure material compatibility, different grades of rubber can be selected for the drainage belt material. However for sulfuric and hydrochloric acid, standard rubber materials such as natural rubber, SBR, and NBR are incompatible to various degrees. Like the vacuum drum filter, higher grades of materials must be considered for the rubber belt such as EPDM (Moss Rubber, 2005). Using these higher grade materials add significant cost to what is already most expensive piece of the rubber belt filter.

Indexing Belt Filter

In the indexing belt filter, an endless filter cloth is backed by a stationary perforated filter bed. During operation the belt moves only periodically, step-wise down the filter bed. When the belt is stationary, vacuum is applied through the filter media to allow filtration. When the belt is indexing the vacuum is vented to allow motion across the stationary bed.

Like the rubber belt filter, the indexing belt filter is well utilized in fast settling slurries. Intensive washing can be performed like the rubber belt filter. There is a slight advantage for the indexing belt filter when it comes to countercurrent wash schemes as the wash additions and filtrates are better separated with the indexing design. Also with the indexing filter since the belt is stationary for a period of time, a press can be utilized to squeeze the solids while simultaneously blowing air through the cake to remove all remaining free moisture. The drawback to the indexing design is that side time introduced during the atmospheric pressure belt movement which reduces the filtration capacity compared to a continuous rubber belt filter design. For that reason the indexing belt filter it is better suited for medium to slower filtering slurries compared to ideal continuous rubber belt filter applications.

When considering the use of this design for hazardous slurries, the indexing belt filter is a perfect example of flexibility in both environmental considerations and available materials of construction. Because the filter bed is stationary and the cloth can be changed from the discharge end of the filter, a full hood is able to be built over the entire surface of the filter without side gaps (Image 1). This is a significant benefit when designing the filter for sulfuric or hydrochloric acid slurries as it completely isolates any interaction with the product. Through the use of blower a slight negative pressure can be maintained over the filter at all times. Alternatively the indexing design can also provide for a full gas tight enclosure which is beneficial for filtration when explosion proofing must be considered. The gas tight enclosure is able to be nitrogen purged and maintained using a nitrogen recirculation skid.

When considering materials of construction, the indexing design is able to be built from a multitude of polymers which commonly include polyethylene, polypropylene, or PVDF. These materials are able to be used for 100% of the wetted surfaces of the filter. The chemical resistance against sulfuric and hydrochloric acid with this selection of polymers is excellent at all concentrations ("Chemical Compatibility Database", 2016). Additionally the material cost of these polymers compared to high grade alloys such as Alloy 20 and Hastelloy is much less. In fact, the use of polypropylene is even cheaper than construction using 316 stainless steel. The main limitation for using polymer constructed belt filters is temperature resistance. For a polypropylene belt filter, the maximum operating temperature is 175^{0}F – 194^{0}F (80^{0}C-90^{0}C) and 212^{0}F – 230^{0}F (100^{0}C-110^{0}C) for PVDF. However, for most vacuum filter applications this is not an issue as temperatures are often below the boiling point of water.

Image 1. BHS polypropylene indexing belt filter with oscillating slurry feed, fume hood and pressing/blowing device

Case Study – Sulfuric Acid Filtration

A company producing solid chemicals used in paper coatings was using a process in which product dissolved in sulfuric acid (H_2SO_4) is precipitated using a neutralization step via NaOH. This company was looking to move away from NaOH neutralization as the price of NaOH had increased significantly over the past year. An alternative production method was known in which the concentration of sulfuric acid was diluted with water to approx. 25wt%. The dilution then allowed the product to precipitate from solution. With this method any residual sulfuric acid must be removed from the solids as all downstream equipment was constructed of stainless steel. Additionally the precipitated solids are fragile so no intense agitating or centrifugation was possible. Benchtop scale testing using a 20cm^2 pocket leaf filter

(PLF) was performed to determine the appropriate technology, sizing, wash requirements and filter cloth to meet the application requirements.

PROCESS DATA

Slurry Throughput:	12,500 l/hr	(55 gpm)
Slurry Density:	1.3 g/mL	(30% H_2SO_4)
Suspended Solids:	13.5 wt%	
Dry Solids Throughput:	2,200 kg/h	
H_2SO_4 Concentration:	22-30 %	
Slurry Temperature:	max. 140°F	(60°C)

Through benchtop scale testing for the vacuum belt filter with the 20cm^2 PLF the following data was measured:

Filter Cloth:	80μm polypropylene
Test Filter Area (A_f):	0.002m^2 (20cm^2)
Slurry Feed:	425g
Filtration Time:	11 seconds
Mother Filtrate Quantity:	175 mL
Wash Volume:	235 mL
Wash Time:	55 seconds
Wash Filtrate:	238 mL
Drying time:	20 seconds
Heel Filtrate:	10mL
Cake Height:	50 mm
Wet Cake Weight:	185 g
Residual Cake Moisture:	65 wt%
Dry Cake Weight (m):	65 g

It was determined through preliminary calculation that a BF100 with 1.0 meter wide active belt width and an area of A_{zone} = 0.75m^2 per zone was appropriately sized. The cycle time and zone arrangement scale up for the BF100 at a filter cake height at 50mm was performed as follows:

Solids loading per stroke:

$$\frac{m}{A_F} * A_{Zone} = \frac{0.065kg}{0.002m^2} * 0.75\frac{m^2}{Stroke} = 24.38\frac{kg}{Stroke}$$

Number of strokes per hour:

$$N = \frac{2200kg/h}{24.38kg/Stroke} = 90Strokes/h$$

Duration of one cycle:

$$t_{Cycle} = \frac{3600\frac{s}{h}}{90\frac{Strokes}{h}} = 40\frac{\sec}{stroke}$$

With an indexing time of 5 sec with no applied vacuum, the active processing time per cycle is 35 sec. The belt filter zone arrangement based on an active time of 35 seconds is recommended as follows:

Table 1. Indexing Belt Filter BF100 Zone Arrangement

Process Steps	Test Times	Theoretical No. of Zones	Actual No. of Zones
Feed	-	-	1
Filtration	11 sec	0.31	1
Wash	55 sec	1.57	2
Drying	20 sec	0.57	1
Total	-	-	5 zones

With a total number of 5 zones required at $0.75m^2$ each, a belt filter BF100-038 with $3.75m^2$ in filter area was recommended to process 2,200 kg/h dry solids from the sulfuric acid solution (Image 2). A feed zone at the beginning of the filter with no vacuum was also recommended to allow for a more uniform solids distribution.

Image 2. Polypropylene belt filter with fume hood built for the filtration, washing, and drying of the sulfuric acid slurry.

Conclusion

When considering new technology or a new production processes it is always important to verify that the specified materials match the chemical and temperature resistance requirements for each stream. While the technology in the market place for vacuum filtration is often well known, when it comes to dealing with hazardous slurries such as those containing sulfuric or hydrochloric acid the options should be carefully considered. Production areas containing acids should be fully enclosed to prevent splashing or spills in an upset condition, and to capture any hazardous acid vapors. Special attention must be made to specify compatible materials as stainless and carbon steels are often incompatible with most sulfuric and hydrochloric acid concentrations. However specifying compatible specialty alloys such as Alloy 20 and Hastelloy for wetted parts becomes very expensive quickly. With traditional filters such as the rotary drum filter or the continuous rubber belt filter providing containment and identifying economic compatible materials can be a near impossible task. With the indexing vacuum belt filter, fully enclosed or even gas tight operation is easily achievable. More importantly, through the use of polymer wetted parts such as polypropylene or PVDF the cost of the filter is able to be kept low, even lower than 316 stainless steel construction. Polymer constructed belt filters in hazardous acid environments have a proven track record of safe and reliable operation with low costs of purchase.

References

1. ACGIH: Documentation of the Threshold Limit Values (TLVs) and Biological Exposure Indices (BEIs) - Hydrogen Chloride. 2003.
2. "Chemical Compatibility Database." Chemical Compatibility Data Base. Cole-Parmer, n.d. Web. 19 Sept. 2016.
3. Fontana, M.G., "Corrosion Series," Ind. Chem. Eng., *44* March 1952, P89A
4. Grocki, John. "Relative Cost 10-12." Pressure Vessel and Tank Fabrication Using Stainless Steels. Proc. of Steel World Expo, Houston. STI/SPFA, Oct. 2012. Web.
5. Moss Rubber and Equipment Corp. Chemical Resistance Chart. Benicia: Moss Rubber and Equipment, 2005. Print.
6. National Institute for Occupational Safety and Health. (1997). *NIOSH pocket guide to chemical hazards*. Cincinnati, Ohio: U.S. Dept. of Health and Human Services, Public Health Service, Centers for Disease Control and Prevention, National Institute for Occupational Safety and Health.

472208 Evaluation of the Cactus Based-Mucilage As an Alternative Natural Dispersant on Toxicity, Effectiveness, and Surface Tension of Oil-in-Water Emulsion

Abstract:

Cactus based-mucilage has the advantage of being biodegradable and can be obtained through sustainable agriculture. This functional natural surfactant was extracted from the *Opuntia ficus indica* cactus plant, and the extraction process yielded two types of mucilage: the non-gelling extract (NE) and the gelling extract (GE). The properties of the cactus plant-based mucilage were studied on the toxicity, surface tension, droplet size and stability of dispersed crude oil in water emulsions (O/W emulsions). The toxicity of the mucilage and conventional dispersants (Corexit 9500A) were evaluated by a standard EPA toxicity test using Daphnia magna colonies exposed to both NE and GE mucilage extracts in concentrations ranging from 0 to 2000 mg/L for 48 hours. It has shown that mucilage can be classified as practically non-toxic to the Daphnia magna colonies with a LC_{50} above 2000 mg/L. The LC_{50} of Corexit 9500A was below 2 mg/L, which is moderately toxic to the Daphnia magna colonies. Cactus mucilage extracts and conventional dispersants has been evaluated by testing the surface tension and measuring the droplet size of O/W emulsions under different concentrations, including three concentrations of oil (3, 6, 30% v/v) and several dispersant to oil ratios (1:1, 1:3, 1:10). Synthetic seawater and surrogate oil from BP were used. The results of the dispersion obtained using mucilage extracts were compared with using Corexit 9500A. The surface tensions of emulsions with cactus mucilage were similar as with Corexit 9500A. The average droplet size in the systems with cactus mucilage (1.5% of NE has generated 5 μm droplets) were smaller when compared with the systems using the Corexit 9500A (1.5% of Corixit 9500A has formed 6.2 μm droplets). Smaller droplet size implies higher emulsion stability, and may improve the biodegradation of oil. An improved laboratory dispersant testing protocol developed by EPA, baffled flask test, was performed to measure the dispersion effectiveness at DOR 1:20, 1:50, and 1:100. The effectiveness of DOR at 1:20 and 1:50 were around 35%. Cactus mucilage is readily available through sustainable agriculture, it is inexpensive to

extract, and has exhibited a long shelf live. Therefore, cactus mucilage can be an alternative technology to mitigate the damage that oil may cause to the aquatic ecosystem and minimize undesired effects associated with the use of synthetic dispersants in oil spills.

462488 Optimization of UF Backwash with Real-Time Control of Coagulant Dosing

Abstract:

Over the past two decades, reverse osmosis (RO) has emerged as the leading technology for seawater and brackish water desalination. However, membrane fouling remains a major challenge for robust RO operation. Hence, effective RO feed pre-treatment is vital to long-term operation of RO desalination plants. Amongst existing RO pre-treatment technologies, ultrafiltration (UF) has proven to be effective in producing consistently high quality filtrate for RO feed. One of the major advantages of UF membrane pre-filtration is the ability to reverse flow through an UF module (a process known as backwash) and thereby removing the foulant layer. After UF backwash, a portion of the foulant may remain on the membrane (i.e., not removed by backwash) as irreversible foulant layer. Irreversible fouling is typically removed through chemical washing which can significantly increase RO feed pretreatment. UF filtration performance can be increased via coagulant dosing of the raw feed water. Studies have also shown that UF backwash effectiveness improves as coagulant dose increases up to a certain threshold, beyond which UF backwash effectiveness no longer improves via additional coagulant dosing. Optimization of coagulant dosing is therefore critical in order to reach both optimal UF and backwash effectiveness while reducing coagulant consumption. However, real-time coagulant dose optimization is a major challenge when feed water quality is temporally variable both in the short term as well as seasonally. Under such conditions coagulant dose optimization via static jar tests is inappropriate. Therefore, in the present work, a novel method is introduced to allow real-time adjustment of coagulant dose so as to ensure effective UF backwash even under conditions of changing water feed quality. In the present method, the change in initial UF filtration resistance (post backwash) is used as a metric of UF backwash effectiveness. The above information is utilized by a coagulant dose controller designed to utilize the above metric to determine whether the coagulant is being underdosed or overdosed, and correspondingly adjust the coagulant dose to the UF feed. The controller was implemented in an integrated UF-RO seawater desalination pilot plant (capable of 18,000 GPD production of fresh water). The controller performance was evaluated, in a field study in California, with respect to its performance, including during a storm event in which raw seawater feed quality rapidly worsened, and its ability to reduce coagulant consumption. Results of the study demonstrated that the present approach enabled system operation that was able to adapt to worsening feed water quality conditions while reducing coagulant dose by up to ~29% relative to operation without real-time coagulant dose optimization.

451292 Performance Analysis of Texas Eagle Ford Shale Oil Hydro Fracturing Produced Water Treatment Process

Abstract:

Performance Analysis of Texas Eagle Ford Shale Oil Hydro Fracturing Produced Water Treatment Process

Chris Skrivanos, Nico Dwarica, Cameron McKay, John Allen Floyd and Mahbub Uddin

Trinity University, San Antonio, Texas

Extended Abstract

With the rise of horizontal drilling, hydraulic fracturing has become increasingly common and it is estimated that 90 percent of new wells drilled in the United States are hydraulically fractured. Hydraulic fracturing involves the injection of a very large quantity of high pressure liquid to fracture the rock formation and pry open the fracture. This fracturing allows natural gas/oil to flow freely from the formation into the well for collection. The components used in the hydraulic fracturing process consist primarily of water, sand, salts, gel, diluted acid, surfactant corrosion inhibitor, scale inhibitor, and other additives to aid in the hydro fracturing process. During the hydro fracturing operation when the pressure used to inject the fracturing fluid into the well is released, some of the fluid returns to the surface; this fluid is called "flowback" water. Once gas production begins at the well, all wastewater emerging from the well is called "produced water". Both types of wastewater – flowback and production phase water – contain potentially harmful constituents. These constituents can be broadly grouped into several principle categories: salts, organic hydrocarbons, metals, chemical additives and some naturally occurring radioactive material. Because of these constituents, shale oil/gas wastewater should not be released into the environment without adequate treatment. The purpose of this paper is to report the performance analysis of a water treatment process of Texas Eagle Ford Shale oil hydro fracturing produced. The treatment process consists of four primary unit operations: Adsorption column, Reverse osmosis, Ion-Exchange and Pump. The produced water is first pumped through the adsorption column, then a second high-pressure pump which then forces the water through the reverse osmosis membrane. The permeate of the membrane then flows through two ion exchange columns and received in a collection bin.

Adsorption is the adhesion of atoms, ions, or molecules from a fluid or dissolved solid to an insoluble surface and is used to remove suspended organics like oil and grease as well as dissolved organics like benzene and toluene. Our adsorption column consists of a cartridge and housing unit rated for operating pressures of 90 psi and temperatures no higher than 125 °F. The cartridge was 10" tall and accepted 10" x 4.5" cartridges. Reverse osmosis filters work on the principle of membrane divergence. An extremely high pressure (800 psi in our process) is applied at the housing entrance that forces the fluid through the fine membrane. Due to the pressure spike, particles that are small enough pass through the fine holes in the filter transfer, while the larger particles are rejected and thus removed from the membrane. In our application, the reverse osmosis membrane rejected the ions and other larger particles like metals, minerals, and organics, allowing only water and smaller particles to remain. The membrane we used is a FilmTec Sea Water Membrane designed to handle increased salinity (greater than 10,000 ppm). The reverse osmosis unit operation used had a flow ratio of permeate to retetante of 73 to 27, which was determined by a series of mass balances across the unit operation. We used a mixed ion exchange in order to remove any remaining positive or negative ions. Our process consists of two ion exchange columns in parallel, both of which were 20" vertical columns housed in separate units. The process uses two pumps. The first pump transferred the water from the tank to the adsorption column at a low pressure. The second pump primed the system to the operating pressure, 800 psi, to transfer the produced water through the reverse osmosis membrane and the ion exchange.

Each water sample analyzed was obtained from one of 4 major positions. Position 1: the untreated water, Position 2: after the activated carbon operation, Position 3: after the reverse osmosis operation and Position 4: after the ion exchange operation. Produced water processed by the system once is defined as first iteration treated water. Produced water processed by the system twice is defined as second iteration treated water. The first iteration sample collection began after flowing untreated produced water through the system for 60 seconds. The second iteration was run identically to the first iteration except produced water that had previously been treated by the system was used instead of untreated produced water. Once collected, samples were labeled according to the iteration, position and the trial number of extraction.

In order to assess the effectiveness of each unit operation and iteration in the wastewater treatment system, a series of experiments were designed to monitor the flow ratio, total dissolved solids present in solution, pH, and change in organic content. Inductively Coupled Plasma Optical Emission Spectrometry (ICP-OES) was implemented as a means to measure the metallic and salt content present in the water at various purification steps. The initial metallic content was determined to be 184.45 ppm, and the initial salt concentration was determined to be 2262.26 ppm for the produced water. A simple pH test using a probe was conducted on samples to monitor the change in acidity of the water as it passed through each unit operation, and the initial pH of the produced water was determined to be 4.25. Gas chromatography measurements were implemented in order to monitor the changing levels of total organic compounds present in the water.

Analysis of Results

First iteration ICP-OES results showed that the initial salt content of the water was 2262.26 ppm and the heavy metal content was 184.45 ppm. The carbon adsorption column removed 99.01 percent of the initial salt content and 98.5 percent of the initial heavy metallic content. The reverse osmosis membrane removed an additional 0.14 percent and 0.16 percent of the initial salt and heavy metal content respectively. After the ion exchange column approximately 100 percent of the total salt content was removed and 99.84 percent of the total heavy metals were removed, leaving a heavy metal concentration of 0.29 ppm and the salt concentration of essentially 0 ppm. Analysis of the second iteration treatment indicated the carbon adsorption column added 0.37 ppm of heavy metal and 2.09 ppm of salt to the water. The reverse osmosis membrane removed 0.29 ppm of heavy metals and added 0.85 ppm of salt content. The ion exchange column removed 0.04 ppm of heavy metals and 1.55 ppm of salts, bringing the final concentration of the second iteration treated water to 0.32 ppm of heavy metals and 1.39 ppm of salts. The first iteration pH results showed that the produced water had an initial pH of 4.25. The carbon adsorption column increased the pH to 8.26, the reverse osmosis membrane lowered the pH to 7.24, and the ion exchange column lowered the pH to the final value of 5.31. Second iteration treated water pH results showed that the carbon adsorption column increased the pH from 5.31 to 8.27. The reverse osmosis membrane decreased the pH to 7.24. The ion exchange column brought the pH to a final value of 6.49. The produced water contained a plethora of unknown organic constituents and it would be unpractical to identify each individual constituent to generate a calibration curve. So instead of measuring the changing concentrations of each constituent, the GC measurements were used to observe the change in the overall organic content in the water, relative to position 1 of the first iteration which is representative of the initial organic content of the produced water. The first iteration GC results showed that the carbon adsorption column removed 48.4 percent of the initial organic content, the reverse osmosis membrane removed an additional 28.7 percent of the initial organic content, and the ion exchange column removed 6.4 percent of the total organic content; the system overall removed 83.5 percent of the initial organic content.

Conclusions and Recommendations

The produced water treated in our process in a single iteration met the EPA standards for acceptable levels of metallic and salt content for reclaimed water, however the pH is at the lower end of the standard. The first iteration treatment reduced the metallic content by 99.84 percent, reduced the salt content by essentially 100 percent, raised the pH from 4.25 to 5.31 and reduced organic content by 83.5 percent. With a second iteration, the treated water pH is within the acceptable range for reclaimed water. The second iteration was shown to be unnecessary since the metallic and salt concentration changed negligibly, and the organic content increased. The pH was raised to acceptable levels, but the addition of a controlled injection pH adjustment unit operation would allow the pH to be raised without the need for a second iteration, making the produced water treatment process more efficient. This change to the process would generate treated water that meets the EPA drinking water standard. Our analysis indicates that the wastewater treatment processes has applications in commercial and residential areas where hydraulic fracturing is common, to treat wastewater that meets EPA standards.

461844 Macroalgae and Chitosan-Macroalgae Biocomposite As Potential Adsorbents of Water-Soluble Hydrocarbons: Effect of pH, Organic Matter and Ionic Strength

Abstract:

Macroalgae and Chitosan-macroalgae biocomposite as potential adsorbents of water-soluble hydrocarbons: Effect of pH, organic matter and ionic strength.

Abstract

Nowadays the recovery of low-molecular aromatic hydrocarbons (HC) released into natural bodies of water continues to be a challenging task. These contaminants bring sever consequences to the environment and to the human health. The oil spill cleanup strategies are primarily design to deal with the heavy fractions accumulated in the water surface. Unfortunately, very limited information is available regarding the treatment of dissolved fractions. Biosorption on macroalgae (Ma) biomass seem to be a potential alternative to overcome disadvantages associated to high-priced costs and complexity of activated carbon production. Three seaweeds of a representative sample of brown, green and red macroalgae (*Macrocystis pyrifera*, *Ulva expansa* and *Acanthophora spicifera*, respectively) were evaluated to remove benzene and toluene from water, which are the most soluble hydrocarbons. Toward this objective, the influences of temperature and ionic strength were also determined. Raw biomasses were characterized by different physical and chemical techniques to assess their potential as biosorbents and the mechanisms involved. The organic and inorganic dissolved fractions were also considered. The characterization techniques mainly identified carbohydrates, lignin and proteins. *M. pyrifera* biomass registered the highest removal capacities: about 112 and 28 $mg \cdot g^{-1}$ for benzene and toluene, respectively, probably attributed to alginate and lignin content. It was found that the hydrocarbons biosorption affinity was not affect up to an ionic strength of 0.6 M, due to the biosorbents pore occlusion by water clusters. For

benzene and toluene nonpolar solutes, the biosorption mechanism is an addition of simultaneous interactions between the sorbents and the chemical constituents of the cell wall, presumably by London (dispersion) forces onto â€“OH and â€“NH active sites from polysaccharides and proteins and hydrophobic interactions from lignin and lipid fractions.

Despite these advantageous properties, the physical characteristics (small particle size, low strength and density) of such biomaterials are not viable for a continuous process operation and make biomass difficult to apply. Therefore the development of innovative low-cost techniques of immobilized biosorbents with special attention to increase their effectiveness in the biosorption process is mandatory. The biopolymer-biopolymer interactions between positive and negative macromolecules can enhance the compatibility and stability of biosorbents, and these new structures are called polyelectrolytes complexes (PEC). The linear amino polycationic compound of chitosan (Ch) possess remarkable quantity of amine and hydroxyl groups that interacts with negatively charged substances with excellent chelation behavior and high affinity toward pollutants like metals and dyes. On the other hand pectin (Pe) is a negatively charged polysaccharide composed mostly by esterified D-galacturonic acid residues linked together by Î±-(1-4) chain. Pectine molecule is also used in the food industry for its well-known thickening, stabilization and gelling properties.

Â The main objective of this research was to explore the performance of Ma and PEC complexes of different proportions of Ma/Ch/Pe on biosorption capacity of the two main soluble hydrocarbons (benzene and toluene) in water. The biocomposite synthesis was optimized by application of the factorial design and response surface methodology. Moreover, a through chemical and physical characterization analysis, by textural properties, potentiometric titrations, elemental content, chemical stability, KBr-FT-IR and TGA analysis, were performed to explain the adsorption mechanisms.

458236 Concentrating 17 -Ethinyl Estradiol (EE2) Using PES Hollow-Fibre Membranes: Experimental Exploration and Molecular Modelling

Abstract:

The 17Î±-ethinyl estradiol (EE2) is a synthetic estrogen derived from 17Î²-estradiol (E2), the major endogenous estrogen in humans. EE2 is widely used as an active ingredient for birth control and other hormonal treatments. EE2 is excreted into the sewage and the environment mainly via urine. Wastewater treatment plants are usually ineffective in removing EE2 from water which results in the release of EE2 into the environment. The concentration of EE2 in surface water bodies are typically at sub-ppb levels. Even at such low concentration, several studies have reported negative environmental effects of EE2 including feminization of aquatic animals. At present, it is not known if EE2, at such low concentration, has any adverse effect on human or other mammals over short and long term exposure. Developing novel strategies for the efficient removal and treatment of EE2 from water has become an imperative in water reclamation and reuse. Here, we report a novel approach whereas a widely used polyethersulfone (PES) membrane was used to concentrate EE2 through adsorption and solvent elution. The concentrated EE2 can then be further treated with high efficiency using approaches such as biodegradation. Molecular mechanisms underlying the interaction between EE2 and PES membrane was elucidated using molecular dynamic simulation.

454263 Natural Organic Matter Fouling Behaviors on Superwetting Nanofiltration Membranes

Abstract:

Natural organic matter fouling behaviors on superwetting nanofiltration membranes

Linglong Shan, Hongwei Fan, Hongxia Guo, Shulan Ji and Guojun Zhang

Beijing Key Laboratory for Green Catalysis and Separation, College of Environmental and Energy Engineering, Beijing University of Technology, Beijing, China. Email: shanlinglong@126.com

In natural water, natural organic matter (NOM) components have been identified as precursors of disinfection by-products. Nanofiltration has been widely recognized as a promising technology to remove NOM from natural water. Currently, the most challenging aspect is the often inevitable occurrence of membrane fouling, which is caused by NOM accumulation on the membrane surface, as in NOM removal in drinking water treatment and production. NOM fouling is considered to be governed by the interface relationship between NOM and the membrane surface in aqueous solution. Much work has been carried out, but the effect of hydrophilicity/-phobicity of NOM and the membrane surface on fouling behavior is still a controversial issue. [1-7] The controversy originates from the unilateral consideration of membrane wettability or foulant hydrophilicity/-phobicity, when in fact both need to be considered. In our study, both superhydrophilic and superhydrophobic membranes with stable structure have been successfully obtained using two facile hybrid approaches (see Fig. 1). The fouling behavior of NOM on superwetting nanofiltration membranes has been extensively investigated using humic acid (HA: model foulant) and real natural water. The extended Derjaguin-Landau-Verwey-Overbeek (XDLVO) approach and nanoindentor scratch tests suggested that the superhydrophilic membrane had the strongest repulsion force to HA due to the highest positive ¦¤G^{TOT} value and the lowest critical load. Excitation emission matrix analyses of natural water also indicated that the superhydrophilic membrane showed resistance to fouling by hydrophobic substances and therefore high removal thereof. Conversely, the superhydrophobic membrane showed resistance to fouling by hydrophilic substances and therefore high removal thereof (see Figs. 2 and 3). Long-term operation suggested that the superhydrophilic membrane had high stability due to its anti-NOM fouling capacity, which is crucial for practical application. Based on the different anti-fouling properties of the studied superwetting membranes, a combination of superhydrophilic and superhydrophobic membranes was examined to further improve the removal of both hydrophobic and hydrophilic pollutants. The R_{UV254} and R_{DOC} could be increased to 83.6% and 73.3%, respectively, by a combination of superhydrophilic and superhydrophobic membranes. In view of the hydrophilic and hydrophobic NOM fouling behaviors on superwetting membrane surfaces, the superwetting strategy can be regarded as a very important approach for obtaining a membrane with both high NOM rejection and good anti-NOM fouling properties.

Keywords: Superhydrophilic, Superhydrophilic, Nanofiltration, NOM, Fouling

References

[1] L. Shan, H. Fan, H. Guo, S. Ji, G. Zhang, Natural organic matter fouling behaviors on superwetting nanofiltration membranes, Water res., 93 (2016) 121-132.

[2] L. Shan, H. Guo, Z. Qin, N. Wang, S. Ji, G. Zhang, Z. Zhang, Covalent crosslinked polyelectrolyte complex membrane with high negative charges towards anti-natural organic matter fouling nanofiltration, RSC Adv., 5 (2015) 11515-11523.

[3] M. Ng, S. Liu, C.W.K. Chow, M. Drikas, R. Amal, M. Lim, Understanding effects of water characteristics on natural organic matter treatability by PACI and a novel PACI-chitosan soagulants, J. Hazard. Mater., 263 (2013) 718-725.

[4] J.A. Nilson, F.A. DiGiano, Influence of NOM composition on nanofiltration, J. Am. Water Works Assoc., 88(5) (1996) 53-66.

[5] H. Yamamura, K. Okimoto, K. Kinura, W. Yoshimasa, Hydrophilic fraction of natural organic matter causing irreversible fouling of microfiltration and ultrafiltration membranes, Water Res., 54 (2014) 123-136.

[6] L. Zhang, G. Chowdhury, C. Feng, T. Matsuura, R. Narbaitz, Effect of surface-modifying macromolecules and membrane morphology on fouling of polyethersulfone ultrafiltration membranes, J. Appl. Polym. Sci., 88 (2003), 3132-3138.

[7] A. Tiraferri, Y. Kang, E.P. Giannelis, M. Elimelech, Superhydrophilic thin-film composite forward osmosis membranes for organic fouling control: fouling behavior and antifouling mechanisms, Environ. Sci. Technol., 46 (2012) 11135-11144.

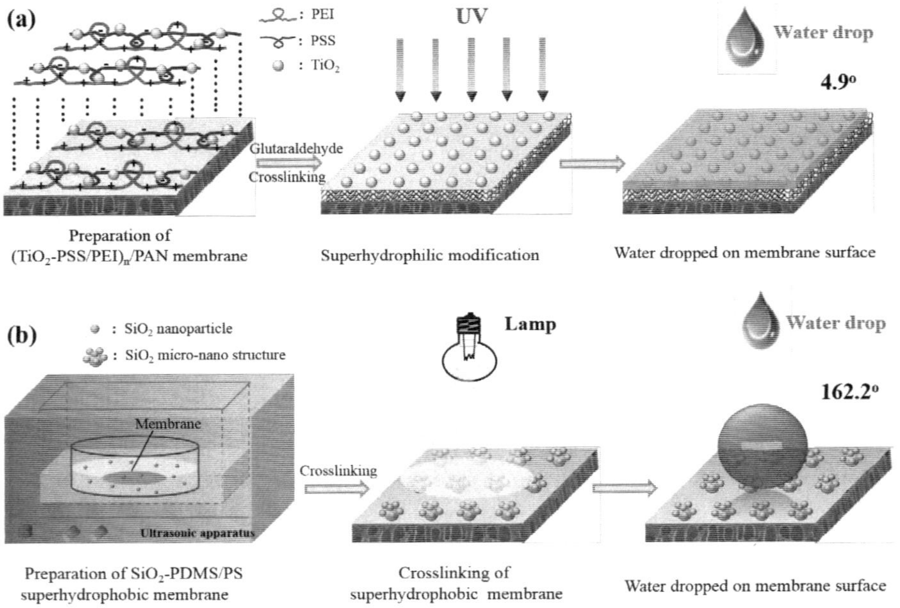

Figure 1. (a) Schematic illustration of the preparation of the superhydrophilic membrane and its surface wettability; (b) schematic illustration of the preparation of the superhydrophobic membrane and its surface wettability.

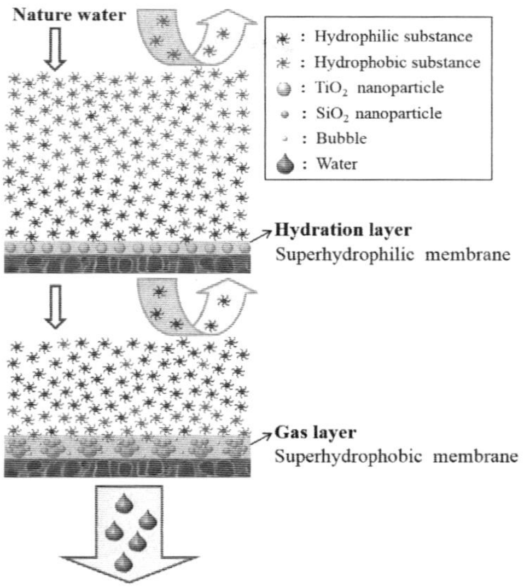

Figure 2. Schematic illustration of different anti-fouling properties of the studied superwetting membranes. Superhydrophilic membrane showed resistance to fouling by hydrophobic substances, while the superhydrophobic membrane showed resistance to fouling by hydrophilic substances.

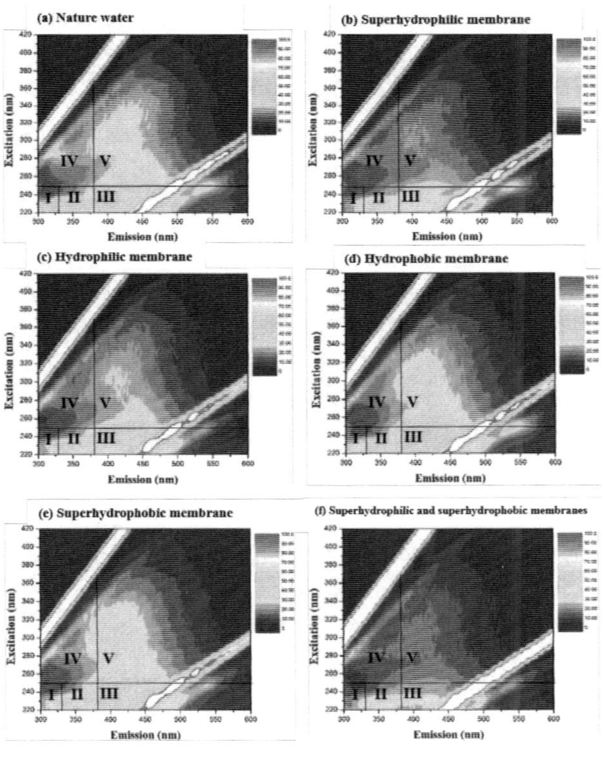

Figure 3. EEMs for (a) natural water; (b) natural water nanofiltrated by superhydrophilic membrane; (c) natural water nanofiltrated by hydrophilic membrane; (d) natural water nanofiltrated by hydrophobic membrane; (e) natural water nanofiltrated by superhydrophobic membrane; (f) natural water nanofiltrated by superhydrophilic membrane-superhydrophobic membrane.

Experimental and Simulation Analysis for Optimization of Solar Desalination System Using Humidification-Dehumidification Method: Open and Close Air Cycle

***Nickyar Ghadirinejad* [1], *Farshad Farshchi Tabrizi* [2], *Mohammmad Mehdi Zarei* [3], *Taher Chegini* [4], *Roberto Moreno-Atanasio* [5]**

[1] Chemical Engineering Department, University of Sistan and Baluchestan, Zahedn, Iran
[2] Chemical Engineering Department, Shiraz University, Shiraz, Iran
[3] Chemical Engineering Department, Khazar University, Baku, Azerbaijan
[4] Civil Engineering Department, Khazar University, Baku, Azerbaijan
[5] Chemical Engineering Department, Newcastle University, Callaghan, Australia

Abstract:

Over the last decades, desalination of brackish water and seawater has been studied extensively for producing fresh water. Desalination technologies are divided into two major categories: Thermal and Membrane. Humidification-Dehumidification (HDH) which is the subject of the present research, is one of the thermal technologies. It is inspired by the rain cycle and is based on evaporation and consecutive condensation.

In a series of experiments, the processes of production of fresh water has been investigated under various operational parameters such as temperature and flowrate of cooling water, humidifier intake from water boiler, air flowrate and height of packed bed column. Subsequently, an appropriate mathematical model of the system has been developed and validated using the collected data. Furthermore, based on the developed model a series of simulations has been carried out in ASPEN Plus in order to obtain optimum physical parameters of various parts of the desalination system.

The experimental results showed that under constant heat flux and optimum air flowrate, three parameters increase the production of distilled water the most: increasing flowrate of cooling water, decreasing flowrate of water inlet and increasing temperature. Moreover, although increasing flowrate of cooling water leads to an augmentation in the distilled water flowrate, but this effect is mitigated at high rates.

Furthermore, based on the obtained overall heat transfer coefficients of the humidifier and condenser, the most reliable Height Equivalent to a Theoretical Plate (HETP) model of random packing is found to be Mangers & Ponter. The estimation error of the model in predicting the height of packed bed was less than 4%. Based on the results of this model, closed air system increases the efficiency of the production of distilled water by 15% in comparison to open air. This increase is the result of prevention of wasting of air humidity in the condenser outlet.

Finally, optimum parameters of operational units are obtained by applying a Differential Evolution (DE) strategy. The optimum values of flowrate and temperature of cooling water are 0.0298 m^3/hr, 13 °C, respectively. The best value of flowrate of closed air is found to be 0.008075 m^3/hr and of water intake to be 0.021497 m^3/hr. These values produced fresh water with flowrate of 0.3777 m^3/hr.

Keywords: Solar Desalination, Dehumidification, Humidification, Heat Transfer

1- Introduction:

The rapid growth of industries and population, rivers and lakes pollution problems caused by industrial wastewater and flowing municipal wastewater into freshwater resources have caused a shortage of fresh water [1]. However, it should be noted that the only sources of never-ending water reservoir are oceans; water salinity reduction to overcome the shortage of fresh water, is a more applicable solution [2]. Desalination includes many different methods and required energy for this procedure also encompasses a wide range. The heat produced by fossil fuels and/or the excess heat-released of nuclear reactions, renewable energies such as ocean waves or solar energy, all can supply heat power for water desalinisation plants. Desalination technologies are divided into two categories: Thermal and membrane methods. Water is heated in thermal processes and after evaporation, fresh water is produced via condensation. In membrane methods, thin permeable membranes are used to separate feed into both fresh water flow and more saline flow [3].

In recent years, Humidification-Dehumidification (HDH) system, as a new method has been investigated; through this system, air flows as the intemediate fluid. In fact, this method is a combination of membrane processes and thermal distillation. It operates based on the mass diffusion principles and uses cooling air due to water evaporation and humidification. Fresh water is resulted via dehumidification (condensation of humid air). The obvious benefit of HDH is to be one of the most cost-effective desalination mechanism regarding to terms of low temperature and pressure [4]. The investigated method includes solar collector/s, humidification tower/s (evaporation unit), dehumidification tower (condensation unit) which even to achieve higher efficiency, it could include heat recovery unit as well. In order to increase the efficiency of heat transfer, at first, water is passed through the condenser coil, by heat absorption from hot air, water is pre-heated. Afterward, the pre-heated water is led into a solar heater and an additional heater (auxiliary heater) subsequetly to reach a higher temperature. In the next step, hot water is showered into the humidification tower.

2- Closed-air, open-water (CAOW) water-heated systems:

Alluded humidification tower including random packings/ Bubble Columns/ Wetted-wall Columns [5], showered with hot water, gets totally moist using counter-current air flow from the botton; subsequently, after heat absorption, hot air is guided through the condenser to produce condnesate-water (fresh water). In closed air cycle, after passing the condenser, the air is recycled into the system. HDH method has been studied by Al-Hallaj et al. [6,7], Ben-Bacha et al. [8], Garg et al. [9] ،Nafey et al. [10] ،Nawayseh et al. [11], [12] Muller-Holst.

In order to achieve optimum operational conditions and achieve the best yield, many research projects have been studied [13,14,15,16,17] and various mathematical models have been proposed considering measurement of mass and heat transfer coefficients [13], impacts of input parameters into the HDH system, mass and energy balance equations, Consideration of waste heat during HDH process and finding a solution due to heat recovery [18]. To simulate a system including both a flat-plate water-heater and a double-pass solar air heater, assuming some simplified assumptions, Yildirim and Solmus [19] proposed an optimized mathematical model by using fourth order Runge-Kutta solution method. Cost-estimation of produced fresh water via an elaborated, multiple solar desalination system [20] and via a novel solar HDH system [21] has been investigated.

3-HDH Apparatus:

In the study, the humidifier consisted of emedded random packing (Pall ring type with 25 mm nominal diameter). Water was showered from the top and air wsa flown counter-currently. The coil-condensor was a single 8-meter-long cupper pipe. For circulation of air in side the HDH system, two fans were used, without any direct contact between motors and humid air.

As it shown in figure 1, intake salt water in low temperature (as feed water into the HDH system) entered into the condenser's shell, and after relative pre-heating subsequently entered into the solar heat collector. During experimental research, to achieve higher temperatures an electrically additional heater was used. Heated saline water was showered inside the humidifier. Water flow rate was regulated by passing through some valves and rotameters to reach desired amounts. Air-flow rates adjusted through entire HDH system using fans' regulators. By directly counter-current contact between hot saline water and air in random-packed bed, the air temperature got augmented. Consequently, it tended to embrace greater capacity to absorb more moisture. Passing through the dehumidifier, fresh water was by-passed and collected; outlet air was recycled in closed air system and was blowed from the bottom into the humidifier.

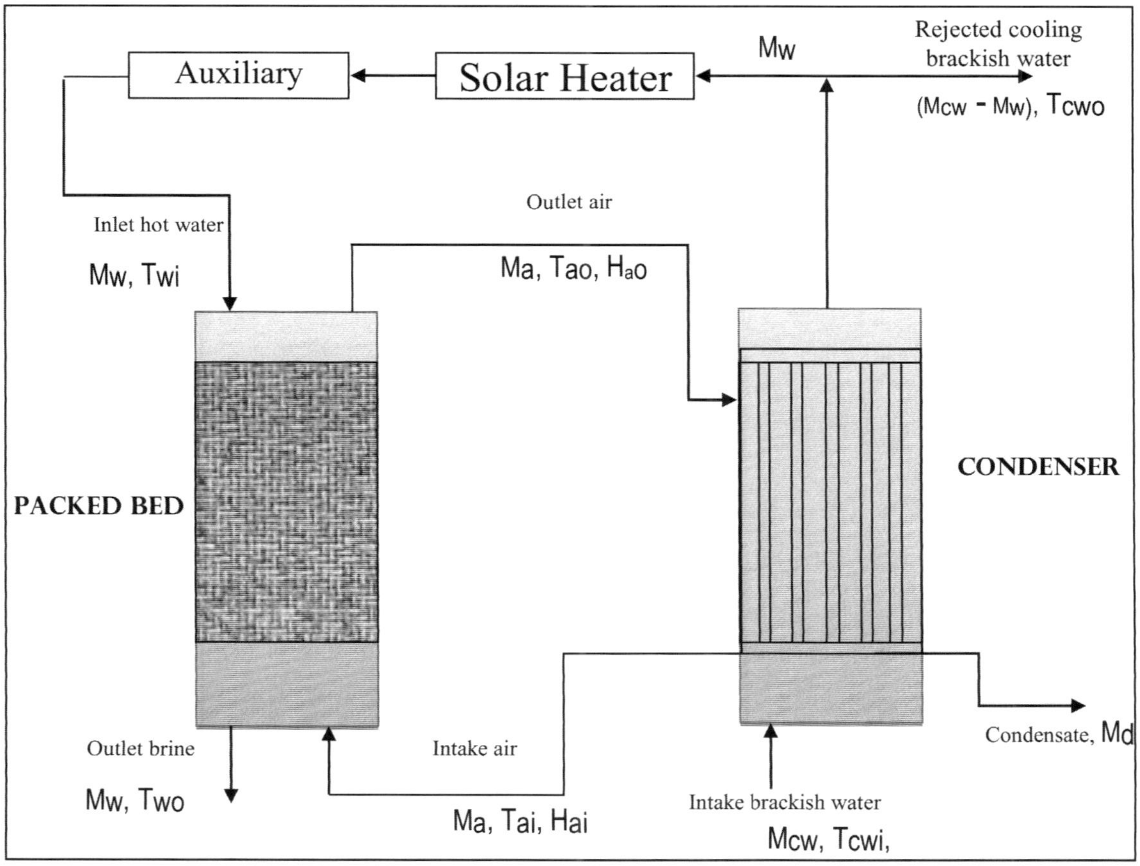

Fig. 1: Layout of HDH system

For experimental data collection, the system was ran continuously for 15 to 20 minutes in the same mode, reaching considered steady-state temperatures and humidity amount in each test; then tests were performed. Each of the experiments were repeated at least three times to ensure their correction.

4- Modeling of humidification-dehumidification system:

In order to determination of diameter and Height Equivalent to a Theoretical Plate (HETP), based on the obtained overall heat transfer coefficients of for the random-packed-bed humidifier and condenser [22], Considering given data about Rasching ring and Pall ring packings [5] following models have been investigated: Shulman et al. [23], Shulman & de Grouff [24], Van Krevelen & Hoftijzer [25,26], Mangers & Ponter [27], Shi & Mersmann [28], Billet and Schultes [29], Onda et al. [30].

The estimation HTEP model of **Mangers & Ponter**:

$$k_L = \frac{(3.22\times10^{8})Sc_L^{0.5}D_L}{a_e(1-\cos\theta)}\left(\frac{\rho_L u_L}{\mu_L}\right)^{\alpha}\left(\frac{\rho_L^2\, g\, h_{pe}}{\mu_L^2}\right)^{0.27} \quad (1)$$

As α in the above equation is calculated by:

$$\alpha = 0.49\left[(1-\cos\theta)^{0.6}\left(\frac{\rho_L\sigma_L^3}{\mu_L^4 g}\right)^{0.2}\right]^{0.108} \quad (2)$$

$$k_G = 0.2\,\frac{D_G}{d_c}\left[\frac{\rho_L u_L}{u_L a_p}\right]^{0.8} Sc_G^{\frac{1}{3}} \quad (3)$$

Additionaly, for comparison the results obtained from the mathematical modeling and practical experiments, both physical and chemical operational behaviors of HDH units -at steady state conditions- have been investigated by modeling in ASPEN Plus. Initial feed's conditions to HDH system, constituted 3884.8 ppm salinity, Ca^{2+}, Mg^{2+}, K^{+}, Cl^{-}, HCO_3^-, SO_4^{2-}, NO_3^- Na^{+} aqueous solutions were imported to ASPEN Plus model. Enick's Redlich-Kwong-UNIFAC equations [31] and Pall ring packings were selected for the mentioned processes.

In preliminary calculation, the capacity of existing packing in the dumidifier was assumed 10 HETP. Regrading experimental results, output data from run of each stage in ASPEN Plus were compared.

4-1- Modeling of HDH system in open and close air cycle:

Effects of intake water flow rate and its temperature entering the humidifier, air and cooling water flow rates, and effect of packed-bed height on the yield of desalinated water amounts have been studied by developing mass and energy transfer balances for overally and even on every sinlge unit in the HDH system.

$$Mcw.Cp.Tcwi - (Mcw - Mw).Cp.Tcwo + Q_{Heater} = Mw.Cp.Two + Md.Cp.Ta \quad (4)$$

Saturated-air, as a main simplifying assumption was considered to solve formulated equations. On the contrary, unlike precedent models, the term of air solubility in outlet saline water, as well as unsteady air and water flow rates recycling the entire HDH system was considered [32].

To improve derived equations and to reach more realistic modeling in the range of $0 \leq S_p \leq 16000$ ppm and $10 \leq T \leq 180$ °C, correlations of seawater density (ρ_{sw}), specific heat (C_p), seawater thermal conductivity (K_{sw}) [32] were applied:

$$\rho_{sw} = 10^3(A_1.F_1 + A_2.F_2 + A_3.F_3 + A_4.F_4) \quad (5)$$

Which A_n and F_n are specific variables proposed by Isdale and Morris [33].

Evaporation latent heat (λ) (which is T in K degree)

$$\lambda = 2501.89 - 2.407 \times T + 1.1922 \times 10^{-3} \times T^2 - 1.586 \times 10^{-5} \times T^3 \quad (6)$$

Specific heat coefficient of air (T in K degree)

$$C_{pa} = 1.03409 - 2.84887 \times 10^{-4}.T + 7.816818 \times 10^{-7}.T^2 - 4.970786 \times 10^{-10}.T^3 + 1.077024 \times 10^{-13}.T^4 \quad (7)$$

To calculate air density with respect to Al-Sahali & Ettouney's proposal [13] for HDH desalination plant, air assumed as an ideal gas.

$$\rho_a = \frac{P.M_{wei}}{R.T} \quad (8)$$

Thermal conductivity coefficient of air:

$$K_a = -2.276501 \times 10^{-3} + 1.2598485 \times 10^{-4}.T - 1.4815235 \times 10^{-7}.T^2 + 1.73550646 \times 10^{-10}.T^3 - 1.066657 \times 10^{-13}.T^4 + 2.47663035 \times 10^{-17}.T^5 \quad (9)$$

In order to calculate the amount of dissolved air in the water (dependent on its pressure and temerature) in the humidifier [34] Henry's constants for air solubility in water (K_{sol}) were considered too.

Newton's method was used to solve mentioned non-linear equations, using some most probable intial guesses via experimental data. For optimization of outlet fresh water dependent on Q, Mw, Ma, Tcwi and Mcw values, Differential Evolution (DE) strategy was applied to find optimum values. DE strategy [35], one of genetic algorithm methods, via successive tries to generate correct interpolations was coded in MATLAB software.

In the same method, calculations of open-air cycle in the HDH system were performed by rewriting similar equations as above.

5- Results and discussion:

Keeping constant given thermal energy to the inlet water (including solar heater and additional heater), several sets of experimental data were collected in various conditions. By changing the initial independent parameters, trends of intermediate variables and the final variable (amount of produced fresh water) were evaluated.

5-1 Effect of inlet water flowrate and its temperature on deslination production rate:

At constant thermal energy, inlet water flow rate and its temperature were realized as two dependent vaiables. In this series of tests, parameters of air flow, cooling water flow and packed-bed height were kept constant; and consequently, effects of two mentioned parameters of inlet water entering to the humidifier on the yield of HDH system were investigated (Fig.2). As it is shown, at constant conditions, by increasing the temperature of inlet water, production rate of fresh water increases. It is noticible that during experiments, relative humidty of circulating air was checked continuously and it was near to saturation state (97.33%).

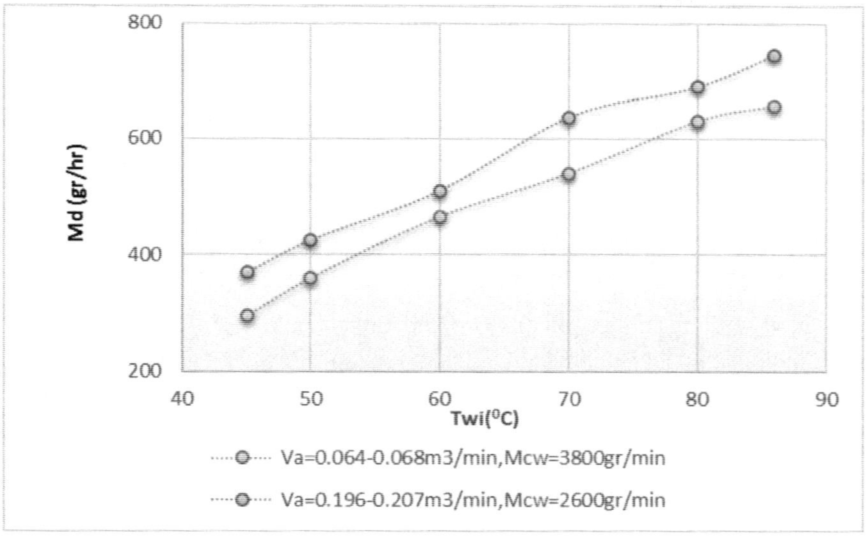

Fig. 2: Effect of temperature of inlet water on produced fresh water

5-2 Survey on air flow rate, considering effects of cooling water flowrate on yield of the HDH system:

As it presented in Figure 3 for three flow rates of circulated-air, the temperature of the exhausted air from the tower is a dependent variable of cooling water flow rate. At constant air flow rate, increasing in water temperature leads to augmentation of air temperature and consequently fresh water production.

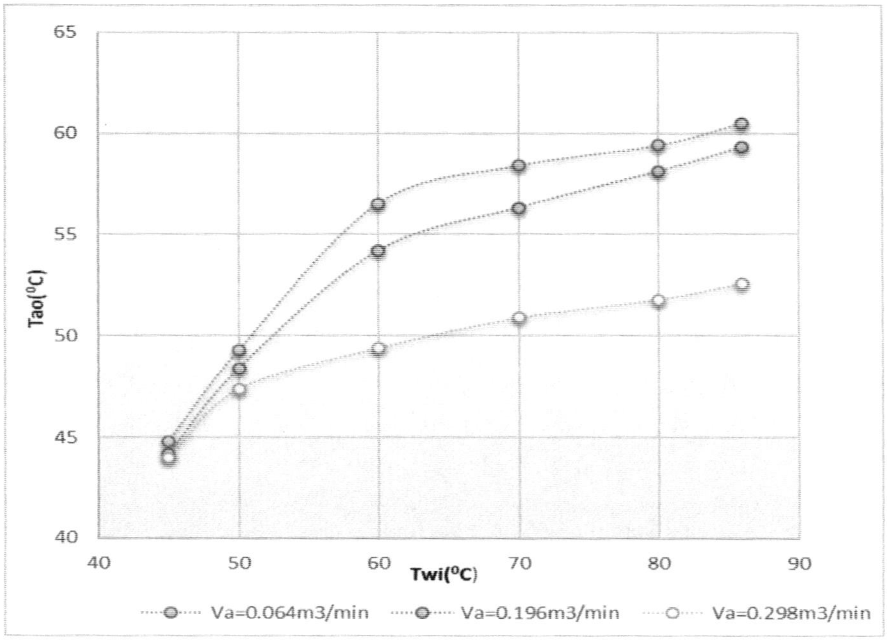

Fig. 3: effects of changes in Twi on Tao

In the present HDH design, to avoid wasting moisture, transfered by exhaust air flow from the condenser, and to spend less energy, close air cycle was contrived. It is noteworthy that the validity of this assumption has been proved by comparing output-results of close and open air models. Experimental tests were conducted in three different ranges (0.064-0.068 $\frac{m^3}{min}$, 0.196-0.207 $\frac{m^3}{min}$ and 0.298-0.310 $\frac{m^3}{min}$). Regarding Fig. 3, 4 and 5, range of 0.196-0.207 $\frac{m^3}{min}$ was determined as the optimal one for the air flow. Obviously, increase of air velocity causes decline of outlet air temperature in the HDH system. Besides, it leads to reduction in air settelment time in the humidifier and finally shrinkage of distilled water.

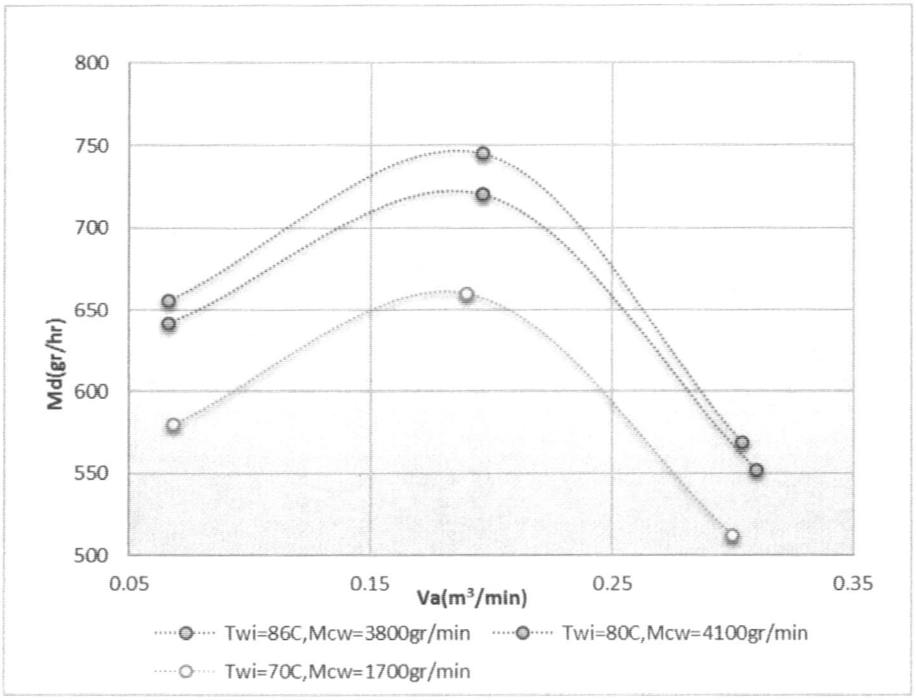

Fig. 4: Trends of M_d in three different air velocities at 70, 80 and 86 °C.

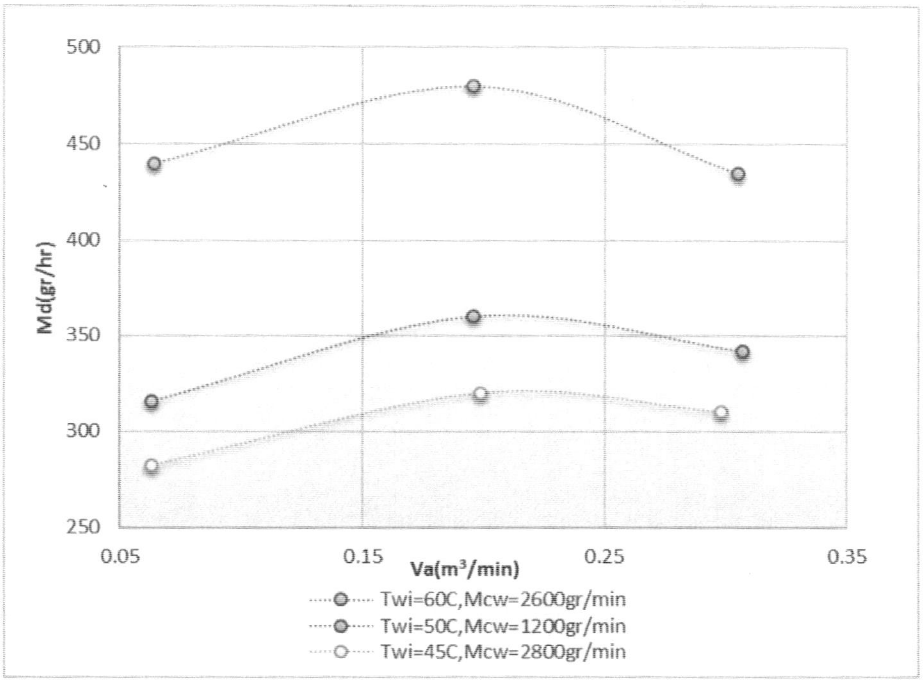

Fig. 5: Trends of M_d in three different air velocities at 45, 50 and 60 °C.

5-1 Distinguishing between results of open and clos air cycle in the investigated HDH system:

This comparibility is shown in table 1. It shows 15% higher productivity of fresh water in close air cycle than production rate in open air cycle. The most probalbe justification for this conlcusion could be waste-heat recovery and lack of moisture loss in the exhaust air from the condenser and recycling to the processes.

Table 1: comparability of 4 samples for open and closed air cycle

Closed cycle				**Open cycle**			
Twi (°C)	Tao (°C)	Tcwo (°C)	Md (m3/hr)	Twi (°C)	Tao (°C)	Tcwo (°C)	Md (m3/hr)
71.6	60.4	32.3	1.70E-04	70.4	56.2	32.1	1.01E-04
91.3	66.5	32.6	1.95E-04	88.4	61.4	31.1	1.50E-04
40.4	41.6	24.9	1.28E-05	37.6	35.1	25.6	1.23E-05
80.2	44.6	33.8	1.71E-04	79.4	39.0	33.2	1.27E-04

5-2 HETP investigation and HDH optimization:

To achieve optimal HETP for humidification tower, several experiments have been executed (Fig. 6). At constant practice conditions, using k_L and k_G equations, the most reliable HETP model has been proposed; Mangers & Ponter's model was determined as the most precise one (Table 2).

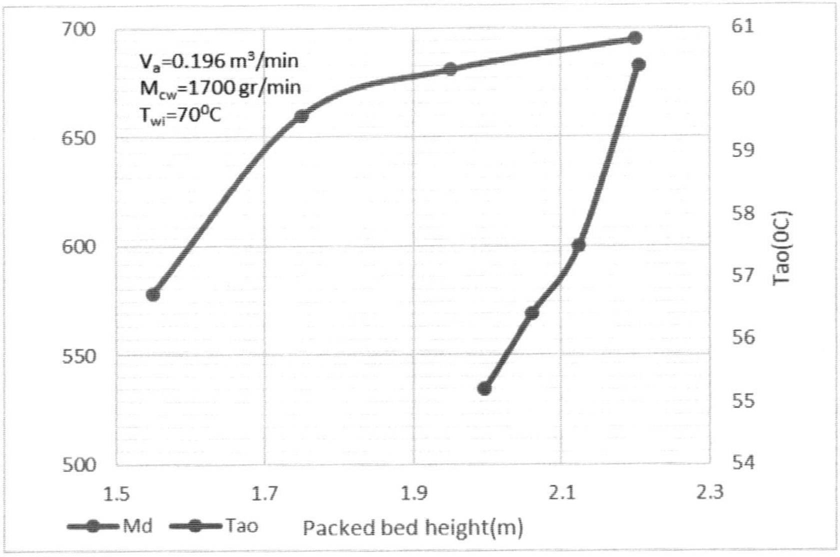

Fig. 6: effects of Packed bed height on M_d ($\frac{gr}{hr}$)

Table 2: estimating HETP models

Model	HETP (m)
Shulman et al.	0.042601
Shulman & de Grouff	0.047809
Van Krevelen & Hoftijzer	0.394546
Mangers & Ponter	0.511026
Shi & Mersmann	0.437404
Billet and Schultes	0.407311
Onda et al.	0.660333

Furthermore, due to admit validity of collected data modeling and simulation in ASPEN Plus have been done. At first, 10-equivalent-step was presumed. After every run of functions implemented in ASPEN Plus and comparison between outcome data and experimental values, it was observed that a tower including two or three steps can help us to meet our expected moisture balance (one single run is shown in table 3).

Table 3: ASPEN Plus simulation for 5-equivalent-setps of HETP

Exp. Cond.	Stages	Twi (°C)	Two (°C)	Tao (°C)	Tcwo (°C)	Md (mol/s)	Mw (mol/s)	Md/Mw
Mw=0.00305 (m3/hr)	5	82.31	42.23	66.48	32.9	0.04489	0.6095	0.07365
Ma=0.00108	4	82.17	43.77	65.9	32.9	0.04334	0.6095	0.07112
(m3/hr) Tcwi=28(°C)	3	81.95	46.18	64.92	32.9	0.04089	0.6095	0.06709
Tai=32.9(°C)	2	81.56	50.26	62.97	32.9	0.03647	0.6095	0.05985
	REAL	80.4	36.2	58.1	30.2			0.06011

According to the Eckert's proposed coefficients [36] and the use of Pall ring packing with 25 mm nominal diameter, the effective height of humidifier was suggested about 170 cm.

Using Differential Evolution (DE) strategy and via differing the number of dependent variables (such as air flow rate, T_{cwi}, M_w and finally M_{cw}) individually and also simultaneously, oprimized model was correlated successfully. Practical values which were resulted optimum $M_d=0.3777 \frac{m^3}{hr}$ were $M_{cw}=0.0298 \frac{m^3}{hr}$ at $T_{cwi}=13$ °C, wiht $M_a=0.008075 \frac{m^3}{hr}$ and $M_w= 0.021497 \frac{m^3}{hr}$.

6- Conclusion:

One of the most applicable methods for desalination brackish water is humidification-dehumidification system. This thermal method has been studied on a pilot system. After experimental data collecetion due to data validity proof and also finding an optimum fresh water rate, two mathematical models, one on open air cycle and another one on close air cycle have been derived. Furthermore, HETP for the humidifier was studied by comparing experimental data resulted from various tower's height and data from simulation of HDH system in ASPEN Plus; Estimation error of the HETP modelfor Pall Ring packed bed was less than 4%. Yield in closed air system was shown 15% better than open air system.

Convincingly, using of DE strategy, optimal parameters to obtain the choisest yield (0.3777 m^3/hr) have been provided: 0.0298 m^3/hr, 13 °C respectively for flow rate and temperature of cooling water, 0.008075 m^3/hr air flow rate (in closed air system) and 0.021497 m^3/hr for water inake.

Nomenclatures:

D (m)	diameter	C_{pa} $(\frac{KJ}{Kg.K})$	Specific heat coefficient of air
G(kg/s)	Gas mass flow rate	λ $(\frac{KJ}{Kg})$	Evaporation latent heat
L (kg/s)	Liquid mass flow rate	K_a $(\frac{W}{m.K})$	Thermal conductivity coefficient of air
M (gr/min)	Mass flow rate	μ $(\frac{kg}{m.s})$	*viscosity*
Ma (kg/s)	Intake air mass flow rate at the bottom of humidification tower	ρ $(\frac{Kg}{m^3})$	*density*
Mcw (kg/s)	Mass flow rate of saline water Inlet to cooling tower	ρ_a $(\frac{Kg}{m^3})$	Air density
Md (kg/s)	Fresh water (condensed water) mass flow rate	σ	*Thickness*
Mw (kg/s)	Mass flow rate of rejected water from condenser	***Subscripts***	
Q (w)	Heat received /transferred	*a*	*air*
R.H. (%)	Relative Humidity percentages	*ai*	*Inlet air*
T (K)	Temperature	*ao*	*Outlet air*
Tai *(K)*	Intake air temperature at the top of Humidifier	*cw*	*Cooling water*
Tao *(K)*	Outlet air temperature at the top of humidifier	*cwi*	*Inlet Cooling Water*
Tcwi *(K)*	Cooling water inlet temperature	*cwo*	*Outlet Cooling water*
Tcwo *(K)*	Water temperature of rejected water from condenser	*d*	*Distilled water*
Twi *(K)*	Water temperature of inlet water to humidifier	*G*	*Gas*
Two *(K)*	Water temperature of outlet brine water at the bottom of humidifier	*L*	*Liquid*
V_a $(\frac{m3}{min})$	Volumetric flow rate of air	*w*	*water*
R $(\frac{Pa.m^3}{Kmol.K})$	Universal Gas Constant	*wi*	*Inlet water to Humidifier*
M_{wei}	*Molecular weight*	*wo*	*Outlet water from Humidifier*

References:

[1] Malik, M., Tiweri, G.N., Kumer, A., Sodha, M.S., Solar distillation, New York, pergamon, pp. 3-4,1985

[2] Kalogirou, S.A., Solar thermal Collectors and applications, Progress in Energy and Combustion science, Vol. 30, pp. 231-295, 2004.

[3] Watson, I.C., Morin, O.J., Henthorne, L., Desalting handbook for planners, Aqua Resources International Inc., 2003.

[4] Qiblawy, H.M., Banat, F., solar thermal desalination technologies, Desalination, Vol. 220, pp. 633-644, 2008.

[5] Treybal R.E. Mass transfer operations, 3rd ed., NY: McGraw-Hill; 1980.

[6] Al-Hallaj S, Farid MM, Tamimi AR. Solar desalination with humidification–dehumidification cycle: performance of the unit. Desalination 1998;120:273–80.

[7] Farid MM, Al-Hallaj AW. Solar desalination with humidification–dehumidification cycle. Desalination 1996;106:427–9.

[8] Ben-Bacha H, Damak T, Bouzguenda M. Experimental validation of the distillation module of a desalination station using the SMCEC principle. Renewable Energy 2003;28:2335–54.

[9] Garg HP, Adhikari RS, Kumar R. Experimental design and computer simulation of multi-effect humidification (MEH)–dehumidification solar distillation. Desalination 2002;153:81–6.

[10] Nafey AS, Fath HES, El-Helaby SO, Soliman AM. Solar desalination using humidification–dehumidification processes. Part II. An experimental investigation. Energy Conversion Management 2004;45(7–8):1263–77.

[11] Nawayseh NK, Farid MM, Al-Hallaj S, Tamimi AR. Solar desalination based on humidification process. Part I. Evaluating the heat and mass transfer coefficients. Energy Conversion Management 1999;40:1423–39.

[12] Muller-Holst H. Solar thermal desalination using the multiple effect humidification (MEH) method. In: Book chapter, solar desalination for the 21st century; 2007. pp. 215–25.

[13] Al-Sahali M, Ettouney HM. Humidification dehumidification desalination process: Design and performance evaluation, Chemical Engineering Journal, 143, 257–26, 2008.

[14] Farid M.M., Parekh S., Selman J.R., Al-Hallaj S., Solar desalination with humidification–dehumidification cycle: mathematical modeling of the unit. Desalination; 151:153–64, 2002.

[15] Yamali C, Solmus I. Theoretical investigation of a humidificationdehumidification desalination system configured by a double-pass flat plate solar air heater. Desalination 205, 163–177, 2007.

[16] Zamen M, Amidpour M, Soufari SM. Cost optimization of a solar humidification–dehumidification desalination unit using mathematical programming. Desalination 239, 92–99, 2009.

[17] Dai Y.J, Wang RZ, Zhang HF. Parametric analysis to improve the performance of a solar desalination unit with humidification and dehumidification. Desalination; 142:107–18, 2002.

[18] W.F. He, L.N. Xu, D. Han. Parametric analysis of an air-heated humidification-dehumidification (HDH) desalination system with waste heat recovery. Desalination 398 (2016) 30–38.

[19] Cihan Yıldırım,Ismail Solmus. A parametric study on a humidification–dehumidification (HDH) desalination unit powered by solar air and water heaters. Energy Conversion and Management 86 (2014) 568–575.

[20] S.W. Sharshir, Guilong Peng, Nuo Yang, Mohamed A. Eltawil, Mohamed Kamal Ahmed Ali, A.E. Kabeel. A hybrid desalination system using humidification-dehumidification and solar stills integrated with evacuated solar water heater. Energy Conversion and Management 124 (2016) 287–296.

[21] Nabil A.S. Elminshawy, Farooq R. Siddiqui, Mohammad F. Addas. Experimental and analytical study on productivity augmentation of a novel solar humidification–dehumidification (HDH) system. Desalination 365 (2015) 36–45.

[22] Wang GQ, Yuan XG, Yu KT, Review of mass transfer correlation for packed columns. Ind. Eng. Chem. Res. 2005, 44, 8715-8729.

[23]Shulman, HL, Ullrich CF, Proulx AZ, Zimmerman JO, Performance of Packed Columns. 2. Wetted and Effective Interfacial Areas, Gas- and Liquid-Phase Mass Transfer Rates.AIChE J. 1955, 1, 253.

[24]Shulman, H. L.; de Gouff, J. J. Mass Transfer Coefficients and Interfacial Areas for 1-Inch Raschig Rings. Ind. Eng. Chem.1952, 44, 1915.

[25] Van Krevelen, D. W.; Hoftijzer, P. J. Studies of Gas Absorption. 1. Liquid Film Resistance to Gas Absorption in Scrubbers. Recl. Trav. Chim. Pays-Bas. 1947, 66, 49.

[26] Van Krevelen, D. W.; Hoftijzer, P. J. Kinetics of Simultaneous Absorption and Chemical Reaction. Chem. Eng. Prog. 1948, 44, 529.

[27]Mangers, R. J.; Ponter, A. B. Effect of Viscosity on Liquid Film Resistance to Mass Transfer in a Packed Column. Ind. Eng. Chem. Process Des. Dev. 1980, 19, 530.

[28]Shi, M. G.; Mersmann, A. Effective Interfacial Area in Packed Columns. Ger. Chem. Eng. 1985, 8, 87.

[29] Billet, R.; Schultes, M. Advantage in Correlating Packing Column Performance. Inst. Chem. Eng. Symp. Ser. 1992, 128, B129.

[30] Onda, K.; Takeuchi, H.; Okumoto, Y. Mass Transfer Coefficients between Gas and Liquid Phases in Packed Columns. J Chem. Eng. Jpn. 1968, 1, 56.

[31] Eckert, J. S.: Chem. Eng., 82, 70. Apr. 14, 1975.

[32] Mostafa H. Sharqawy, John H. Lienhard, Syed M. Zubair. Thermophysical properties of seawater: a review of existing correlations and data. Journal of Desalination and Water Treatment. 16 (2010) 354–380.

[33] J.D. Isdale and R. Morris, Physical properties of sea water solutions: density, Desalination, 10(4) (1972) 329–339.

[34] Francis L. Smith and Allan H. Harvey (September 2007). "Avoid Common Pitfalls When Using Henry's Law". CEP (Chemical Engineering Progress). ISSN 0360-7275.

[35] Feoktistov, V. (2006). Differential Evolution: In Search of Solutions. Springer. ISBN 978-0-387-36895-5.

[36] H. Bahmanyar, Mass Transfer. Tehran University's Jahad Daneshgahi Publication, 2003.

462318 Improved Olefin Polymerization Process with Reduced Energy Consumption

Abstract:

Improved Olefin Polymerization Process With Reduced Energy Consumption

Carrying out olefin polymerization reaction in solution has some distinct advantages such as a homogenous environment to facilitate consistent fast reaction; solvent acting as natural heat sink to absorb heat of polymerization, and fast product grade transition. On the other hand, the need to remove and recover large amount of solvent used in solution polymerization reaction, and the requirement to meet target solvent residues in the final product, greatly increases the capital expenditure and energy consumption of the designed asset.
The conventional method to remove solvent in a solution polyolefin process is by heating up the polymerization reactor effluent to desired temperature, then flashing off the solvent through vapor liquid equilibrium (VLE) using multiple devotalization stages at different operating pressures, which normally require deep vacuum in the final vapor liquid separator. The amount of solvent that needs to be processed by the devotalization and solvent recovery section of the plant dictates the maximum production throughput, and leads to higher energy cost than other alternative olefin polymerization process.

The approach of applying Lower Critical Solution Temperature (LCST) behavior of polymer solution2, and subsequently inducing a "front end" liquid-liquid separator of polymer solution followed by one or two finishing vapor liquid separator, has been explored in the recent years. Under carefully designed temperature, pressure and solvent environment, a homogenous polymer solution may be split into two liquid phases: a polymer rich phase and a solvent rich phase. This almost isothermal separation process1 has some unique characteristics that can be advantageous to designing an energy efficient solution polyolefin manufacturing process. However, achieving sufficient and reproducible liquid-liquid separation can be difficult if the solvent composition does not possess the needed high volatility, or the system temperature is not substantially higher than the Lower Critical End Point (LCEP) of the reactor effluent.

This presentation explored the key thermodynamic aspects of polymer solution Liquid-Liquid Equilibrium (LLE), and described a unique approach to achieve effective liquid-liquid separation. Unlike the conventional process of using solvent with a fixed boiling point, It applies a mixed solvent composition of one high boiling point solvent, and a low boiling point solvent. The low boiling solvent acts as the anti-solvent that significantly alters the LCST behavior of the polymer solution and reduces the LCEP of the system by 40 ~ 60 °C. This mixed solvent approach enables effective polymer-solvent separation inside the liquid-liquid separator without the need to put out energy to heat up the polymer solution stream exiting the reactor. In addition, the amount of low boiling solvent (anti-solvent) can be adjusted and used as a control knob to ensure a consistent liquid-liquid separation outcome (i.e., similar polymer concentration in the polymer rich phase). The amount of low boiling solvent can also be used to adjust for changes in compositions or reactor conditions (e.g. different product design or buildup of inerts).

This new liquid-liquid separation based Olefin Polymerization Process3 brings a 35% reduction of the total plant energy input as compared to conventional technology using vapor liquid separator alone, because heat is added only to the polymer-rich phase that is downstream of the Liquid-Liquid Separator (LLS). In addition, the cooling requirement for the solvent-rich phase downstream from the LLS is also lower as no heat was used to heat up the reactor effluent. Moreover, this new design also brings capital savings as it enables the use of lower cost adiabatic reactors operating at lower polymer concentration.

Reference:

1. TG. Gutowski, NP Suh, C Cangialose, and GM Berube, "A Low-Energy Solvent Separation Method," Polym Eng & Sci 23:230-237, 1983
2. Freeman, P.I. and J.S. Rowlinson, Lower critical points in polymer solutions. Polymer, 1960. 1: p. 20-26.
3. P. Jog, Midland, R. Swindoll, N. Mead, P. Jain, A. Wang, J. Guzman, United States Patent 9018328, Polymerization process for olefin-based polymers.

472392 On the Dynamics of Non-Coalescing Microscale Bubbles in Turbulent Flows for Intensification of Gas-Liquid Mass Transfer

Abstract:

In many gas–liquid contactors or reactors, mass transfer from the gas phase to the liquid phase is a major player in the reactor's overall production and/or economics. The mass transfer rate can be enhanced either by a) increasing interfacial area *a*, b) intensifying the mass transfer coefficient k_L, or c) increasing the solubility of gas via higher pressure or complexing reagents. All three of these beneficial phenomena may be leveraged simultaneously by use of micron-scale bubbles. Therefore, to intensify reactor mass transfer rate, we studied the behavior of microbubbles (< 200 micron) in a 100L bubble column and in a 500 mL bench top column. Specifically, the stability of microbubbles against coalescence was studied under additions of surfactant (sodium dodecyl sulfate (SDS) [$CH_3(CH_2)_{11}OSO^{-3}$ Na^+]) and salt (KCl), both of which are known to decrease bubble coalescence (Parmar & Majumder 2014). According to previous literature, adding salt to the liquid phase in the presence of SDS has two different effects on the bubble surface tension and the hydrostatic repulsion between bubbles (Prosser & Franses, 2000): a) Na^+ ions allow closer packing of surfactant molecules at the gas-liquid interface, which diminishes surface tension because surface tension is inversely proportional to the density of surfactant molecules at the surface, and b) decreases the repulsion between bubbles because the electrostatic Debye length decreases. Depending on the salt concentration, the stability can be changed from stable to unstable. Low concentrations of SDS (0-35) ppm w/w and KCl (0-75) ppm w/w were added to the solution. We observed that the presence of SDS up to 35 ppm w/w and KCl up to 75 ppm w/w did not alter surface tension by more than a few percent. However large change was observed in mean bubble size, k_La, and bubble coalescence rate. It was hypothesized that the stability of microbubbles was not from the slight decrease in the surface tension but from an increase in zeta potential. Zeta potential (ζ) was measured by high-speed video of the time-dependence of bubble motion inside a 50 cm tall bubble column with an oscillating electric field applied across the bubble column, similar to a method used previously by Takahashi (2005). Bubble size and coalescence rate were also measured. Results showed a significant increase in ζ-potential of bubbles as SDS and KCl concentrations were increased. An energy balance confirms ζ-potential to be the leading factor for decreased coalescence rates and decreased bubble size. In a test reactor, k_La and *a*values increased both by 50% as SDS and salt concentrations was increased, likely due to an increase of interfacial area.

References:

Parmar, R. & Majumder, S. K. 2014 Hydrodynamics of microbubble suspension flow in pipes. Ind. Eng. Chem. Res. 53, 9, 3689–3701.

Prosser, A.J. & Franses, E.I., 2000 Adsorption and surface tension of ionic surfactants at the air–water interface: review and evaluation of equilibrium models. Colloids and Surfaces A: Physicochemical and Engineering Aspects. 178, 1–40.

Takahashi, M. 2005 ζ Potential of Microbubbles in Aqueous Solutions: Electrical Properties of the Gas-Water Interface. J. Phys. Chem. B. 109, 21858-21864.

468032 Mathematical Modelling of Liquid â€“Liquid Extraction in the Slug Flow Regime in a Microchannel

Abstract:

Slug flow is a commonly observed flow regime in the microchannel. Mixing in the slug flow regime is enhanced by internal circulations induced by shearing due to wall. This helps improve mass transfer in this flow regime.

In this work, we exploit the low Reynold's number characteristic of the flow and seek a numerical solution to understand the structure of the vortex patterns formed in the two phases in the slug flow regime. Further, we study liquid-liquid extraction in the system and determine the improvement in mass transfer performance. A Lagrangian approach where we analyze the behaviour of a unit cell in the channel in a moving reference frame is adopted for the analysis. The system is analysed for two cases when there is (i) no film surrounding the slug (ii) a thin film surrounding the slug. The 2D governing equations for fluid flow are solved using two approaches: a) a stream function formulation based on finite differences, b) primitive variable formulation with the Chebyshev collocation method. Results obtained from the two methods are validated with each other. Then, the species balance equation is solved numerically using a finite volume code. The extraction performance is analysed in terms of extraction efficiency and mass transfer coefficient. It is believed that the enhancement in mass transfer is due to the mixing induced by the strong internal circulation set up due to the channel wall. The model is validated with published experimental results.

464382 Mass Transfer Enhancement as a Function ofÂ the Oscillatory Baffled Reactor Design

Abstract:

The oscillatory baffled reactor (OBR) is an intensified design of plug flow reactor (PFR). It is able to achieve a high degree of plug flow at very low flow rate due to the interaction between the oscillatory flow and baffles which improves mixing inside the OBR. It was found that superimposing fluid oscillation in a baffled column enhanced the mass transfer significantly comparing with many devices, for instance, bubble column, internal and external loop column, and stirred tank. This is due to the highly turbulent flow pattern that generated in the OBR as a result of the oscillatory motion which leading to frequent bubble breakage, a subsequent reduction in the Sauter mean diameter, an increase in bubbles retention, when trapped in recirculating vortices, and the gas-liquid contacting area

Recently, oscillatory baffled column (OBC), has used for the same purpose, mass transfer, and presented a significant change in which has been increased up to sixfold comparing with a bubble column. This was due to the combanation effect of baffles and oscillatory motion that increase the radial mixing (diffusion). Also gas-holdup affected by the oscillation frequency and amplitude.

In this study, the volumetric mass transfer coefficient ($k_L a$) in air-water system has been calculated and compared using three designs of OBRs, smooth periodic constrictions (SPC) design, helical baffled design, and single orifice design. This work was conducted over a wide range of oscillation condition and aeration rate, volumetric air flow rate. The SPC presented higher values of $k_L a$ than the single orifice baffled reactor and the helical baffled reactor. This indicates that SPC with oscillatory motion provide bigger interfacial contact area between air and water. In addition, the visualisation of bubbles movements inside the three designs of OBR has shown that the bubbles collision and breakage in SPC was more severe. Therefore the bubble size in SPC was smaller than those that achieved in the single

orifice OBR then in the helical OBR.

455522 Isolation of Carboxylic Acids from Dilute Aqueous Solutions – Intensifying Approach By Conversion in Emulsified Reactive Solvents

Abstract:

The focus on the production of biobased bulk chemicals as well as the progress in the biorefinery require innovative downstream processing of the product effluents. Examples are provided by different cellulose pulping processes with varying yields of carboxylic acids (e.g. acetic acid or formic acid). The effluents from such processes are usually characterized by low concentration of the constituents in aqueous solution. Conventional separation techniques quickly reach economic limits. The requirements for downstream processes capable of dealing with such feed specification, must suffice effective separation to ensure economic utilization of the valuable constituents.

The intensification of conventional separation techniques by combining with chemical reaction provides high potential in overcoming these obstacles. Two approaches, reactive distillation for the isolation of carboxylic acids and chemical conversion of the target constituents in both phases of a liquid-liquid extraction process, may offer a feasible option. Reactive conversion of choice for isolation of carboxylic acids is the esterification with an aliphatic alcohol, whereby the market value of the carboxylic acids can be significantly increased and the significant change of physical properties favorably influences the separation process.

As the partition behavior of the reactants is supportive when using higher aliphatic alcohols (e.g. *n* butanol or *1*-octanol), the use of these alcohols, either dissolved in corresponding esters or hydrocarbon based solvents has been investigated. Catalysis and elevated temperature are provided to increase the reaction rate of the esterification reaction. *4* dodecylbenzenesulfonic acid (4DBSA), a strong anionic surfactant and highly active homogeneous catalyst, has been used in the experimental validation in order to highly increase the contact between the reactants and the catalyst by emulsification. After having finished the reaction the emulsion is separated from the aqueous phase by settling and split by coalescence. A continuous process has been designed by altering conversion, phase separation, emulsion splitting and isolation of the product. Exemplarily 80% separation efficiency was obtained for acetic acid separation from aqueous feed by esterification with octanol in the emulsified solvent phase in single stage isolation.

460057 Bioconversion of Natural Gas to Higher Value Chemicals: Engineering Considerations for Commercialization

Abstract:

The low cost and abundant supply of natural gas have been motivators for the development of many technologies that convert this low cost carbon feedstock into compounds of greater value. Natural gas bioconversion is a developmental technology which applies synthetic biology to a naturally methane consuming bacteria, a methanotroph, to program it to produce higher value materials of interest. Intrexon has formed joint ventures to commercialize methanotrophic bioprocesses for biofuels and high value chemical intermediates. Biocatalytic processes which achieve high yields under milder processing conditions allow for reduced capital and operation expenditures while also offering greater economic viability. Currently, the production of both isobutanol and farnesene has been demonstrated in lab scale fermentation. To successfully bring this technology to commercial scale, consideration must be given not only to optimizing the metabolic pathway engineered into the methanotroph; but to key issues in the fermentor design, downstream processing and overall plant economics.

At any scale, aerobic natural gas fermentation must overcome the challenge of gas to liquid transfer of low solubility gases. Low gas to liquid mass transfer reduces the process productivity and yield, which in turn increases fermentor cost, process operating cost and may add additional equipment necessary to recycle gases. There are many fermentor designs currently available to increase the overall mass transfer coefficient of these gases; however, capital and operating expenditures of the reactor become a critical metric for evaluating commercialization feasibility. Unique fermentor configurations can greatly increase capital cost by raising the cost of fabrication and decreasing the ability to receive competitive bids. These unique fermentors also may not create a homogenous well-mixed reactor, which increases the cost of controls and instrumentation.

Downstream process selection for commercialization is a balance of capital cost, operating cost and the risk of cutting-edge over well-established techniques. Low fermentation broth concentrations of product increase the size and energy requirement of downstream equipment. Cutting-edge technologies can increase the efficiency of separation and reduce energy inputs, but these technologies can be capitally intensive and unproven at commercial scales. Bioprocesses also create metabolic byproducts which need to be evaluated for separation and purification to meet specifications for sale.

Favorable plant economics are necessary to move forward with commercialization of a technology. Scale is carefully selected based on a comparison of the production rate, fermentation productivity and yield targets to revenue, capital and operating costs. The plant location is also an important selection factor as costs are affected by utility rates in the area and/or co-location of the plant with shared infrastructure.

Traditional engineering workflows, such as a front-end loading strategy, were used to shape the commercial scale process design. The biologic, technical and economic commercial feasibilities of the end-to-end natural gas to isobutanol process were critical to Intrexonâ€™s plant design. Laboratory scale learnings were applied to select technology that was commercially feasible at demonstration scale. This demonstration scale design was scaled down to pilot scale, where the operation will yield learnings into refinement of at-scale design, equipment selection, empirical data on heating and cooling, and controls strategies.

474172 Be an Engineering Detective--Technical Assessments for Process Licensing

Abstract:

Licensing of chemical process, equipment, and catalyst technologies is an important part of the chemical manufacturing business. For each technology there can be multiple providers to select from. We will explore methods of how to evaluate them. Questions about how to find the "best" technology, or "lowest cost" technology are nontrivial to answer and much work needs to be done to get accurate and confident answers. Throughout the process we try to exercise risk reduction to potentially influence and improve the benefits of the plant and to identify any "Unproven" technology.

470115 Tablet Compression Scale-up Risk Assessment Using a Model Based Approach

Abstract:

Scale-up of processes involves uncertainties from many sources: material properties, equipment design, operation parameters, etc. The use of several models, even simple models, can constrain risks and predict failures before they occur. This talk will focus on the use of modeling in conjunction with small scale experimentation as a predictive methodology to estimate and reduce risks due to the scale-up of the tablet compaction process.

471157 Best Practices Identified While Transferring Lyophilized and Syringe Biologic Drug Products from Internal Site to a CMO

Abstract:

In this presentation, lessons learned while transferring Lyophilized and Syringe biologic drug products to a new site will be shared. The goal is to go over the critical quality attributes of the selected commercial drug product and identify best practices to maintain those attributes in the new commercial site. During the technology transfer to a second commercial site post-approval, the firm might find opportunities to improve the current manufacturing process and face challenges to implement these changes due to regulatory requirements. This presentation will cover some of those challenges as well.

472624 Minimizing Risk of Upgrading Projects on Existing Assets

Abstract:

As is the case with any upgrading projects, the addition of new units/equipment introduce uncertainties and unforeseen risk to the existing assets. In DNV GL's work processes, methodologies incorporating cutting edge technology have been developed in order to measure risk and optimize performance. One of these methodologies that will be presented in this paper will quantify and minimize the risk of the upgrading project on the existing assets as well as the risk of the existing assets on the new units in order to maximize profit.

The main objective of this work is to apply reliability, availability, and maintainability (RAM) study to optimize the performance (from production standpoint) of the integrated project. Specifically, RAM analysis will be conducted to provide a quantitative assessment of the level of performance that may be expected from the system as a whole over its design life. DNV GL will use TARO software to execute the proposed methodology.

In order to illustrate the applicability of the proposed approach, a case study addressing the major risks of an upgrading projects will be evaluated. This case study will compare the as-is performance to the future performance by evaluating the availability and utilization of all the production-critical units and their impact on production. This analysis will assist in identifying the major units contributing to production losses. In addition, the case study will quantify the impact of the increase in utility system demand (Hydrogen, steam, power...etc.) on production and identify any bottlenecks. Finally, the case study will quantify the impact of holding tanks working volumes on production and existing assets.

The results of this study will improve stakeholder confidence in the proposed design and ensure that the integrated project yields maximum return on investment. The proposed methodology will provide managers with decision support in design optimization and integration, process debottlenecking, benchmarking of performance, and mitigation of unforeseen risks.

Process Development Division

(501g) Managing Technology Transfer Risk through Predictive Sciences

Stamato, H., Bristol-Myers Squibb

Development and transfer of products can be improved and accomplished with a fewer risks by employing a focus on predicting the behaviors of materials and the processes acting on them. Looking at the process at multiple scales of action and examining the underlying material/machine interactions small scale tests and simulations can be used to more accurately predict the performance of the final process and product. The presentation will discuss advances in applying both test and modeling to reduce risk, experimentation and help assure success.

468367 Halophyte Biochar for Water Desalination Brine Concentrate Management

Abstract:

Disposal of concentrate from brackish water desalination is limited by concerns of salt leaching into and contaminating surface and ground waters. These concerns, and measures to prevent leaching, add substantially to concentrate disposal costs with little or no opportunity to recover those costs. One option being explored is the use of concentrate as irrigation water for producing halophyte biomass for forage and other applications. Continuing development of this option requires more knowledge about what happens to the salt in the concentrate once land applied as irrigation water.

The purpose of this project is to determine how much of the salt taken up by the halophyte crops can be protected from leaching by pyrolysis of the biomass, i.e. to determine the potential of sequestering salt in mineral form. Results from this research will be useful for those managing high salinity biomass streams, such as halophyte crop wastes and animal manures, and those seeking local concentrate disposal options with a potential for producing economic and/or environmental returns. The ability to sequester salt in biochar would decrease the negative impacts of land-applying high-salinity materials, including those used in desalination concentrate management systems.

455179 ^{13}c Pathway Analysis for the Role of Formate in Electricity Generation By *Shewanella Oneidensis* MR-1 Using Lactate in Microbial Fuel Cells

Abstract:

Microbial fuel cell (MFC) is a promising technology for wastewater treatment. Basically, the microorganisms sticking on the anode of the MFC utilize various organic compounds from the wastewater as the electron donors to generate electricity, while in the meantime removing the organic compounds in the wastewater. The type of electron donors fed into MFCs affects the electrical performance, and mechanistic understanding of such effects is important to optimize the MFC performance. In this study, we used a model organism in MFCs, *Shewanella oneidensis* MR-1, and ^{13}C pathway analysis to investigate the role of formate in electricity generation and the related microbial metabolism. Our results indicated a synergistic effect of formate and lactate on electricity generation, and extra formate addition on the original lactate resulted in more electrical output than using formate or lactate as a sole electron donor. Based on the ^{13}C tracer analysis, we discovered decoupled cell growth and electricity generation in *S. oneidensis* MR-1 during co-utilization of lactate and formate (i.e., while the lactate was mainly metabolized to support the cell growth, the formate was oxidized to release electrons for higher electricity generation). To our best knowledge, this is the first time that ^{13}C tracer analysis was applied to study microbial metabolism in MFCs and it was demonstrated to be a valuable tool to understand the metabolic pathways affected by electron donors in the selected electrochemically-active microorganisms.

461312 *Advanced Redox-Based Electrochemical Separations for Wastewater Treatment*

Abstract:

Water scarcity is one of the principal challenges for the 21st century, both from a geographical and an economic standpoint. Developing novel, more energetically efficient technologies for wastewater treatment, retreatment and reuse is crucial for long-term water sustainability. One of the principal challenge for water remediation, both in the U.S. and worldwide, is to target contaminants of emerging concern, such as complex organic molecules including pharmaceuticals, chemical precursors, endocrine inhibitors and pesticides.[1] These originate from chemical processes and organic synthesis, the food industry, electronic manufacturing among others. Many of these pollutants disperse in remote locations, such as agricultural locations, in which large-scale purification systems cannot be implemented. Electrochemical methods are particularly attractive for addressing these problems due to their modularity, fast kinetics and integration with renewable energy sources.[2]

We have developed an redox-mediated separation method based on organometallic-functionalized electrodes (e.g. ferrocene or cobaltocene) for the sorption and release of specific ions from the liquid-phase under oxidizing and reducing conditions.[3] Many of the organic contaminants are anionic in nature (e.g. carboxylates, sulfonates or phosphonates), thus selectivity based on chemical interactions with their functional groups is essential in chemical design. Rather than rely solely on charge or size, our immobilized redox-centers targets these compounds through the activation of a specific chemical binding. Moreover, in the charging and discharging steps, electrochemical charge can be stored by ion-selective adsorption onto the porous electrodes, thus dramatically decreasing energetic costs.

This modular system can then be easily implemented at the point-of-exit in chemical manufacturing and wastewater treatment. Here, we present a complete asymmetric redox-system in both batch and flow for (1) water remediation, through the specific separation of anionic and cationic contaminants in water, (2) tandem energy recovery due to pseudocapacitive charge storage and (3) increasing sustainability in chemical processes, through the recovery of expensive synthesis products from complex organic reactions, thus reducing organic waste and decreasing water usage during separation steps.

References

[1] a) Gilliom R.; Hamilton P. *U.S. Geological Survey*. 2014. (USGS fact sheet: http://pubs.usgs.gov/fs/2006/3028/

). b) Doughton, C.G. and Thomas, A. T. (1999). Pharmaceuticals and personal care products in environment: agents of suble change? *EHP* **107** (6), 907-938.

[2] Suss, M. E.; Porada, S.; Sun, X.; Biesheuvel, P. M.; Yoon, J.; Presser, V. *Energy & Environmental Science* **2015**. Zhao, R.; Satpradit, O.; Rijnaarts, H. H.; Biesheuvel, P. M.; van der Wal, A. *Water Res* **2013**, *47*, 1941.

[3] **Su, X.**; Kulik, H; Jamison, T.F.; Hatton, T. A. 2016. Anion-selective redox electrodes: electrochemically-mediated separation with organometallic interfaces. *Advanced Functional Materials*. Advance Article Online. DOI: 10.1002/adfm.201600079

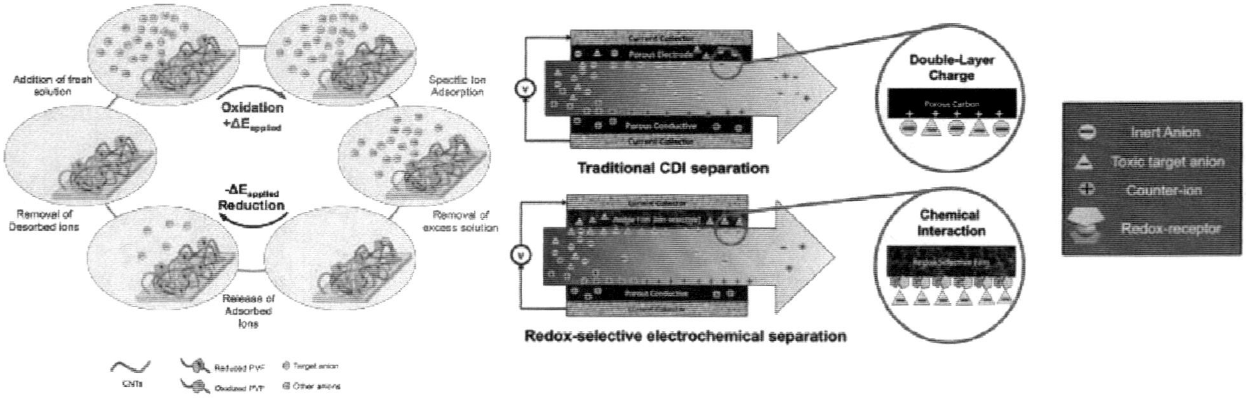

458179 Influence of Elevated Salinity and the Presence of Cytostatic Drugs on Microbial Community in FO Anaerobic Membrane Bioreactors

Abstract:

Wastewater treatment using foward osmotic anaerobic membrane bioreactors (FO-AnMBR) has a number of advantages over traditional wastewater treatment approaches. Previous studies have investigated the overall performance of lab-scale FO-AnMBR in treating synthetic wastewater. During the operation of FO-AnMBR, it is expected that salinity gradually increases, which has also been shown experimentally. In addtion, emerging contaminants such as cytostatic drugs in wastewater would also be accumulated in the FO-AnMBR. Little has been known on whether and how the increased salinity and the presence of cytostatic drugs affect the microbial community in FO-AnMBR. The objective of this study was to investigate the influence of increased salinity and the presence of cytostatic drugs on the structure and function of the microbial community in FO-AnMBR through a metagenomics approach. A lab-scale FO-AnMBR was operated at 25 ^{o}C and fed with synthetic wastewater. Total DNA was extracted from sludge samples taken at pre-determined time points and sequenced using a HiSeq 2500 platform with 250-bp paired-end sequences. Based on the analyses of the sequences, structure and potential functions of the microbial communities were obtained. We found that the increased salinity in FO-AnMBR did not reduce the microbial diversity; however, the relative abunance of methane-producing archaea decreased significantly at high salinity condition, resulting in a reduced methane yield. The general metabolism did not differ significantly under low and high salinity conditions. Intriguingly, genes responsible for cell attachment and biofilm related signalling were enriched at high salinity condition. Transition of the gene clusters for assimilatory and dissimilatory nitrate reduction was also found. The presence of cytostatic drugs only slightly affected the microbial community composition. Biofilm-forming, multidrug resistant and siderophore-producing bacteria were found more dominant after the treatment with drugs.

471970 Batch and Packed-Bed Adsorption Study for the Removal of Dye from Textile Wastewater Using in-House Prepared Silica-Based Nanosorbcats: Kinetics and Computational Modeling

Abstract:

The worldwide shortage of fresh water and the huge competing demands from variety of users stimulates an urgent need for finding alternative cost-effective and environmentally friendly wastewater treatment processes. For instance, synthetic dyes wastewater discharged from the textile and dyestuff industries poses an environmental alert since these organic pollutants are toxic to many aquatic and non-aquatic living organisms. In addition, some of these pollutants are non-biodegradable, and thus, they will exist for a long time in the environment, which may cause a real challenge to the conventional wastewater treatment processes. Accordingly, innovative and environmentally sound techniques are needed. The application of nanoparticle technology as adsorbents and catalysts (nanosorbcats), whether as a standalone or as an enabling technology, in cleaning up wastewater has recently received great attention. This is because of the unique chemical and physical properties of nanoparticles in comparison with their counterparts, which make them superior to the conventional adsorbent/catalysts.

Accordingly, the present study investigated the employment of newly in-house prepared silica-based nanosorbcats functionalized with active species for cleaning up textile wastewater. A facile co-precipitation synthesis route was used to prepare those nanosorbcats, which were characterized by different characterization techniques like XRD, BET, TPD, TPR, HRTEM and zeta potential. The prepared nanosorbcats were employed for the adsorptive removal of cationic and anionic dyes from synthetic and real textile wastewater. Batch adsorption studies were performed for kinetics and equilibrium modeling on the removal of different dyes from wastewater. The operating variables studied were contact time, initial dye concentration, and solution pH. Results showed an optimum pH value, where the maximum removal of dye was achieved. The experimental adsorption data have been modeled using Sips and BET adsorption isotherm models. In addition, a computational modeling on the selected model compounds was carried out to understand the adsorption mechanism on the prepared nanosorbcats. Breakthrough capacities were also investigated in a packed-bed column study. The most successful silica-based nanosorbcats in cleaning up the real textile wastewater in the batch experiments were prepared in extrudate form and used in up-flow packed-bed adsorption experiments. The effects of the influent concentration, flow rate, and bed height were investigated. Breakthrough time and adsorption capacity of the packed-bed increased with increasing bed height, whereas decreased with the increase in influent initial concentration and flow rate. The reliable results of this study will be elaborated and discussed further during the presentation.

463610 Optimization of Operational Parameters for Application of Zirconium/Polyvinyl Alcohol Modified Flat-Sheet Polyvinylidene Fluoride Membrane As Better Adsorptive Material in Treatment of Organic Arsenic Containing Wastewater

Abstract:

Organic arsenic in waters has been a global concern due to its toxicity. In this study, a novel polyvinyl alcohol (PVA)/zirconium modified polyvinylidene fluoride (PVDF) membrane was applied to remove organic arsenic from water. The impregnation of zirconium ions within the modified membrane was attributed to the coordination reactions among the zirconium ions, ether and hydroxyl groups. The synthesized membrane worked better at the acidic conditions and achieved the optimal uptake for both monomethylarsonic (MMA) and dimethylarsinic (DMA) at pH 2.0. The adsorption isotherm study demonstrated that the adsorption of both organic arsenic species was controlled by the mono-layer adsorption process, and the maximum adsorption capacities of 73.04 and 29.78 mg/g for MMA and DMA can be achieved at pH 2.0. The presence of humic acid had a negligible impact on the uptake of organic arsenic, whereas varying impacts on the arsenic adsorption were observed due to the presence of coexisting anions such as fluoride, phosphate, carbonate and silicate. The filtration study was also carried out to evaluate the performance of the modified membrane on the removal of organic arsenic. Based on the XPS analyses, the exchange between chloride ions on the membrane surface and organic arsenic species should be responsible for the removal of both MMA and DMA.

450795 Realization of Sustainable and Cost-Effective Cellulosic Ethanol Production By Deploying Enzyme and Advanced Yeast Technologies in an Integrated Approach

Abstract:

Ethanol is considered the most potential next generation automotive fuel because it is carbon-neutral and could be produced from renewable resources like lignocellulosic biomass. Key to economic viability of cellulosic ethanol is cheap production and application of different enzymes hydrolyzing lignocellulosic carbohydrates into fermentable sugars and efficient conversion of the derived monomeric sugars into ethanol at high productivities and yields using advanced yeast strains. DSM is a frontrunner in the development of bioconversion and fermentation technologies for the production of cellulosic ethanol, enabling the industry to diversify from starch crops to lignocellulosic agricultural residues, unlocking the full industrial potential of this cheap, abundantly available and sustainable type of feedstock.

Within DSM, research projects aim at developing both commercially competitive cellulolytic enzymes for the liquefaction and saccharification processes, and genetically engineered yeast strains that ferment hexoses and pentoses to ethanol. DSM's enzyme cocktail is uniquely active at high temperature and low pH, offering many benefits and product differentiation. Further strain and process development, enzyme cocktail ratio optimization, protein and genetic engineering have led to significant improvements in productivity, efficiency and costs. Fermentation of hexoses and pentoses from lignocellulosic hydrolysates by DSM's yeast strains has been significantly improved by applying both rational and evolutionary engineering strategies, resulting in C6- and C5-sugars fermenting *Saccharomyces cerevisiae* strains being commercially available. Additionally, next generation advanced yeast strains, which have accelerated pentose utilization, improved robustness, as well as the ability to convert inhibitors and specific biorefinery waste products into additional ethanol, are currently finding their way to commercial implementation.

In this context, it is of great importance to fully integrate the design, construction and operation of the aforementioned technologies into a biorefinery for biofuels production. In this presentation, we will discuss the latest developments and how those fit within the platform of on-site manufacturing of enzymes and yeast fermentation.

468107 Alternative Transport Fuels and Their Production Using Surplus Renewable Electricity, Water and CO_2

Abstract:

Electrical power generated from wind energy fluctuates strongly. As put forward by the EU Commission, in 2050 80 % of the electricity demand should be covered by renewables. Therefore, the integration of fluctuating electricity generation from renewables plays a fundamental role in Europe's and especially Germany's future energy system. Additionally for 2050, the EU Commission postulates a reduction of anthropogenic CO_2 emissions of about 80-95 %, compared to 1990. In Germany, the transport sector was accountable for approximately 20 % of the domestic CO_2 emissions in 2014, while contributing 30 % to the domestic energy demand. One reason for this high share is that 90 % of the transport sector is run on fossil fuels. Hence, to achieve the goals for 2050, non-fossil alternative fuels need to be found. A popular option is the further integration of fuel cell cars. However, especially for heavy duty vehicles and planes, there is long term need for liquid fuels such as diesel and kerosene. Another promising option is the idea of Power-to-Fuel (PTF): the synthesis of fuels based on sustainably produced hydrogen and usage of CO_2 as a feedstock. The energy required to run those processes can be provided by surplus electricity from renewable sources. This is energy which cannot be fed into the grid when there is no demand on the consumer side.

If an alternative renewable fuel is able to penetrate the market, two main requirements need to be met: its potential for an implementation strategy into an existing market as well as a technically feasible and economic production. Put differently, a renewable fuel should not require a significant alteration of the fuel infrastructure, thereby allowing a gradual integration in the existing infrastructure. Additionally, a range of different fuels for different usages need to be offered. Just as the fossil fuel market, the renewable fuel market needs diversification.

At the Institute of Energy and Climate Research (IEK-3) of Forschungszentrum Jülich, intensive research is being carried out on a comparable overview of the several products and process routes of PTF. In this contribution, investigations on two different routes from synthesis gas to an alternative renewable fuel will be outlined. The respective products of the two outlined processes are Fischer-Tropsch (FT)-diesel and polyoxymethylene dimethylene (POMDME). Both products can be blended with conventional diesel and with each other because they have appropriate compatibility. Therefore, an implementation strategy into the existing fuel market is practicable.

Despite an equal product application, the comparison of both technologies is widely diversified. At the molecular level, FT-diesel is closely related to conventional diesel; this applies less to POMDME. POMDME also has a considerably lower heating value; however, it has great potential as a fuel due to its beneficial combustion properties.

Although nowadays FT is not used for PTF in large scale, state-of-the-art FT-processes have been used industrially for many decades. In contrast to FT, owing to low product demand, processes for POMDME have not yet been developed on an industrial scale. Both processes have various options for the interconnection of system components and for complete process control. The system's interconnections as well as the process parameters have influence on raw material and energy demand and also product quality, co- and byproducts. Just as the product specifications, the process analysis results will play an important role in the fuel implementation strategy.

462845 Co-Products from the Catalytic Fast Pyrolysis of Biomass

Abstract:

Catalytic fast pyrolysis (CFP) of biomass is a potentially viable process for producing drop-in, renewable hydrocarbon transportation fuels and the economics may be improved by providing alternative pathways to valuable co-products. This provides flexibility to a biorefinery to adjust conditions to improve economics or to valorize waste low value streams. As an example, CFP using zeolite catalysts such as HZSM-5 produces aromatic molecules with 1-4 rings. The single ring products are excellent additives for gasoline because of their high octane numbers and their high heats of combustion. However, the larger aromatics have boiling points too high for gasoline and low cetane numbers, eliminating them as potential diesel additives. Hydrotreating these molecules is required to convert them to a fuel, but they may have value as co-products. Likewise oxygenated intermediates and products, such as phenols, formed during CFP may be high value products that can also help the economics of hydrocarbon fuels. These oxygen functionalities are often difficult to add to hydrocarbon molecules extracted from petroleum and producing these molecules from biomass may be intrinsically more environmentally benign. In addition, leaving oxygen in the molecules may improve the yields of biofuels and bioproducts, which would again improve the economics. We will discuss studies to improve selectivity of co-products and efforts to separate them and prepare polymers.
Catalytic fast pyrolysis (CFP) of biomass is a potentially viable process for producing drop-in, renewable hydrocarbon transportation fuels and the economics may be improved by providing alternative pathways to valuable co-products. This provides flexibility to a biorefinery to adjust conditions to improve economics or to valorize waste low value streams. As an example, CFP using zeolite catalysts such as HZSM-5 produces aromatic molecules with 1-4 rings. The single ring products are excellent additives for gasoline because of their high octane numbers and their high heats of combustion. However, the larger aromatics have boiling points too high for gasoline and low cetane numbers, eliminating them as potential diesel additives. Hydrotreating these molecules is required to convert them to a fuel, but they may have value as co-products. Likewise oxygenated intermediates and products, such as phenols, formed during CFP may be high value products that can also help the economics of hydrocarbon fuels. These oxygen functionalities are often difficult to add to hydrocarbon molecules extracted from petroleum and producing these molecules from biomass may be intrinsically more environmentally benign. In addition, leaving oxygen in the molecules may improve the yields of biofuels and bioproducts, which would again improve the economics. We will discuss studies to improve selectivity of co-products and efforts to separate them and prepare polymers.

480817 Application of Molecular Characterization to Bituminous Crude Oil to Study Asphaltene Precipitation

Abstract:

Application of molecular characterization to bituminous crude oil to study asphaltene precipitation

Md Rashedul Islam, Yifan Hao, Meng Wang, Toni Kirkes, and Chau-Chyun Chen

Department of Chemical Engineering, Texas Tech University

Abstract

Bituminous crude oils are vulnerable to asphaltene precipitation when blended with paraffinic samples. Precipitation behavior depends on the nature and content of the paraffins in the blend. A thermodynamic modeling of asphaltene precipitation from crude oil is important as part of the larger effort to understand the compatibility of bitumen or heavy crude oils with the paraffinic ones. Recently, asphaltene precipitations are predicted from wide spectrum of solvents by aggregation thermodynamic framework [1] developed from Yen-Mullins model [2]. However, precipitation prediction from crude oils is still very difficult because of lack of reliable techniques to represent petroleum fluids with real molecules. In this work, a molecule based characterization approach [3] is utilized to represents Cold Lake bitumen with real hydrocarbon molecules. Additionally, asphaltene precipitations are predicted form mixing n-alkanes with Cold Lake bitumen.

References

[1] M. Wang, Y. Hao, M. R. Islam, and C.-C. Chen, "Aggregation thermodynamics for asphaltene precipitation," *AIChE Journal,* vol. 62, pp. 1254-1264, 2016.

[2] O. C. Mullins, "The Modified Yen Model†," *Energy & Fuels,* vol. 24, pp. 2179-2207, 2010.

[3] C.-C. Chen and H. Que, "Method of Characterizing Chemical Composition Of Crude Oil For Petroleum Processing," US Patent Application Number: 2013/0185044, July 18,2013.

464593 Rapid Analysis of Syngas Using Micro GC Fusion

Abstract:

Rapid Analysis of Syngas using Micro GC Fusion

In 2012, the World Bank estimated that 140 billion cubic meters (bcm) of the 3370 bcm of natural gas produced worldwide was flared. Flaring burns surplus gas and is detrimental to the environment, contributing over 350 million tons of CO_2 in annual emissions. An alternative to flaring is to convert the associated natural gas into synthesis gas (syngas), which is an intermediate gas used to produce synthetic liquid fuel. Converting syngas to liquid chemicals through the Fischer-Tropsch (FT) process presents an economical and greener alternative to flaring. Syngas is composed predominantly of hydrogen and carbon monoxide, though it may also contain carbon dioxide, nitrogen, methane, ethane, ethylene, or H_2S in the high ppm to percent level. After catalytic conversion, determining the syngas composition is a critical step before using the gas in any downstream process. To ensure efficiency, the analysis of syngas must be analyzed quickly and accurately. Building on proven microelectromechanical systems (MEMS) technology, Micro GC Fusion is capable of providing rapid analysis of typical syngas components in a compact instrument. A method equipped with temperature programming provides an area %RSD of less than 1.0% repeatability. This accurate and repeatable analysis of syngas is performed in less than 50 seconds.

469260 A Regulatory Perspective for Manufacture of Complex Drug Substances

Abstract:

The quality of chemically synthesized, small-molecule drug substances usually can be characterized by routine analytical testing. In contrast, for drug substances that are polymeric molecules (e.g. peptides), heterogeneous mixtures (e.g. heparins, glatiramer acetate, etc.), or molecular complexes (e.g. iron sucrose), the structure and impurity profile are more complex than small-molecule drugs, and generally cannot be characterized easily by analytical means. As a result of this complexity, quality control of these complex drug substances cannot solely rely on analytical testing. The identity, impurity profile, and content of complex drug substances are highly dependent on their manufacturing process as well as raw material and in-process controls. This presentation will discuss a regulatory perspective for manufacture of complex drug substances in support of both new drug and generic drug applications.

473006 CQA Control Strategy for a Synthetic Peptide Therapeutic Candidate Etelcalcetide

Abstract:

Application of innovative continuous improvement strategies to ensure consistency and robustness of pharmaceutical drug substance (DS) manufacturing processes can add significant value over a productâ€™s lifecycle. A case study from etelcalcetide, a synthetic peptide therapeutic candidate, will be presented highlighting Amgenâ€™s modality-independent approach to process design. Application of QbD and QRM principles were fundamental in establishing a solid understanding of the impact of variability on DS critical quality attributes (CQAs) and overall process performance. Holistic frameworks to proactively identify, assess, and mitigate CMC risks was leveraged during the process design phase. Examples of specific DS CQAs will be used to show how establishment of appropriate controls required a combination of material input controls, procedural controls, in-process testing, and release testing.

457444 Development and Implementation of a Mathematical Model As a Key Component of a Palladium Control Strategy for a Semi-Continuous Process to Produce a Small Molecule Drug Substance

Abstract:

The manufacture of a small molecule drug substance employs a coupling reaction catalyzed with soluble palladium in the final synthetic step of the sequence. The control strategy around palladium removal involves selective adsorption of the palladium onto a functionalized silica resin in a continuous fashion followed by a continuous deprotection reaction and the selective crystallization of the drug substance. During the registration stability campaign, at-line and off-line samples were collected. To minimize cost associated with the selective adsorbent, and to ensure a robust control strategy for palladium, a predictive model was developed to aid in the appropriate design, scaling and setup of the manufacturing equipment. Use of the model and the selected control strategy for palladium, alleviated the need for continuous sampling for palladium during processing and enabled a robust operation during validation with significant production cost reduction.

467794 Monitoring the Trace Levels of Free Sulfate Release to Terminate a Depolymerization Reaction Using ATR-FTIR

Abstract:

The synthesis of a pharmaceutical product involves a critical depolymerisation reaction which determines the final product quality attributes. Hence, termination of this reaction at the desired end point is utmost important to achieve a consistent product with required specifications. However, monitoring this reaction is challenging due to low chemical specificity of the reaction species involved. In this work, we developed a novel PAT method based on ATR-FTIR combined with chemometrics to monitor this reaction. To achieve this, we first identified the reaction mechanism and the rate-controlling step and monitored the reaction progress by tracking the free sulfate release which forms as a by-product during the reaction. The PAT method has been validated at lab and kilo scales and subsequently transferred to the plant and successfully implemented at manufacturing scale.

474335 FDA Review Chemist's Participation in Pre-Approval Inspections: Case Study from Drug Substance Facility

Abstract:

In 2014, FDA initiated a new approach to incorporate review chemist's expertise in the inspection of drug substance/product manufacturing facilities. A pilot was launched with the objective to evaluate the feasibility, effectiveness, and efficiency of Office of Pharmaceutical Quality's vision for the integrated team-based quality assessment of A/NDAs. A risk analysis was conducted jointly by the compliance officer, drug substance, drug product, and facility reviewers to determine the need for a pre-approval inspection (PAI) and prioritize the issues to be covered on the inspection. The integrated team-based approach resulted in improved knowledge sharing among the reviewers and compliance officers, and in better identification and evaluation of high-risk areas during the application review. These high-risk areas identified by the review team were covered during the PAI. Collaboration between the review chemist and investigators helped uncover significant issues and resulted in the issuance of deficiencies to the firm.

Based on outcomes of the Agency's pilot integrated PAI program, other inspections were conducted using this paradigm including the recent PAI of a complex drug substance manufacturing facility. Participation on this inspection provided the reviewer with an opportunity to examine the equipment and process closely, to gain a better understanding of the development and optimization of manufacturing process and controls, and in turn allowed the inspection team to identify the CMC-related observations. Furthermore, the firm provided positive feedback regarding the presence of a review chemist on inspection deemed particularly helpful to better understand and improve regulatory submission of future applications. In summary, implementation of this new integrated approach for conducting a PAI has enhanced the efficiency and quality of review process, and improved communication between FDA and industry.

474643 High-Temperature Ethane Dehydrogenation in Microporous Zeolite Membrane Reactor

Abstract:

Ethene is the largest base chemical for the chemical industry and produced either by cracking or by dehydrogenation of light alkanes. The increasing demand for ethene has stimulated substantial research into the development of new processes to reduce energy consumption. Catalytic ethane dehydrogenation using a membrane reactor is an attractive solution because the cracking equilibrium can be shifted in favor of ethylene by selective removal of hydrogen.

In this study, we report the intensification of ethane dehydrogenation reaction in packed-bed membrane reactors operating with a Pt/Al_2O_3 catalyst. The effects of tubular MFI zeolite membranes on ethane conversion and ethene selectivity are investigated. The rate of H_2 membrane permeation significantly enhances the ethane conversion. MFI membrane reactors allow the equilibrium limit of ethane conversion to be surpassed at high temperatures. It is demonstrated that medium-pore MFI membranes with moderate H_2/C_2H_6 selectivity can be effective in enhancing ethane conversion at high operation temperature by timely removal of H_2 through the membranes.

464166 Analysis of a Cyclic Redox Process for Oxidative Dehydrogenation of Ethane

Abstract:

Ethylene is commercially produced from ethane via high-temperature pyrolysis in the presence of diluting steam a.k.a. steam cracking. The process requires a high temperature (>1100 K) and a large steam generation load, making it highly energy intensive (17-21 GJ/tonne ethylene). The process also emits significant amount of NO_x and CO_2 and requires periodic shutdowns for coke burnout. Moreover, single-pass ethylene yield is limited by thermodynamic equilibrium. Oxidative dehydrogenation (ODH) of ethane, where ethane is partially oxidized to ethylene and water, represents an alternative route. The partial burning of the product hydrogen provides the required heat while shifting the equilibrium towards the product side. As a result, higher single-pass ethylene yield can potentially be achieved with reduced energy consumptions. The requirements for highly-selective catalyst, co-feed of combustible oxygen and ethane mixture, as well as cost associated with oxygen separation represents key challenges for such a process. A chemical looping based ODH (CL-ODH) scheme addresses the abovementioned challenges. In this process, the required oxygen is supplied by lattice oxygen from a redox catalyst composed of mixed metal oxides. The reduced catalyst is oxidized with air in a separate reactor, where it is regenerated and then recirculated to the ODH reactor. This allows built-in air separation with minimal parasitic energy loss and better temperature control. Overall, CL-ODH transforms the upstream of a traditional steam cracking process and has the potential to enhance ethylene production with significant energy savings.

Mn-based oxides have been shown to be effective as redox catalysts for the ODH of ethane. Selective combustion of hydrogen to water, over the redox catalyst, shifts the equilibrium towards ethylene and also provides the heat required for the endothermic dehydrogenation reaction. In this work, the ODH of ethane is modeled using ASPEN Plus® and is compared with a conventional stream cracking process. Results show that CL-ODH with 85% ethane conversion provides over 70% reduction in the overall energy demand with >75% reduction in the overall CO_2 emissions. The exothermic nature of the regenerator and elimination of the steam requirement lead to major reductions in the upstream energy consumption. Combustion of hydrogen further reduces the downstream separation loads by >20%, with >25% drop in the compression work. For every tonne of ethylene produced, steam cracking requires an external fuel input equivalent to 1.5 GJ_{Th}, whereas the CL-ODH results in surplus fuel energy in excess of 6 GJ_{Th}. Increase in single-pass ethane conversion further increases the energy savings. Preliminary results from CHEMKIN-PRO® are in close agreement with literature values in terms of product compositions for ethane cracking. Addition of surface kinetics of the redox catalyst towards hydrogen combustion provides further understanding of the dependence of product output on the process conditions.

474954 Novel Method to Isolate Polymeric Slurry Based Systems in a Continuous Process

Abstract:

During polymerization reactions, if the final polymer is not soluble in the solvent, a slurry or precipitate is formed as the molecular weight builds up. Typically, after the reaction is completed, a series of unit operation such as centrifugation followed by drying is used to isolate dry polymer. Slurry processes have inherent challenges with maintaining stable flow control for continuous operation; powder carryover with vapor; powder handling and other risks associated with potentially explosive dusts.

To describe the new process, we have used the example of a polymer that forms a slurry in the solvent most suitable to carry out its reaction. It was identified that above a certain temperature higher than the boiling point of the solvent, the polymer was soluble in the solvent. In order to form a solution, the reaction mixture would need to be maintained at an elevated temperature and corresponding vapor pressure. By having a polymer solution instead of a slurry, it was possible to use standard thermomechanical separators such as an agitated thin film evaporator (ATFE) or devolatization extruders to isolate the polymer in the continuous process. It was observed that the polymer would crash out of the solution if the temperature or pressure dropped below a certain limit. With this in mind, an overview of the risk mitigation activities undertaken to test the feasibility of reaction and isolation of the polymer in an ATFE operating under pressure is provided. A similar approach could be taken for isolation of other polymers, which are not soluble in a solvent below its boiling point, but are soluble above the boiling point and can be achieved under elevated pressure conditions.

455485 Scale-up of Oscillatory Baffled Reactors (OBRs)

Abstract:

Scale-up of oscillatory helically baffled reactors (OHBRs)

Safaa M.R. Ahmed, Anh N. Phan, Adam P. Harvey

School of Chemical Engineering and Advanced Materials, Newcastle University

Newcastle upon Tyne - NE1 7RU

Mobile:+44 7774101215

E-mail: s.m.r.ahmed@newcastle.ac.uk

Keywords: Oscillatory baffled reactor, scale-up, residence time distribution

Abstract

An Oscillatory Baffled Reactor (OBR) is an intensified design of continuous plug flow reactor (PFR) in which plug flow behaviour can be achieved at very low net flows (laminar flow regime). OBRs consist of tubes with periodically spaced baffles of various designs (orifice, helical, integral etc. baffles). There is a net flow through the reactor, and a superimposed oscillatory flow. The oscillatory flow interacts with the baffles to produce flow structures (usually vortices) that provide mixing. The mixing in the OBR is therefore independent of the net flow. As a result, the OBR's niche application is to operate "long" reactions in continuous mode. This is usually impractical in conventional tubular reactors.

Scale-up in conventional reactors, i.e. stirred tank reactors, is unpredictable due to the non-uniform mixing at large scale, leading to large variations in concentration, temperatures etc. This means that optimum conditions obtained from laboratory scales cannot be directly used at large scales, therefore process development/ product-to-market time would increase. However, scale-up of OBRs should be more predictable, as the flow structures can be reproduced across length scales.

Recent studies on oscillatory helically baffled reactors (OHBRs) at small scales (millilitre volume) found that the helical baffled design could provide high degree of plug flow across a wide operating window due to the combined effect of vortex formation and swirl flow. It was found that the behaviour of residence time distribution remained the same at all tested scales (10mm diameter to 25mm diameter, corresponding to volume 0.078L to volume 0.834L) when the similarity in geometric and dynamic parameters was maintained. The degree of plug flow was quantified in terms of number of tanks-in-series (N). At a fixed geometry, a scale-up correlation was established and validated over a range of operating conditions such as oscillatory conditions (Strouhal number, *St*, and oscillatory Reynolds number, Re_o) and the velocity ratio of oscillatory flow and net flow (ψ).

$$\frac{N}{N_o} = 40\, Str^{-0.3}\, Re_n^{\ -1}\, \psi e^{-0.1\psi} + 1$$

Where, $N_o = 0.25\, Re_n$

456248 Flow Microreactor Synthesis of ZIF-8 Particles with Controlled Size, Shape, and Adsorption Properties

Abstract:

Significant progress has been made in research and development of metal-organic frameworks (MOFs), which consist of metal ions and organic ligands. Their chief features include large surface area, highly ordered porous network, and structural flexibility. The flexibility leads to a structural transition of the host framework from a non-porous to a porous phase that causes a step-wise increase in guest molecules at a threshold pressure, which is called â€œgateâ€ adsorption. Because of this unique adsorption property, soft MOFs are promising for various applications not only to gas storage and separation, but also to molecular sensors. To exploit full potential of soft MOFs, their particle size and shape must be monodispersed because their unique adsorption property can depend on their size and shape. In the present study, we synthesized zeolitic imiazolate framework-8 (ZIF-8), which is a subclass of soft MOFs and is composed of Zn ions and 2-methylimidazole linkers. We mixed aqueous $Zn(NO_3)_2$ and 2-methylimidazole solutions in a central-collision type microreactor, which utilizes shear forces at the collision of two inlet fluids to realize rapid and homogeneous mixing. The mixed solution from the outlet was kept in a glass vessel for an hour. The nucleation starts upon mixing the solutions in the microreactor, followed by the particle growth in the glass vessel. Typical particle size is around 300 nm, and resultant particle size and shape distributions synthesized with the microreactor were more uniform than the case of a batch reactor. This would be due to a high mixing performance of the microreactor that can divide the nucleation process and the particle growth process. Moreover, we synthesized ZIF-8 particles for varied ratio of 2-MeIM to Zn^{2+} (R), concentration of Zn^{2+} ($[Zn^{2+}]$) and temperature. The average size of ZIF-8 decreases with the R, and a higher $[Zn^{2+}]$ results in smaller size at a fixed R, which is because of a higher degree of supersaturation at higher R and $[Zn^{2+}]$ conditions. Prepared ZIF-8 particles have BET surface area of 1500 m^2/g and exhibit the gate adsorption behaviour with different transition pressures depending on the particle size and shape. We also confirmed the microreactor synthesis is applicable to a different type of soft MOFs (ELM-12). In this manner, our flow process using the microreactor is demonstrated to be a simple but versatile technique for the synthesis of monodispersed MOF particles.

457239 Intensification of Batch-Distillation Processes for Production of Alcoholic Beverages

Abstract:

During production of alcoholic beverages, especially the alcoholic fermentation of pomaceous fruit, formation of methanol is an issue that has to be considered. Since the middle lamella which binds adjacent fruit cells, consists of methylated polygalacturonic acid, i.e. pectin, the formation of methanol is unavoidable during natural fruit specific enzymatic digestion. Furthermore pectin methyl esterase is commonly added to the fruit mash in order to gain an enhanced yield of ethanol. While methanol in principle is not toxic to human beings, the effect of metabolites formaldehyde and formic acid can cause severe injuries to health, e.g. blindness or acidosis and can even cause death. Thus the separation of methanol from distilled alcoholic beverages has to be explicitly taken into account.

Against the widely shared notion it is not possible to separate methanol by distillation using common batch distillation equipment, no matter if multi-stage columns or single-stage pot stills are used. Experimental investigations as well as dynamic simulations of single stage batch distillation with Aspen Hysys showed matching results. In fact methanol is distributed between foreshots, heart cut and tailings. Further the ratio of methanol to ethanol is increasing with cumulative distillation time. In addition high regulatory limits in the European Union for methanol concentrations in pomaceous fruit spirits of 1350 mg/100 ml pure ethanol compared to other spirits, e.g. wine brandies are often exceeded even though the high values are not set due to harmlessness to health but insufficient separation performance of standard batch distillation equipment. Hence the production of distilled pomaceous fruit spirits with less methanol content leads to the necessity of intensifying distillation processes with focus on new technical approaches.

The main goal of the current research is to improve the separation of methanol in the foreshots, using an extension of common lyne arms. For this purpose a pot still is equipped with an optional bypass which provides the possibility to route the vapor via an additional randomly packed lyne arm at the beginning of distillation to increase the separation efficiency. Further the bypass lyne arm enables enhancement of internal reflux, realized with a coaxial heat exchanger. In this manner methanol accumulates in the foreshots and may be separated from the heart cut sufficiently.

453255 Droplet Formation of H2SO4/Alkane System in a T-Junction Microchannel: Gravity Effect

Abstract:

A special system of concentrated sulfuric acid (H2SO4) and n-hexane was used to study the droplet formation in a glass T-junction microchannel with H2SO4 as the continuous phase. The effects of capillary number, flow ratio, and viscosity ratio on the droplet formation were investigated. The effect of gravity was explored by changing the flow direction in the microchannel. Results showed that the formation of transition flow pattern from squeezing to dripping is much easier for this special system compared with common aqueous/organic systems. This phenomenon is due to the considerably higher viscosity of H2SO4 than that of common aqueous phase and the higher density difference of the system compared with those of common systems. In addition to capillary number and flow ratio, gravity evidently affects the formation of droplets and flow patterns. The droplet size is smaller than that during the horizontal flow when the flow direction is consistent with gravity. By contrast, flow direction contrary to that of gravity results in larger droplet size than that at horizontal flow. This phenomenon provides guidance on the operation of these special systems in microchannels. Finally, mathematical models of droplet size at different flow patterns have been established, and these models can predict droplet size very well. This study could be helpful to extend the application of microreactors to new working systems.

464667
Flare Minimization for Simultaneous Turnaround Operations of Two Olefin Plants

Abstract:

Flare minimization for olefin plant turnaround operations such as start-up and shut-
down is a double-
win practice, which brings not only industrial economic benefits, but also local environm
ental credits. By far, almost all of flare minimization studies focus on single units or singl
e plant normal or abnormal operations. No systematic studies have been reported on flare
minimization coupling abnormal operations of two chemical plants. Conceivably, if mate
rial and energy exchange between two chemical plants could be effectively integrated and
scheduled, flare emissions from one plant might be adopted by another plant, so that flar
e minimization could be achieved to its maximum extent. In this paper, a systematic meth
od has been developed to study the flare minimization for simultaneous turnaround operat
ions of two olefin plants: one plant is experiencing start-
up while another plant is conducting shut-down operations.

When one olefin plant undergoes its shut-
down operation, its flare minimization strategy is conducted as: (i) keep the minium num
ber of furnace running by shuting down extra furnaces; meanwhile, reduce liquid invento
ries in tanks, drums, and columns; (ii) shut down the remaining furnaces and keep CGC (
crack gas compressor) running with vapor recycles; meanwhile, vaporize liquid remains i
n tanks, drums, and columns; (iii) shut down CGC after recovering liquids and vapor inve
ntory as much as possible; (iv) perform decommissioning for each facilities including left
liquid discharge, vapor discharge, and nitrogen purge. For another plant performing its st
art-
up operation, the general operating procedure includes: (i) perform nitrogen replacement
with hydrocarbons; (ii) input liquids to columns to build up lthe vapor -
liquid balance in various vessels; meanwhile columns are engaged in the full-
reflux mode; (iii) start CGC with vapor recycles; meanwhile, precooling the chilling train
and warm up columns; (iv) start up furances gradually; (v) gear up each column operatio
n to their normal status.

In this paper, flare minimization opportunities coupling shutdown operating steps (ii) thro
ugh (iv) together with start-
up operating steps (ii) and (iii) are exploried. To study the feasibility of this coupling turn
around operation and calculate flare emission sources, plant-
wide dynamic simulations for these two plants have been conducted. The developed duo-
plant flare minimization strategy has been virtually tested and demonstrated. The obtaine
d flare minimization strategies and simulation results provide an indepth understanding o
n synergies of two plant abnormal operations for flare minimization.

469916 Consequence Assessment for Event Tree Analysis in Transitions Using Dynamic Simulation

Abstract:

Incidents that occur in industrial facilities can lead to the release of hazardous materials, uncontrolled fire and explosions, economic losses and environmental contamination. Consequence assessment is an examination of the undesirable impacts of an accident. Fuzzy logic combined with Event Tree Analysis (ETA) is used for quantitative risk assessment (QRA).

In this work dynamic simulation was used to assess the consequences of event tree branches for two examples of equipment switching transitions in chemical plants. The first example was switching between two solid desiccant dryers in an ethylene plant. The other example was the switching of a reflux pump with its spare in a reformate stabilizer column. Both examples have standard operating procedures (SOP) commonly accomplished manually by operators. Numerous operator induced fault cases were considered. These SOPs and fault cases were simulated in Aspen HYSYS V8.0 using the event scheduler feature. The consequences were identified by monitoring variables during dynamic simulation of each case. Thus, the fuzzy ETA algorithm was used to evaluate the risk of equipment switching transition faults while dynamic simulation was used to assess the consequence.

AIChE Annual Conference 2016

Division: Process Development

Session: Process Research and Development for Industrial Sustainability

Title: Effect of chemical price uncertainty on project viability: an economic evaluation

Authors: Gaurav Arora, Mohammad Shafiei, Mike Huckman

Abstract: A case study is presented for evaluating economic viability of a chemical process. The discussion includes process design choices, cost estimates, and economic evaluation of the process. The economics are found to be very sensitive to the feed and product prices. Three specific scenarios for process are analyzed in detail – at current chemical price, low feed-high product price and high feed-low product price. While these scenarios are helpful in understanding the cost, revenue breakdown, and economics of the project, a better insight is obtained by analyzing the effect of chemical prices over the last 15 years on the economic viability of the project. Margin required to meet an IRR of 15% is calculated followed by a market price analysis to calculate the probability of this occurring in the last 15 years.

Keywords: Economic Analysis

468268 Molecular-Level Kinetic Model Informed Life Cycle Analysis of Green Diesel Production

Abstract:

The drive to reduce our reliance on petrodiesel coupled with the advent of new environmental policies and economics has prompted the production of green diesel as an alternative, sustainable fuel. In this study, a molecular-level kinetic model was used to investigate the impact of changing feedstock composition and reaction conditions on the life cycle analysis of green diesel. The kinetic model simulates a fixed bed hydrotreating unit that includes the detailed triglyceride and free fatty acid hydroprocessing chemistry to produce paraffin-rich diesel range fuels. Linear free-energy relationships and literature data were used to reduce and tune the parameters of the LHHW rate laws allowing for product cloud point and cetane number property calculations. The product green diesel has excellent combustion properties but poor cold-flow properties, so hydrocracking severity was used to regulate end use and value of the green diesel. The process inputs and outputs were used in Argonne National Laboratoryâ€™s Greenhouse Gases, Regulated Emissions, and Energy Use in Transportation (GREET) model to evaluate greenhouse gas emissions of green diesel compared to petrodiesel. Scenario testing was performed to study the relative emission reduction for the different reaction conditions.

470164 Application of the Advanced Distillation Curve Method to Characeterize Two Alternative Transportation Fuels Prepared from the Pyrolysis of Waste Plastic

Abstract:

The properties of alternative fuels must be evaluated to determine their suitability for use in existing engines, either as blends with conventional petroleum products or as stand-alone substitutions (that is, drop in replacements). Alternative fuels made from waste feedstocks are especially attractive since they address the dual purpose of fuel production and mitigation of environmental pollution environmental concerns. In this study, two alternative liquid fuels generated by pyrolysis of waste polyethylene plastic bags were assessed using a composition-explicit distillation approach, which provides a rich data matrix that includes including fuel volatility, composition, hydrocarbon classification, and energy content in each distillate volume fraction. The data collected using this approach provide valuable information for the development of equations of state to describe the thermodynamic behavior of these complex fluids. The analysis showed that, despite significant differences in composition, the distillation curves and associated data of the alternative fuels show promising similarity to their conventional counterparts.

469215 Application of Real-Component Based Molecular Characterization on Petroleum Refinery Simulation

Abstract:

Abstract

As light to medium sweet crude oils keep depleting while demand of oil products are increasing, heavier and sourer crude has become a growing component to global oil market. Accompany with that, extreme bulk properties (including high molecular weights, high boiling points, large densities, high viscosities, etc.) are expected. However, current experimental approaches are still limited in characterizing heavy crudes in two aspects: 1) chemical structures of hydrocarbons with carbon number up to merely 12 can be accurately analyzed; 2) distillation cuts can only be obtained up to 750 K of normal boiling point, even with deep vacuum fractionation apparatus[1]. With limited experimental data available, less than 50 wt. % of individual heavy oil or bitumen can be analyzed, which renders simple model extrapolation over the whole crude highly unreliable.

The present work compares performance of petroleum refining simulation between two different methods: 1) the traditional API correlations commonly used in industry, 2) a real-component based molecular characterization method[2]derived from thermodynamically-consistent correlation of available crude oil data on chemical and physical properties. Compared with the former method, the later one not only gives reasonable predictions for properties of crude oil including heavy cuts, but also provides chemical structures and compositions of model constituent components over the whole crude oil, which serves as a cornerstone of high fidelity refining simulation.

Reference

1. Sánchez-Lemus, M. C., et al. (2016). "Physical properties of heavy oil distillation cuts." Fuel 180: 457-472.

2. Chen, C.-C. and H. Que (2013). "Method of characterizing chemical composition of crude oil for petroleum refinery processing". US Patent Application 20130185044 A1.

460128 Economic Design of Bioconversion Process: Natural Gas to Isobutanol

Abstract:

The low cost and abundant supply of natural gas have been motivators for the development of many technologies that convert this low cost carbon feedstock into compounds of greater value. Natural gas bioconversion is a developmental technology which applies synthetic biology to a naturally methane consuming bacteria, a methanotroph, to program it to produce higher value materials of interest. Intrexon has formed joint ventures to commercialize methanotrophic bioprocesses for biofuels and high value chemical intermediates.

Using a methanotrophic bioprocess to produce fuels and chemicals has several advantages other than the low cost and minimal processing of feedstock. Natural gas bioconversion offers scaling opportunities beyond traditional gas-to-liquid (GTL) facilities, as well as biomass feedstock facilities. Theoretical yields of methanotroph/methane exceed that of yeast/sugar approaches by a wide margin. Biocatalytic processes that achieve high yields under milder processing conditions allow for reduced capital and operation expenditures, offering greater economic viability.

Intrexon's flagship molecule of interest in the methanotroph platform is isobutanol, a fuel for gasoline blending which has several positive attributes. Isobutanol is a clean burning "drop-in" fuel which can be blended with gasoline and is compatible with the existing petroleum infrastructure. Currently, provisions allow for bio-butanol blending up to 12.5% with gasoline and ASTM specifications exist for sale as fuel. Combustion of isobutanol has cleaner emissions as compared to gasoline by reducing hydrocarbon, carbon monoxide, nitrous oxide and sulfur dioxide production. Isobutanol also has advantages over first generation biofuels, such as ethanol, because of its higher energy density.

Currently, the production of both isobutanol and farnesene has been demonstrated in lab scale fermentation. To successfully bring this technology to commercial scale, consideration must be given not only to optimizing the metabolic pathway engineered into the methanotroph, but to key issues in the fermentor design, downstream processing and overall plant economics. Traditional engineering workflows, such as a front-end loading strategy, were used to shape the commercial scale process design. The biologic, technical and economic commercial feasibilities of the end-to-end natural gas to isobutanol process were critical to Intrexon's plant design. Laboratory scale learnings were applied to select technology that was commercially feasible at demonstration scale. This demonstration scale design was scaled down to pilot scale, where the operation will yield learnings into refinement of at-scale design, equipment selection, empirical data on heating and cooling, and controls strategies.

463172 Techno-Economic Analysis of Indirect, Direct, and Hybrid Coal-Biomass to Liquids (CBTL) Plants with CO_2 Capture and Storage (CCS)

Abstract:

Indirect coal liquefaction (ICL) and direct coal liquefaction (DCL) technologies are commercially available for producing alternative transportation fuels. However, both processes are plagued with high CO_2 emission. Because the syncrude produced in the ICL and DCL reactors cannot satisfy the current specification of transportation fuels, additional upgrading technologies are required. For example, syncrude from the ICL process has low octane number for naphtha cut but high cetane number for diesel cut, while syncrude from the DCL process has high octane number for naphtha cut but low cetane number for diesel. Therefore the hybrid coal liquefaction (HCL), a combination of ICL and DCL technologies, can reduce the penalty of downstream syncrude upgrading unit by optimal blending. Addition of biomass and application of carbon capture and storage (CCS) technologies are two possible solutions to reduce the carbon footprint, but would lead to higher operating cost and capital investment. Several studies have been conducted for indirect coal-biomass to liquids (CBTL) plants with CCS, while few studies have been conducted for direct and hybrid CBTL processes at the systems level. With this motivation, we will present a techno-economic study conducted in Aspen Process Economic Analyzer® (APEA®) environment for indirect, direct, and hybrid CBTL plants with CCS using high fidelity process models developed in Aspen Plus® and Excel.

In the indirect CBTL plant, coal and biomass are first gasified to syngas. Then the syngas is converted to syncrude via Fischer-Tropsch (FT) synthesis. CO_2 is captured from both raw syngas and FT vapor product. In the direct CBTL plant, coal and biomass are directly converted into syncrude in the catalytic two-stage liquefaction (CTSL) unit by adding hydrogen produced from gasification of coal/biomass/liquefaction residue or reforming of shale gas. Significant amount of CO_2 that is generated in the hydrogen production unit(s) is captured to satisfy the target extent of CO_2 capture. In the hybrid CBTL plant, pre-processed coal and biomass are sent to either syngas production unit or the CTSL unit. Produced syngas is sent either to FT unit or hydrogen production unit. Naphtha and diesel products from the FT unit and the CTSL unit are blended to reduce the syncrude upgrading penalty. Different CCS technologies are considered and optimized for the indirect, direct and hybrid CBTL plant depending on the sources of CO_2 containing stream and corresponding CO_2 partial pressure.

The focus of this presentation will be on optimal synthesis of the hybrid plant. In particular, following aspects will be presented: (1) development of plant-wide process model and rigorous economic model in Aspen Plus®, Aspen Exchanger Design and Rating®, Excel, and APEA®, (2) sensitivity studies to analyze the impact of key design parameters (i.e. biomass/coal ratio, extent of CCS, CCS technologies, blending ratio of indirect and direct syncrude in the hybrid route) and investment parameters (i.e. price of coal and biomass, project life, plant contingency and plant capacity) on key economic measures, such as net present value (NPV), internal rate of return (IRR) and break-even oil price (BEOP), (3) analysis of the potential environmental credits due to use of biomass and CCS for all liquefaction approaches, (4) comparisons and analysis of trade-offs of indirect, direct, and hybrid CBTL plants.

465496 Finished Fuel Blending Models for Assessing Integration of Biomass-Derived Products with Petroleum Refinery Products

Abstract:

As the bioenergy industry focuses on the near-term development of conversion pathways from biomass feedstocks to infrastructure-compatible fuel blendstocks, analysis capabilities and models are essential for assessing the values of biomass-derived blendstocks based on its fuel properties and the constraining parameters for blending them with petroleum-derived products. Funding from DOEâ€™s Bioenergy Technologies Office (BETO) has enabled the analysis team to develop both linear programming models (Aspen PIMS) and process models (Aspen Plus and Aspen HYSYS) to meet the new analysis needs. In addition to developing these analysis tools, the team has also utilized published experimental blending data to validate the results of the models and to improve blend property predictions. These models are already playing a vital role at NREL in understanding the potential to integrate fuel products from biomass conversion pathways under BETOâ€™s research and development portfolio.

This presentation will focus on (1) an overview of the analysis approaches and process models developed under this effort, (2) the benefits for utilizing the tools for assessing biomass-derived blendstocks, (3) examples of blending analysis results, (4) comparison of results from blending models relative to experimental data and (5) future work for this effort in assessing blending of oxygenated compounds.

462192 Pharmaceutical Crystallization: A Manufacturing Process Perspective

Abstract:

Typically a number of critical quality attributes (CQAs) such as crystal structure and crystal component identification, purity, crystallinity, crystal size distribution (CSD), polymorphism and impurity, and stability are evaluated to assess the robustness, controllability, repeatability, and consistency of a pharmaceutical crystallization process. There are different ways to achieve in-depth process understanding and to realize rationale process control. This presentation provides several FDA research case studies to demonstrate the technical feasibility of real time process monitoring approaches for better understanding, design and control of pharmaceutical crystallization processes and product characterization. The case studies are: 1) Combined real-time PAT monitoring and process chemometrics for mapping the state of a pharmaceutical crystallization process; 2) Combined real-time PAT process monitoring and first-principle modeling for elucidating the nucleation mechanisms of a dynamic pharmaceutical crystallization process; 3) Combined real-time PAT process monitoring, Design of Experiments (DOE), and General Linear Modeling (GLM) to establish a hybrid approach for process characterization and process design space development; and 4) Integrated PAT approach for nucleation induction time measurement. This presentation will briefly address some of the current opportunities in pharmaceutical crystallization, including switching from batch to continuous crystallization and several promising, novel technologies in the area of crystallinity monitoring, from a process engineering perspective.

456484 Cooling Crystallization within Micromixers: Mathematical Modeling, Theoretical Analysis, and Experimental Validation

Abstract:

A key objective in crystallization is to control crystal size distribution (CSD), to improve product quality (e.g., pharmaceutical and biopharmaceutical) and process efficiency. Micromixers have been an effective technology to enhance the control of nucleation, which is the key step towards final product CSD in continuous [1] and semi-continuous crystallization processes [2]. Common micromixers combine solution and anti-solvent streams such as in a dual-impinging-jet (DIJ) micromixer (e.g., [3]–[5]).

Rather than anti-solvent crystallization as commonly used, a DIJ mixer has been demonstrated to combine hot and cold saturated solutions of the same solute/solvent to generate small uniform seed crystals [2], which appears to be the first reported cooling DIJ nucleation. The product crystals were less than 10 microns in length, placing them in the correct size range for direct application in inhalers [6]. Theoretically, if the mixing was perfect (that is, concentration and temperature were completely mixed at the molecular scale in the mixer), the average supersaturation level would be too low to nucleate crystals in a cooling DIJ mixer. The surprising result of nucleating crystals by combining hot and cold saturated solutions in a DIJ mixer motivated the theoretical analysis of the system, which showed that nucleation was enabled by the much faster energy transfer than mass transfer rates near the interface. This difference in rates enabled the temperature of the hot solution to drop to approximately the average temperature of the two solutions before its solution concentration had significantly changed, resulting in a supersaturation sufficiently high to nucleate crystals.

The theoretical analysis is supported with process simulations. The two-dimensional energy and solute mass balances are solved using COMSOL input with an analytical solution for the velocity field, to generate spatial distributions of temperature, concentration, and supersaturation near the hot-cold interface within a cooling DIJ mixer. In the most important spatial region of importance for characterization of nucleation, the two-dimensional fields are shown to be very close to analytical solutions derived from a one-dimensional approximation of the energy and molar balances. This simplification applied the derivation of design criteria to facilitate quick assessment of whether any particular pharmaceutical-solute combination will nucleate crystals in a cooling DIJ mixer, based on the physicochemical properties. These criteria could save time and material by avoiding or reducing trial-and-error experiments, which is helpful at the early stage of pharmaceutical process development. The process simulations are then expanded to include the full solution of the spatially-varying population balance model [5] to calculate the crystal size distributions throughout the spatial domain. The mathematical models and methodologies are validated for the crystallization of L-asparagine monohydrate (LAM) in aqueous solution.

References

[1] M. Jiang, Z. Zhu, E. Jimenez, C. D. Papageorgiou, J. Waetzig, A. Hardy, M. Langston, and R. D. Braatz, "Continuous-flow tubular crystallization in slugs spontaneously induced by hydrodynamics," *Cryst. Growth Des.*, vol. 14, no. 2, pp. 851–860, 2014.

[2] M. Jiang, M. H. Wong, Z. Zhu, J. Zhang, L. Zhou, K. Wang, A. N. Ford Versypt, T. Si, L. M. Hasenberg, Y. E. Li, and R. D. Braatz, "Towards achieving a flattop crystal size distribution by continuous seeding and controlled growth," *Chem. Eng. Sci.*, vol. 77, pp. 2–9, 2012.

[3] B. K. Johnson and R. K. Prud'homme, "Chemical Processing and Micromixing in Confined Impinging Jets," *AIChE J.*, vol. 49, no. 9, pp. 2264–2282, 2003.

[4] M. Midler Jr., E. L. Paul, E. F. Whittington, M. Futran, P. D. Liu, J. Hsu, and S. H. Pan, "Crystallization method to improve crystal structure and size," US5314506 A, 24-May-1994.

[5] X. Y. Woo, R. B. H. Tan, and R. D. Braatz, "Precise tailoring of the crystal size distribution by controlled growth and continuous seeding from impinging jet crystallizers," *CrystEngComm*, vol. 13, no. 6, pp. 2006–2014, 2011.

[6] T. G. D. Capstick and I. J. Clifton, "Inhaler technique and training in people with chronic obstructive pulmonary disease and asthma," *Expert Rev. Respir. Med.*, vol. 6, no. 1, pp. 91–103, Feb. 2012.

452797 Real Time PAT Monitoring of the Solvent-Mediated Transformation of Ciclesonide Acetone Solvate into Methanol or Cyclohexane Solvate

Abstract:

Due to structural differences, pseudopolymorphs of a given drug substance exhibit different physicochemical properties, which impacts drug dissolution, bioavailability, and efficacy. Therefore polymorphic transformation has been a vital aspect of pharmaceutical crystallization process engineering. In this work, three solvates of ciclesonide were characterized by Differential Scanning Calorimetry ï¼ˆDSCï¼‰, Thermogravimetry (TG), and X-ray Powder Diffraction (XRPD) analysis. The driving force of solvates transformation process was investigated via measuring solubility of ciclesonide solvates in different mixed solvent systems. To follow the crystallization process kinetics and understand the phase transformation mechanisms, , several Process Analytical Technology (PAT) techniques such as Fourier Transform Infrared Spectroscopy (FTIR), Raman Spectrum (RAMAN), Particle Video Microscope (PVM) and Focused Beam Reflectance Measurement (FBRM) were employed to monitor the phase transformation from acetone solvate into methanol or cyclohexane solvate in real time. In addition, the effects of temperature, solvent composition, and stirring rate on phase transformation process were discussed. Furthermore, it was demonstrated that real time PAT process monitoring can help to realize better process control and achieve desired crystal product quality attributes.

472470 Crystal Morphology and Size Control in Pharmaceutical Industry

Abstract:

Crystal Morphology and Size Control in Pharmaceutical Industry

Junbo Gong[1,2*], Yan Wang[1,2], Jingkang Wang[1,2]

[1] National Engineering Research Center of Industry Crystallization Technology, School of Chemical Engineering and Technology, Tianjin University, Tianjin, China;

[2] The Co-Innovation Center of Chemistry and Chemical Engineering of Tianjin, Tianjin University, Tianjin, China;

E-mail address: junbo_gong@tju.edu.cn * wangyan57@tju.edu.cn

Abstract

Crystal shape and size engineering is an integral part of pharmaceutical crystallization. It aims to design and control crystal habit and particle size distribution (PSD) of drug substance. This program is of great significance due to the strong impact of product quality, including bioavailability, dissolving rate, flowability, etc. Downstream particle handling is also affected, such as filtration, drying and compaction. The factors influencing crystal morphology and size come from two groups: the one is the operating condition, which includes supersaturation, temperature, the hydrodynamics of the crystallizer (e.g., stirring rate, mixing conditions) and external field (e.g., magnetic field, ultrasound field), and the other is chemical composition of the solution which consists of solvents and additives. In order to control crystal shape and PSD, experiments should be designed and carried out to monitor crystal growth process. The influence on crystallization thermodynamics and kinetics should also be investigated. During the research,

microscopy techniques, like polarizing optical microscope, scanning electron microscopy (SEM) and atomic force microscope (AFM) will be applied to analyze crystal shape. Particle sizing techniques are also used to count for PSD, like the laser diffraction (LD) and focused beam reflectance measurement (FBRM). On the other hand, the use of additives has attracted increasing attention in the drug manufacturing process. The presence of additives could substantially affect the solid forms and morphology. High throughput screening (HTS) techniques to additives should be developed to accelerate the screening efficiency. The effects of additives on crystal shape may be resulted from the specificity of interaction between the additive and the solute molecule, charge distribution, hydrophilic-hydrophobic property, stereoselectivity as well as diffusion rate and competitive adsorption, etc. With fast advances made in computational techniques, modeling and prediction of crystal morphology can be realized in the absence or presence of additives. They not only provide powerful tools to design and screen effective additives, but lead to further understanding of crystal growth mechanisms.

Fig. 1. SEM of the drug substance without (a) with (b) the modification of the crystallization environment

472381 Liquid - Liquid Phase Separation during the Cooling Crystallization of Pyraclostrobin

Abstract:

Liquid–Liquid Phase Separation during the Cooling Crystallization of Pyraclostrobin

Crystallization, as a method of isolation and purification, is widely used in fine chemicals, food and pharmaceutical industries. It is a powerful mean to improve the properties of particle, such as changing crystal form, narrowing crystal size distribution (CSD), modifying crystal habit and so on. However, sometimes droplets like oil drops are generated instead of crystals are appeared from the supersaturated solution during the cooling and/or solventing-out crystallization process. This phenomenon is always termed as liquid¨Cliquid demixing, liquid¨Cliquid-phase separation (LLPS) or oiling out.

The main purpose of this investigation was to investigate the cooling crystallization process of pyraclostrobin in isopropanol/cyclohexane solvent mixtures. It was found that liquid–liquid phase separation will occurred under some certain conditions. In order to understand the mechanism of the liquid–liquid separation and the influence of cooling rate on oiling out, the phase diagram and cloud point curves at different cooling rate were experimentally measured with the aid of in situ tools, such as PVM, FBRM, ATR-FTIR and Raman. The phase diagram includes the solubility curve, the liquid–liquid separation curve, supersolubility curves, and clear point. It was found from the phase diagram that a normal single-phase crystallization would change to a oiling-out crystallization once the concentration of pyraclostrobin exceeded a certain value during crystallization. Contrasting with cloud point curves of different cooling rate, it is obvious that the cooling rate has a significant effect on the metastable state critical oiling out point. The concentration corresponds to the metastable state for critical oiling out point decreases with the increase of cooling rate. In the cooling and heating cyclic process, the solution has experienced five stages. During cooling down the solution from high temperature, oiling out including stable LLPS and metastable LLPS occurred, then nucleation and growth of crystals happened; upon heating up, the crystals dissolved gradually, then the solution with a small amount of particles transformed into transient oil phase and finally it formed a stable and clear solution.

Considering the crystal properties of final products, some experiments designed to investigate the influence of cooling rate and initial concentration. And the result showed that the higher cooling rates, the more serious agglomeration and poorer crystal habit of final products would get. Another result is that initial concentration has little influence on the crystal properties. High quality of products can be obtained by designing route of cooling crystallization according to the phase diagram.

Figures

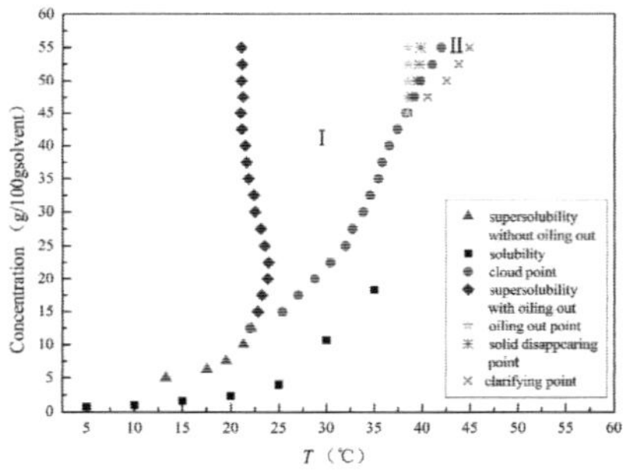

Figure 1 Phase diagram of pyraclostrobin in 10 wt % isopropanol/cyclohexane solvent mixtures.

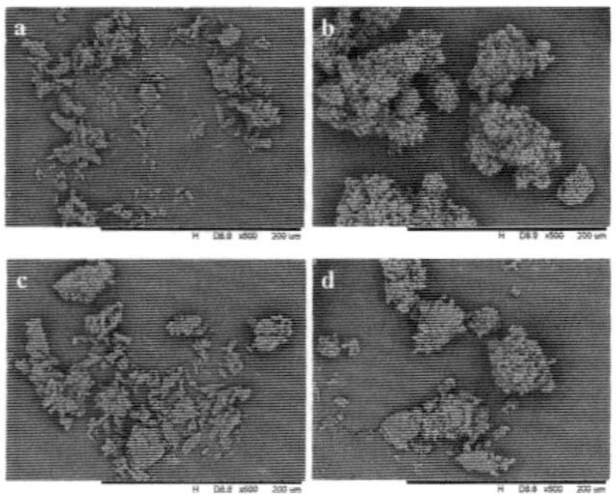

Figure 2 SEM images of pyraclostrobin product obtained from different experiments

a refer to pyraclostrobin products at the cooling rate of 0.1 K/min under concentration of 10g/100g solvent; b refer to pyraclostrobin products at the cooling rate of 0.5 K/min under concentration of 10g/100g solvent; c refer to pyraclostrobin products at the cooling rate of 0.1 K/min under concentration of 15g/100g solvent; d refer to pyraclostrobin products at the cooling rate of 0.5 K/min under concentration of 15g/100g solvent.

472527 Discriminate the Differences Between Surface Water and Channel Water in the Formation of Solvates

Abstract:

Discriminate the differences between Surface water and Channel water in the formation of solvates

Yumin Liu[1,2], Dandan Han[1,2], Jingkang Wang[1,2], Junbo Gong[1,2*]

[1] National Engineering Research Center of Industry Crystallization Technology, School of Chemical Engineering and Technology, Tianjin University, Tianjin, China;

[2] The Co-Innovation Center of Chemistry and Chemical Engineering of Tianjin, Tianjin University, Tianjin, China;

E-mail address: liuyumin@tju.edu.cn junbo_gong@tju.edu.cn *

Abstract

Solvates are defined as Active Pharmaceutical Ingredients (APIs) solids which incorporate one or more solvent molecules in the crystal lattice or channel to form new crystalline substances. Converting the drugs into hydrophilic solvates can help to improve the solubility in water and increase drug bioavailability. Meanwhile, it also can improve the stability of drugs, thus facilitating the formulation and storage of the drugs. There are four binding modes between compounds and solvents, just presented in the following Figure 1. Liquid inclusions and incorporation into the lattice are composed of solvates. With the increased incidence of solvate formation, it is essential to get a comprehensive understanding of their structure, physical stability, and the implication of de-solvation on the properties of the APIs.

Hydrate are the most abundant solvates, owing to the high propensity of water to be entrapped in the lattice of APIs, with its small size and tendency to form multi-directional hydrogen bonds. Pharmaceutical solids may come in contact with water during processing steps, such as crystallization, lyophilization, spray-drying, wet granulation, or aqueous film-coating, and they may be exposed to humid air during storage. Absorbed water molecules may reside on crystal surfaces, in crystal channel, or in crystal lattice structures which can change the packing in the unit cell. Surface water and channel water is quite different from each other. Without a doubt, we can utilize SXRPD to discriminate that differences between them. But in our laboratory, several other offline analytical techniques, such as XRPD, TGA and DVS, were also adopted to distinguish it and achieved remarkable results.

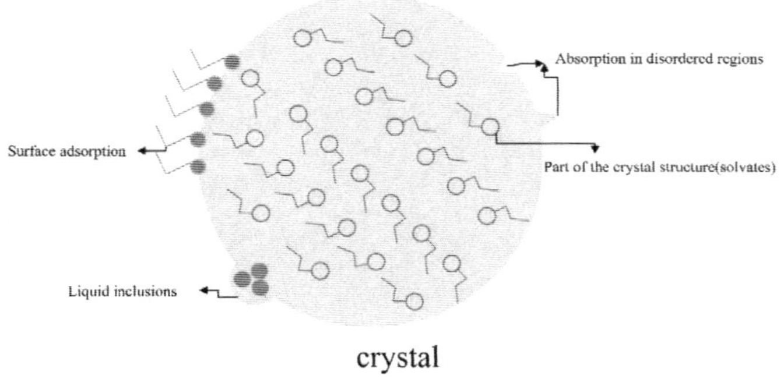

Fig.1 Different principles of solvent associations with crystalline solids

Fig.1 Different principles of solvent associations with crystalline solids

472078 Influence of Ultrasonic Wave on the Crystallization of Lysozyme

Abstract:

Due to the highly sensitivity of proteins tó the crystallization conditions, a slight change in operating parameters might affect the outcome of the crystallization. In this study, the effects of ultrasonic wave on the crystallization of lysozyme were investigated by applying different levels of ultrasound during the purification process of lysozyme from the aqueous solutions of hen egg white through the solvent freeze out (SFO) technology. Tetragonal lysozyme crystals were obtained during the process, which were then verified by X-Ray Powder Diffraction (XRPD). The quality of lysozyme crystal (composition and purity, size and shape, enzyme activity) was measured using SDS-PAGE, polarized light microscopy and spectrophotometer, respectively. It was found that the purity and size of lysozyme crystal were affected significantly by ultrasonic wave, while the enzyme activity of lysozyme crystal werenâ€™t. In addition, the influence mechanism of ultrasonic wave on lysozyme crystallization was discussed. The results obtained by this work will provide essential information for the purification of other proteins.

463163 Techno-Economic Optimization of a Non-Phosgene Polycarbonate Process

Abstract:

Techno-Economic Optimization of a Non-Phosgene Polycarbonate Process

K.M. Bentaher, T.N. Rogers(*)

Michigan Technological University,Department of Chemical Engineering

1400 Townsend Drive, Houghton, MI 49931-1295

(*)Corresponding Author

Abstract

Polycarbonate is an important polymer and one of the fastest growing engineering thermoplastics. It is widely employed in different applications, including automobiles, electric and electronic appliances, CDs and DVDs, and mobile phones. The annual growth rate of the global polycarbonate demand is expected to exceed 5% in the period 2011-2016. The global demand for polycarbonate is likely to exceed 6 million tons by 2020, while the production of polycarbonates is expected to reach about 5 million tons by 2020. The reasons for the wide use and high demand growth rate for polycarbonate are its excellent properties, which include outstanding impact resistance, good transparency, high heat resistance, and dimensional stability. Commercially, polycarbonate is produced in four grades, disk grade, low viscosity grade with high melt index (MI), medium viscosity grade with medium MI, and High Viscosity Grade with low MI.

Until recently, the vast majority of polycarbonates has been manufactured by the phosgene process. This process uses carbon monoxide and chlorine as raw materials to produce phosgene, which is highly toxic and its use is highly restricted as a chemical weapon. The phosgene process suffers from a number of environmental and economic drawbacks due to the use of phosgene as a monomer, the use methylene chloride in the polymerization step, and the need for a large quantity of water to separate the phosgene and chlorine from the polycarbonate product.

In order to overcome these environmental issues, this research project focuses on the development of phosgene-free polycarbonate manufacturing process. One of the green routes to produce polycarbonate without using phosgene is transforming dimethyl carbonate (DMC) to diphenyl carbonate (DPC), which can be further reacted with Bisphenol-A to produce Polycarbonate.

Alternative processes to produce the intermediates required to produce polycarbonate through a non-phosgene route will be evaluated technically, economically and environmentally to develop a base case for the polycarbonate process. After-tax discounted cash flow analysis will be used to evaluate the economic viability of introducing an optimized base case of the final non-phosgene polycarbonate process at a suitable US location.

454726 Optimization of Ultrafiltration and Diafiltration Process for a Shear Sensitive and Aggregation Prone Monoclonal Antibody

Abstract:

A traditional process for monoclonal antibody (mAb) purification ends with an ultrafiltration and diafiltration step. The purpose of Ultrafiltration (UF) is to concentrate and buffer exchange the monoclonal antibody into final formulation. In this case study, the final purification chromatography before the UF step is the hydrophobic interaction chromatography (HIC). The HIC was a prior polishing step to remove process impurities and aggregates, and the target mAb was eluted into a high concentration of sodium sulfate buffer. The HIC chromatography separated impurities from mAb based on hydrophobic interactions of the resin ligand and non-polar regions of the protein. The biochemical properties coupled with the high sodium sulfate concentration in the HIC process conditions resulted in low step yield and increase of aggregates over time under stress conditions. It was expected that during the initial concentration phase of the UF step, the presence of high concentrated sodium sulfate might lead to a higher surface hydrophobicity increasing antibody self-association and binding to the UF membrane. Multiple process parameters were investigated to mitigate the issue: sodium sulfate concentration for UF feed, membrane loading capacities, protein concentration of feed during at diafiltration, and impact of shear force on aggregate formation. Rheometer studies using feed streams with varying sodium sulfate concentration and impurity levels suggested that protein aggregation increased in the presence of high sodium sulfate concentration. Additional studies including investigation of loading capacities and diafiltration feed concentrations were executed to define the ultrafiltration parameters for scale-up. The developed UF process was incorporated into a successful pilot and GMP manufacturing scale to support clinical demands.

465734 Cyclic Process of Catalytic Methane Decomposition and Oxidative Regeneration over Nickel-Carbon Catalyst

Abstract:

Methane was decomposed over the nickel-carbon catalyst for producing hydrogen and carbon nano-onions. The generated carbon nano-onions were gasified by CO_2 and the spent catalyst was regenerated while CO_2 was converted to high-purity CO. The regenerated catalyst was re-used for the methane decomposition and the cyclic operation was repeated for 15 cycles. Both catalytic activities were stably maintained for 15 cycles, which provides a potential for the cycle process. The nickel-carbon catalyst was synthesized from CO_2 and two major green-house gases, methane and CO_2 were converted to hydrogen and CO. Therefore, the cyclic operations over the prepared catalyst can be a practical candidate of the sustainable process.

465738 Efficiency and Cost Comparison Between Dual-Stage a-MdeaÂ and Selexol Processes for Carbon Dioxide Capture in a 300-MW-Class Integrated Gasification Combined Cycle (IGCC) Power Plant

Abstract:

In order to minimize the environmental impact of power plant operations and reduce the CO_2 emissions, carbon capture processes (CCPs) become an important technological issue. In this study, various CCPs for a 300-MW-class integrated coal gasification combined cycle (IGCC) power plant were simulated to evaluate thermal and economic efficiencies. As a first step, the efficiencies of single-stage a-MDEA, Selexol, Pufisol, Rectisol and Cryogenic processes as an acid gas removal unit (AGRU) were evaluated for a screening test. Considering both thermal efficiency based on screening test and the balance between chemical and physical solvent systems, dual-staged a-MDEA and Selexol processes for AGR and CCP were rigorously evaluated in view of power generation, power consumption, exergy destruction, as well as economic analysis. The results indicated that Selexol exhibits better thermal efficiency. And operating cost of Selexol process was also cheaper than that of a-MDEA process for capturing a ton of CO_2. However, the capital cost of Selexol process was more expensive than that of a-MDEA process. Based on the exergy analysis, the detailed exergy flow and destruction were illustrated to evaluate the impact on every constituent component of each CCP. And the development direction to improve the efficiency of each CCP was suggested. The results in the study can be used as a springboard in decision-making and process development for commercial-scale CCPs in view of thermodynamic and economic analyses.

466265 Ionic Liquids As Entrainers for the Separation of Aliphatic/Aromatic Mixtures By Extractive Distillation Systems: Rigorous Model Based Process Optimization

Abstract:

The great energy consumptions are one of the main workhorses that the process industries are addressing in the XXI century. It is well known that the most important consumptions are due to separation processes. Besides, it is common that the more complex operation, the greater energy consumption, which is especially true for the separations of azeotropic mixtures. Azeotropes and close boiling point mixtures are quite usual in chemical industries appearing in solvent, gasoline or monomer industries, making impossible (or very expensive) the separation of individual components of the mixtures by conventional distillation. This is overcome by using, on one hand, material separation agents in operations such as adsorption or liquid-liquid extraction, or by adding an entrainer to a distillation column (azeotropic or extractive distillation). This work is focussed on the application of recently developed solvents, called ionic liquids, as entrainers for the separation of a low differential volatility mixture (n-heptane/toluene). Separation of aromatic hydrocarbons from naphtha has a great interest for the petrochemical industry. On one side, aromatic hydrocarbons represent an extra load for the naphtha crackers, negatively affecting the process economy as additional operative and capital costs. On other side, aromatic hydrocarbons have their own value as raw materials, being furtherly transformed in higher value-added products. Between all technologies studied sulfolane has resulted the most extensively used thanks to an UOP technology, which employs it as extracting solvent. In this process, sulfolane has to be recovered from both the extract and the raffinate after liquid-liquid extraction, which entails high investment and energy costs. Ionic liquids (ILs) are a novel attractive candidate for this separation since Seddon in 1997 reported the great capacity of imidazolium ionic liquids to dissolve benzene. Ionic liquids are molten salts with a melting point below (or near) ambient conditions. They are composed by an anion and an organic cation, so they can be chosen in order to tune their physical properties (tailor made solvents). The application of ionic liquids for the separation of aliphatic and aromatic components is being under research, mainly by experimental measurements of liquid-liquid or vapor-liquid equilibria (isobaric or isothermal measurements) [1]. On the other hand it is also under research the development of accurate models for the correlation and prediction of their properties. In this way, in the last decades new physically founded methods based on quantum chemistry theory have emerged and have been largely applied due to the current computational facilities. That is the case of the COSMO-RS model developed by Klamt [2] or the COSMO-SAC model of Li et al. [3]. These COSMO-based models are a priori methods that predict thermodynamic properties of fluid mixtures by solely using the molecular surface polarity distributions of the individual compounds, obtained by quantum chemical methods. Many thermodynamic properties can be calculated by COSMO-based models, such as vapour pressure, normal boiling points, excess properties and phase equilibria data. Additionally, some COSMO-based molecular descriptors can be used to estimate some properties by quantitative-structure-property-relationships (QSPR) such as viscosity, density or partition coefficients in polymers. In the last years we have successfully integrated COSMO-RS results into process simulations using Aspen Techâ€™s software in order to develop potential new industrial processes with mixtures of components not available in the simulatorsâ€™ databases. We have applied this procedure to biotechnological processes and especially to processes involving ionic liquids (ILs), our main goal being to provide technical and economic information about the potential industrial applications of ILs. All this has the aim of enriching the design/selection of ILs with optimized properties for specific applications.

In this work, we presented the study of how ionic liquids can increase the efficiency of n-heptane/toluene mixtures separation. We have made a screening of potential ionic liquids to be used with the COSMOTherm software and selected a small group of them. Then, following the previously described methodology we have implemented such components in the Aspen Properties databanks by specifying the required properties. Rigorous process simulations of the extractive distillation column can be then developed. IL Mixture design was carried out by defining an optimization problem in Matlab and solved using derivative-free optimization methods. A COM interface was needed in order to connect Matlab with the Aspen Plus software. The objective function implemented to design the entrainer mixture took into account both separation an economic criteria.

Besides, we have correlated the experimental liquid-liquid equilibrium [4] and vapor-liquid equilibrium [5] data of some ionic liquid â€" solvent systems to conventional activity coefficient thermodynamic models (NRTL) in order to evaluate COSMO predictions. Binary interaction parameters were then implemented in Aspen Plus and some process simulations were carried out in order to determine the deviation between the regressive thermodynamic model (NRTL) and the predictions of the COSMO based simulations.

Sothis work is structured as follows: First a short introduction to ionic liquids and to the COSMO based model is shown, then the methodology employed to couple quantum calculations with Aspen Plus and the optimization strategy is described. Third, a detailed comparison of COSMOSAC predictions against regressive thermodynamic models (NRTL) is shown in terms of process simulation results. The next section addresses the optimization problem of the extractive distillation process selecting the best ionic liquid or mixture for the separation proposed and finally the main conclusions of the work are inferred.[1] Navarro, P., Larriba, M., García, J., Rodríguez, F. Vapor-liquid equilibria for n-heptane + (benzene, toluene, p-xylene, or ethylbenzene) + [4empy][Tf2N] (0.3) + [emim][DCA] (0.7) binary ionic liquid mixture. Fluid 540 Phase Equilibria 2016;417:421

[2] Klamt, A. and F. Eckert, COSMO-RS: a novel and efficient method for the a priori prediction of thermophysical data of liquids. Fluid Phase Equilibria, 2000. 172(1): p. 43-72.

[3] Lin, S.-T. and S.I. Sandler, A Priori Phase Equilibrium Prediction from a Segment Contribution Solvation Model. Industrial & Engineering Chemistry Research, 2002. 41(5): p. 899-913.

[4] Larriba, M., Navarro, P., Garcia, J., Rodríguez, F.. Liquid-liquid extraction of btex from reformer gasoline using binary mixtures of [4empy][tf2n] and [emim][dca] ionic liquids. Energy and Fuels 2014;28(10):6666:6676

[5] Navarro, P., Larriba, M., García, J., González, E.J., Rodríguez, F. Vapor-liquid equilibria of n-heptane+ toluene+ [emim][DCA]g system by headspace gas chromatography. Fluid Phase Equilibria 2015;387:209:216.

467193 Process Synthesis of Continuous Downstream Bioprocessing of Recombinant Intracellular Products

Abstract:

Though continuous manufacturing has been implemented in almost every other industry, but biopharmaceutical industry has been reluctant to change from the archaic batch processing model. In the past, the main concern of the biopharmaceutical industry was the regulatory authorities' definition of a batch, but the FDA and European Medical Agency (EMEA) has defined a batch as a specific quantity of a drug that is intended to have uniform character and quality within specified limits, putting more emphasis on drugs meeting specific quality standards and less emphasis on the means of production[1]. Therefore, biopharmaceutical companies should be pursuing integrated continuous biomanufacturing platforms as these methods have been shown to reduce costs (net present value) by 55% relative to conventional batch processing[2]. However, biopharmaceutical companies are still reluctant to pursue integrated continuous biomanufacturing platforms for intracellular products because intracellular cell culture is difficult to achieve continuously, specifically due to the bottlenecks of the downstream processing.

In this research, we explore the development of continuous downstream bioprocessing using process synthesis for recombinant intracellular products. The case study that we are conducting uses the extraction of beta-carotene from recombinant Saccharomyces cerevisiae. The separation and extraction pathway consists of five sections, 1) cell harvesting, 2) cell disruption, 3) cell extraction and recovery, 4) drying and tablet formulation and 5) recovery of media to be reused in the bioreactor. The superstructure is formulated as a mixed-integer nonlinear programming (MINLP) mode using profitability as the objective function based on that the cost of production for a 10% beta-carotene tablet found in literature[3]. The optimal topology for the downstream processing will be integrated with a continuous bioreactor model to give an optimal, continuous production implementation for intracellular product for comparison with traditional batch processes.

1. Jungbauer, Alois. "Continuous Downstream Processing of Biopharmaceuticals." *Trends in Biotechnology* 31.8 (2013): 479–492.
2. Andrew L Zydney, Perspectives on integrated continuous bioprocessing — opportunities and challenges, Current Opinion in Chemical Engineering, Volume 10, November 2015, Pages 8-13, ISSN 2211-3398, http://dx.doi.org/10.1016/j.coche.2015.07.005.
3. Marz, Ulrich. *The Global Market for Carotenoids*. BCC Research, 2015.

467463 Development of Novel Approach to Remove Acid Gas from Natural Gas at Off-Shore Plants

Abstract:

The removal of acid gas from natural gas at offshore plant is critical to meet international environmental and safety regulations. Furthermore, the production rate at offshore oil and gas fields is steadily increasing relative to the production rate at onshore fields. Hence, this study focuses on methods to enhance the overall efficiency of the AGRU (Acid Gas Removal Unit) utilized in the topside of offshore plants by considering various acid gas removal technologies such as adsorption, absorption and membrane-based separation. In particular, this research focuses on absorption using amine-based chemical solvents: a commonly used method in industry, combined with membrane-based separation: a method that is suitable for offshore environment conditions such as rocking motion or permanent tilt. The effects of various different factors are analyzed including the efficiency of various types of amine solvents and configuration of the absorption/membrane systems which are assessed based on their economic feasibility including capital and operating costs, etc. This work investigates and proposes effective designs for the AGRU used in the topside of offshore plants.

Acknowledgement

This work was supported by a research project of "Development of Integrated Interactive Model for Subsea and Topside System to Evaluate the Process Design of Offshore Platform", funded by Ministry of Trade (project No. 10060099).

479159 Production of Theobromine by N-demethylation of Caffeine Using Metabolically Engineered *E. coli*

Abstract:

This study reports bench-scale feasibility study for production and recovery of theobromine from caffeine using metabolically engineered *E. coli*. The N_1-demethylase genes used in this study were originally discovered in *Pseudomonas putida* CBB5. Five strains of *E. coli* engineered with multiple combinations of the two *N*-demethylase genes were compared for growth and activity. Strain pAD1dDD containing one copy of *ndmA* and three copies of *ndmD* in two compatible plasmids, gave the best results (98.5% molar conversion of caffeine to caffeine). The reaction was scaled up to 1.95 L and theobromine was recovered in nearly 100% pure form via preparative chromatography and drying. Final yield was 80% relative to caffeine. Purity of the product was identical to an authentic standard as judged by LC-MS and NMR. This is the first report of theobromine production via a metabolically engineered strain, with the highest yield and purity reported of any biological method. This method also eliminates the harsh reaction conditions and environmental issues associated with the chemical process.

466567 Techno-Economic Analysis of the Ethanol Dehydration Processes: Isothermal Vs. Adiabatic Operation

Abstract:

The present work addresses the dehydration of ethanol over zeolite catalyst, where main products are ethylene and diethyl ether (DEE) while some of hydrocarbons of C3+ are produced by oligomerization, cracking and hydrogenation of ethylene [1,2]. In the suggestion of kinetic mechanism for the dehydration of ethanol, two different pathways are considered; either ethanol is directly converted to ethylene or diethyl ether (DEE) is formed and then, cracked to ethanol and ethylene. For quantitative analysis, reaction rates are developed on the basis of Langmuir-Hinshelwood equation and kinetic parameters are estimated by fitting experimental data under a variety of conditions. Temperature shows positive effects on the conversion, while too high temperature decreases the selectivity of ethylene. The effects of water are also shown that the conversion is decreased with increasing water amount due to the competitive adsorption. In the feed, DEE is also included with varying amount to evaluate the role of each pathway in such a way that the developed model is used to determine the condition for maximizing the yield of ethylene.

The kinetic model was extended to develop pilot-scale reactors under the assumption of both adiabatic and isothermal operation, and then, the separation systems were suggested to produce pure ethylene (purity >99 wt.%). In the case of an isothermal reactor, the effects of the number and size of tubes on the capital expenses of the reactor were evaluated and the most appropriate one was used in the design of the process. Meanwhile, an adiabatic reactor was considered to be composed of several reaction tubes with inter-heating system incorporated between them. Despite the series of reaction tubes, the conversion was below 100%, indicating that additional separation system such as a cryogenic distillation should be employed for the recovery of ethylene from by-products, compared to the isothermal operation.

For preliminary quantitative analysis for the economics of both processes, the optimal size of each reactor was determined, and the purchase and installation costs were estimated using several correlations reported

in the literature. The isothermal operation showed that the operation expenses for the separation were increased with increasing operation temperature, while the capital cost for the reactor was decreased due to the increased reaction rates, resulting in the decrease of total cost in the process. Meanwhile, although the adiabatic case had merits in the design of the reactor tubes, significant increase in the expenses for the separation was observed because of the increased operation cost by additional use of the steam in the reboiler and the cooling media in the condenser. In conclusion, the comparison of two processes based on different operation mode would provide the useful data on the design of the optimal pilot-scale process.

References

[1] Gayubo et al., *Ind. Eng. Chem. Res.* **2001**, 40, 3467–3474.

[2] Galadima and Muraza, *Ind. Eng. Chem.*, **2015**, 31, 1–14.

[3] Chiang and Bhan, *J. Catal.*, **2010**, 271, 251–261.

[4] Phung and Busca, *Chem. Eng. J.*, **2015**, 272, 92–101.

[5] Cameron et al., *Process Design for the Production of Ethylene from Ethanol*, 2012.

[6] Jernberg et al., *Ethanol Dehydration to Green Ethylene*, 2015.

Acknowledgements

This work was supported by the Korea Institute of Energy Technology Evaluation and Planning (KETEP) under the "Energy Efficiency & Resources Programs" (Project No. 20153010092090) of the Ministry of Trade, Industry & Energy (MOTIE), Republic of Korea, and the Human Resources Development of the Korea Institute of Energy Technology Evaluation and Planning (KETEP) grant funded by the Korea government Ministry of Trade, Industry & Energy (No. 20154010200820).

469241 A Combined Computational and Experimental Study of Copolymerization Kinetics for Glycidyl Methacrylate and Tert-Butyl Methacrylate

Abstract:

Methacrylate families are widely used to prepare photoresists in the semiconductor industry.[1] Numerous possible methacrylate polymerization recipes can be formulated to design products with desired properties including molecular weight, molecular weight distribution, and copolymer composition. Demand is growing to develop computational methods that improve the copolymerization product design and help with the product scale-up.[2,3] While the reaction kinetics of simple monomers such as methyl methacrylate (MMA) have been studied extensively,[3-11] more complex methacrylate monomers have been overlooked. In this work, a combined computational and experimental study on copolymerization kinetics of glycidyl methacrylate (GMA) and tert-butyl methacrylate (tBMA) was conducted to investigate the polymerization propagation kinetics. Density functional theory (DFT) was used to investigate the propagation kinetics and to calculate monomer reactivity ratios. A benchmark test of DFT methods was carried out on GMA self-propagation rate coefficients with experimental support.[12] To improve the accuracy in thermochemical calculations, two tiers of treatment of vibrational frequencies were applied and compared, including a harmonic oscillator (HO) model[13] and a one-dimensional hindered rotor (1D-HR) model[14-18] with scale factors. Our results showed that the 1D-HR model generated a more accurate estimate of pre-exponential factors, rate constants, and reactivity ratios than the HO model. The calculated reactivity ratios and rate constants were then used in a polymerization kinetic model to predict the polymerization reaction rate and molecular weight. The predicted molecular weight was in good agreement with the experimental results.

(1) Bauer, W. In *Ullmann's Encyclopedia of Industrial Chemistry*; Wiley-VCH Verlag GmbH & Co. KGaA: 2000.

(2) Bebe, S.; Yu, X. R.; Hutchinson, R. A.; Broadbelt, L. J. *Macromolecular Symposia* 2006, *243*, 179.

(3) Zhang, G. Z.; Konstantinov, I. A.; Arturo, S. G.; Yu, D. C.; Broadbelt, L. J. *J Chem Theory Comput* 2014, *10*, 5668.

(4) Matheson, M. S.; Auer, E. E.; Bevilacqua, E. B.; Hart, E. J. *J Am Chem Soc* 1949, *71*, 497.

(5) Beuermann, S.; Buback, M.; Davis, T. P.; Gilbert, R. G.; Hutchinson, R. A.; Olaj, O. F.; Russell, G. T.; Schweer, J.; vanHerk, A. M. *Macromol Chem Phys* 1997, *198*, 1545.

(6) Simal, F.; Demonceau, A.; Noels, A. F. *Angew Chem Int Edit* 1999, *38*, 538.

(7) von Werne, T.; Patten, T. E. *J Am Chem Soc* 2001, *123*, 7497.

(8) Yu, X. R.; Pfaendtner, J.; Broadbelt, L. J. *J Phys Chem A* 2008, *112*, 6772.

(9) Yu, X. R.; Levine, S. E.; Broadbelt, L. J. *Macromolecules* 2008, *41*, 8242.

(10) Degirmenci, I.; Aviyente, V.; Van Speybroeck, V.; Waroquier, M. *Macromolecules* 2009, *42*, 3033.

(11) Degirmenci, I.; Eren, S.; Aviyente, V.; De Sterck, B.; Hemelsoet, K.; Van Speybroeck, V.; Waroquier, M. *Macromolecules* 2010, *43*, 5602.

(12) Wang, W.; Hutchinson, R. A. *Macromolecules* 2008, *41*, 9011.

(13) Merrick, J. P.; Moran, D.; Radom, L. *J Phys Chem A* 2007, *111*, 11683.

(14) Van Speybroeck, V.; Van Neck, D.; Waroquier, M.; Wauters, S.; Saeys, M.; Marin, G. B. *J Phys Chem A* 2000, *104*, 10939.

(15) Van Speybroeck, V.; Vansteenkiste, P.; Van Neck, D.; Waroquier, M. *Chem Phys Lett* 2005, *402*, 479.

(16) Heuts, J. P. A.; Gilbert, R. G.; Radom, L. *J Phys Chem-Us* 1996, *100*, 18997.

(17) Pfaendtner, J.; Yu, X.; Broadbelt, L. J. *Theor Chem Acc* 2007, *118*, 881.

(18) Pfaendtner, J.; Yu, X.; Broadbelt, L. J.; 1.0 ed., available for use free by contacting Linda J. Broadbelt.

466568 Viscoelasticity and Molecular Mobility of Celecoxib/Pvp/Tpgs Melt: A New Approach for Amorphous Drug Dispersion Assessment during Hot Melt Extrusion Product Design

Abstract:

Innovations at the Product Design and Discovery Interface

Abstract # 466568

Title: Viscoelasticity and molecular mobility of celecoxib/PVP/TPGS melt: A new approach for amorphous drug formulation assessment during hot melt extrusion (HME) product design.

Introduction/Motivation/Objectives:

Formulating drugs (active pharmaceutical ingredients) as amorphous solid dispersions is an alternative for improving the solubility of poorly water soluble drugs. Hot melt extrusion (HME) has demonstrated to be the adequate manufacturing unit process for compounding such amorphous solid dispersions. However, the risk of a potential event of inadequate amorphous content in the extrudate is always present, which is of critical significance when assessing the risk to the quality of such formulations during developmental product design phases.

Celecoxib is a poorly water soluble (BCS class II) non-steroidal anti-inflammatory drug. When compared to other BCS class II compounds, it has relatively low molecular mobility (high viscosity) and crystal growth rate upon cooling from the undercooled melt state [1,2]. The presence of surfactants or polymers has demonstrated to significantly increase or inhibit the radial crystal growth rate of Celecoxib, respectively [3].

We have previously studied the impact on the Celecoxib crystallization by PVP (polymer) and TPGS (surfactant) in terms of molecular mobility. Over all it was demonstrated that the polymer decreases the mobility of the drug, whereas the surfactant acts as a plasticizer. The dynamic viscosity of the Celecoxib with polymer and/or surfactant follows Arrhenius temperature dependence near the melting point (10°C of undercooling), however the VTF model results in higher agreement as the undercooling degree increases.

Even though, viscosity has been assessed for HME processability of melt formulations, it has not been associated to the amorphous content in the extrudate when processed using HME. Herein we present the link between viscoelasticity and amorphous state and content in the extrudate of mixtures of celecoxib, PVP and/or TPGS. We expect these results to complement the knowledge available for when assessing a drug as a candidate for a solid dispersion formulation during developmental stages of a drug product.

Sample Preparation:

Physical samples were prepared by mixing the drug with the corresponding amount of polymer and/or surfactant. Samples were mixed manually following a mixing protocol for homogeneous material distribution, and hence reproducibility.

Binary and ternary mixtures consisted of 50 wt.% TPGS or 50 wt.% PVP dispersed in a 50 wt.% Celecoxib, and ternary mixtures of 25 wt.% PVP, 25 wt.% TPGS and 50 wt.% Celecoxib. Samples where extruded using a Thermo Scientific MiniLab II HAAKE Rheomex CTW5 II HME. Extruded amorphous content and state was characterized using X-Ray Powder Diffraction (XRD), and Raman spectroscopy combined with a microscope for area of sample selection to obtain the spectra. The latter allowed for testing of drug homogeneous content throughout the extruded material.

Methodology/Data Analysis:

Extruded drug samples where analyzed in terms of amorphous content and amorphous state. The results where compared with our previous data on viscoelasticity and molecular mobility of the melt. Together we linked the molecular mobility (viscoelasticity) of the melt with the amorphous content in the extrudate of celecoxib mixed with PVP and/or TPGS.

Results and Analysis:

The samples were extruded using a bench-top HME (Thermo Haake MiniLab II) at a temperature of 140°C and a speed of 25 1/min (rpm). Raman spectroscopy of the extrudate samples resulted in a significantly less amorphous content when celecoxib was combined with PVP only, than when combined with both PVP and TPGS. This is evidenced with a noticeable larger Raman intensity of the celecoxib/PVP/TPGS mixture of each of the most important functional groups of celecoxib (CF_3, SO_2 and NH_2) compared to celecoxib mixed with PVP.

Powder diffraction characterization (XRD) showed a significant reduction in the crystalline state of the drug when mixed with PVP compared to PVP/TPGS, evidenced by the disappearance of the characteristic peaks of Celecoxib between 13 and 25 degrees (2Θ). On the other hand, TPGS increased the intensity of these peaks, and hence the crystalline state.

SEQ Table * ARABIC 1. Comparison between the off-line rheological and Raman data. A decrease in molecular mobility of the melt results in a higher amount of amorphous content in extrudate.

Samples	Off-line rheology			**HME**	Raman
	Molecular mobility (viscosity)*	Viscoelastic behavior*	Viscoelastic behavior* (140°C)		Amorphous content in extrudate*
Celecoxib/PVP	Decreased (viscosity increase)	Liquid to Elastic	Liquid		Increased
Celecoxib/PVP/TPGS	Decreased *(Decreased much less than with PVP alone)*	Liquid to Elastic	Liquid	**140°C, 25 min^{-1}.**	Increased *(Increased much less than with PVP alone)*
Celecoxib/TPGS	Increased (viscosity decrease)	Liquid to Elastic	Liquid		Decreased

*Compared each to pure Celecoxib.

The table summarizes the molecular mobility behavior and the amorphous content of the extrudate. When analyzed together, scientifically justifies the inclusion of viscoelasticity when designing a drug product formulation for HME. A decrease in viscosity results in an increase of the amorphous content of the drug, whereas an increase in molecular mobility results in a decrease in amorphous content in the extrudate. However, it must be highlighted that this correlation is conditional to the materials present in the melt, polymers (PVP) or surfactants (TPGS) as when these are combined less amount of amorphous content may result.

Conclusions:

We have elucidated the relationship between the molecular mobility of celecoxib when in presence of PVP (polymer) and/or TPGS (surfactant), the elastic/liquid (viscoelasticity) behavior, and the resulting amorphous content of the celecoxib in the extrudate. This work provides the proof-of-concept that viscoelasticity characterization combined with analytical techniques in HME could be used to tailor an amorphous drug with reproducible and adequate amorphous content manufactured using HME, being this correlation currently unknown and highly beneficial for when designing an amorphous drug dispersion.

References

[1] Baird, J. A.; Santiago-Quinonez, D.; Rinaldi, C.; Taylor, L. S. Role of Viscosity in Influencing the Glass-Forming Ability of Organic Molecules From the Undercooled Melt State. *Pharm Res* **2011**, *29*, 271–284.

[2] Baird, J.A.; Van Errdenbrugh, B.; Taylor, L.S. A Classification Tendency of Organic Molecules from Undercooled Melts. *J Pharm Sci* **2010**, 99, 3787-3806.

[3] Mosquera-Giraldo, L. I.; Trasi, N. S.; Taylor, L. S. Impact of Surfactants on the Crystal Growth of Amorphous Celecoxib. *International Journal of Pharmaceutics* **2014**, *461*, 251–257.

[4] Trachenko, K. The Vogel-Fulcher-Tammann Law in the Elastic Theory of Glass Transition. *Journal of Non-Crystalline Solids* **2008**, 354, 3903-3906.

[5] Gupta, P.; Chawla, G.; Bansal, A.K. Physical Stability and Solubility Advantage from Amorphous Celecoxib: The Role of Thermodynamic Quantities and Molecular Mobility. *Molecular Pharmaceutics* **2004**, 1, 406-413.

472436 Miniaturized Purification Platform for Automated Screening of Flow Synthesis Using Extraction and Crystallization

Abstract:

R&D for continuous flow chemistry (e.g. end-to-end continuous manufacturing and automated synthesis lab) necessitates a modular and reconfigurable chemical synthesis platform capable of screening of a variety of chemical pathways. Flow microreactors are now commercially available, and more recently, drop-on-demand formulation device, such as ink- and droplet-based pharmaceutical printers, have made it to market. While these devices can be naturally combined to construct a modular end-to-end continuous flow platform, the platform cannot be fully realized without its purification counterpart. Nevertheless, microscale purification platform is not yet well established; Although a number of innovative separation techniques have been reported in the microfluidics literature, including evaporation, extraction, crystallization, filtration and chromatography, most of these processes have been designed as a standalone for lab-on-a-chip applications. An automated and reconfigurable purification device which can seamlessly integrate with other parts of the flow synthesis platform in a flexible and scalable manner is therefore desirable. In this work, a microscale purification platform which can virtually be constructed using off-the-shelf parts is described. Specifically, our device consists of a combination of commercially available parts, such as capillaries, chips, mixers, fittings, etc. as well as custom parts, including tubing heaters and zero-dead volume Y-junctions. In addition, the device support multiple reservoirs for use with aqueous and organic solvents, a Peltier-based heating and cooling system, a couple of process analytical tools, including UV and non-contact Raman spectroscopy and an auto-sampler for sample collection of offline analysis using techniques such as MS and UPLC. The device operation is controlled using a LabView software which integrates, controls and monitors electronic actuators, including syringe pumps, valves, pressure sensors, etc., and executes a sequence of instructions to perform priming, loading and cleaning. The capability of the device is demonstrated via automated screening of continuous and segmented (air and liquid) flow extraction and crystallization under various favorable and unfavorable conditions for the purification of model APIs including diphenhydramine and atropine.

464011 Novel Validation Method Development for a Drug Delivery Device Used during Breastfeeding

Abstract:

4.5 million infants die every year. Many of these deaths can easily be prevented, with existing medications, if hygienic and effective paediatric formulations were available. In low-resource settings, the need for medications, which do not require refrigeration, can easily be accurately dosed, and can be hygienically administered is especially high, due to electricity, potable water, and literacy limitations. Breastfeeding infants are especially in need of appropriate drug and delivery device combinations, considering few drug delivery systems are commercially available specifically for this population, and infant compliance can be low.

We are responding by developing a novel device for administering medications or nutrients to infants, and *in vitro* validation methods for characterising itâ€™s efficacy. The device, when worn by a mother during breastfeeding, releases active pharmaceutical ingredients contained in the device into the milk directly consumed by the infant.Â It therefore has many potential advantages. For example, by suspending the medication or nutrient into the breast milk directly, and since it could be disposable, it overcomes the need for potable water as a suspension or cleaning fluid. It contains a solid dosage form, and therefore does not require refrigeration. Since the device is pre-loaded with medication a mother does not need to be literate to measure out a dose. Furthermore, since it is used during breastfeeding, infants may find it to be a more natural means of receiving medicine than conventional spoons or syringes.

Several validation methods have been used and developed in optimising the device design and potential corresponding dosage forms. These include a breastfeeding simulation apparatus, a novel infant tongue peristalsis simulation apparatus, and a novel tablet disintegration characterisation method using texture analysis.

The breastfeeding simulation apparatus was used to assess the release from the device of several dosage forms containing zinc, used in the treatment for diarrhoeal disease. The apparatus mimics maternal lactation, by flowing human milk through a silicone breast mimic. The milk then passes through the device loaded with the therapeutic insert, and then is sucked by a modified breast pump, at pressures physiologically representative of breastfeeding, into an exit line from which samples are collected with a fraction collector. The generated breast milk samples supplemented with zinc, collected over a simulated breastfeed are then analysed via inductively coupled optical emission spectrometry to test drug release rates from the drug delivery system. Zinc release over the course of 20 minutes, more than the time typically used during breastfeeding, in the simulations for four rapidly disintegrating tablet formulations ranged from 32-51%, indicating further formulation optimisation is needed.

An infant tongue mimic system was developed to simulate *in vitro* the peristaltic motions and associated pressures exerted from an infant tongue during breastfeeding. The system was used to characterise the potential effects of tongue movement on the disintegration of model tablets from the drug delivery system. The tongue mimic system includes a modified peristaltic pump equipped with a novel tongue mimic, a piezoresistive sensor for pressure measurements, a tube for holding the tablet under model breastfeeding flow conditions, and an adjustable palate mimic, allowing for pressure to the tablet to be modified based on clinical data. The release profiles of the model compound, sulforhodamine B, measured at two rotational values possible during breastfeeding for two potential compression values, representative of two potential tongue strengths of infants were statistically different ($P=0.0013$ and $P<0.001$, using 2-way ANOVA with matching for the lower and higher compression values respectively). For breastfeeding rotational rates representing non-nutritive breastfeeding, compression amount was found to be a significant factor ($P=0.0162$). In addition to being useful for characterising dosage forms for the device under the influence of tongue motion during suckling, this system could be useful in characterising other oral dosage forms or products.

Disintegration characterisation in human milk of potential tablet formulations is critical for screening dosage form appropriateness for the device. Therefore, a texture analysis disintegration method was also developed to analytically characterise tablet disintegration in opaque media. This method was developed to overcome visibility challenges which would be associated with use of the USP disintegration test with human milk. In this method, a texture analyser applies a load to a rapidly disintegrating tablet immersed in opaque media, and load data is collected over time as the tablet disintegrates, with fluctuations in the load indicating tablet fracturing. An in-house tablet formulation of a rapidly disintegrating tablet, and a commercially available tablet formulation were characterised in water, bovine, and human milk. These were found by one-way ANOVA to have statistically different disintegration times in each media ($P<0.0001$). This system is useful in characterising other tablet formulations for the device, and may be useful in the characterisation of other materials in opaque media.

Further work aims to continue testing various designs of the drug delivery system, as well as other formulations of corresponding dosage forms, to optimise the release of active pharmaceutical agent from the device.

470420 Photothermal Therapy of Bladder Cancer and Melanoma Via Targeted Single-Walled Carbon Nanotubes

Abstract:

Single-walled carbon nanotubes (SWNTs) are unique in that they strongly absorb near-infrared (NIR) light, while biological systems have very low levels of absorption of NIR light. This project focuses on targeting the SWNTs using conjugation with the human annexin A5 protein. Annexin A5 (molecular weight of 36,000) is a monomeric protein that binds with high affinity to phophatidylserine (PS) in phospholipid bilayers. PS is the most abundant anionic phospholipid of the plasma membrane and is tightly segregated to the internal side of the plasma membrane in most cell types. Anionic phospholipids are largely absent from the external surfaces of resting mammalian cells under normal conditions. PS is expressed on the external surface of tumor cells and also of endothelial cells that line the blood vessels in tumors, but it is not expressed on the outside surface of the vascular endothelium in normal organs. The SWNTs serve as an NIR target resulting in tumor heating and ablation.

This project focuses on the photothermal therapy treatment of two types of cancer that initially occur on surfacesâ €"bladder cancer and melanoma. Approximately 80% of bladder cancer patients have recurrent tumors starting with stage I superficial and quickly progressing towards more muscle invasive and metastatic malignancies with each recurrence. The high incidence of reappearance is believed to be due to the residual tumor left behind in a significant number of patients. The current form of treatment is a transurethral resection for lower grade tumors and cystectomy for higher grade masses. Novel treatments are needed for eliminating all bladder tumor cells with a minimum number of treatments, ideally one. One type of skin cancer, malignant metastatic melanoma (CMM), has a high mortality rate. When CCM becomes metastatic, patient survival rates fall as low as 5%. Over the last half-decade rapid advances in biologic therapeutics and nanotechnology have spawned a new generation of treatment options proving a rich opportunity to investigate combinatorial therapies.

In the study of this therapy for treating bladder cancer, *in vitro* binding tests confirmed a strong binding affinity of SWNT-annexin A5 to MB49 mouse and J82 human bladder cancer cells. Combining SWNT-annexin A5 heating via NIR light confirmed significant cell death as compared to untreated controls for both cell lines. The *in vitro* tests provided statistically significant validation for the potential of this targeted ablation therapy. *In vivo* testing on immune competent mice was conducted to confirm the efficacy of this treatment even further. A biodistribution study followed by FT-Raman analysis confirmed no non-specific accumulation of SWNT-annexin A5 in any organs. Using a 360° diffuser to deliver NIR light to the inside of mouse bladders, NIR power tolerance tests confirmed that no healthy tissue damage occurred at 50 J/cm^2. *In vivo* therapy testing of MB49 cells implanted in the bladder of immune competent mice confirmed a promising treatment modality which is currently being further investigated with more testing using bioluminescence imaging of the cells in combination with fluorescence microscopy.

The work on treating melanoma used B16F10-luciferase mouse melanoma cells injected into immune competent mice. Plasma levels of SWNT-annexin A5 were monitored for up 24 hours after i.v. injection in B6 immune competent albino mice; plasma levels of the conjugate were negligible 1 hour after injection. The minimum injected dose of SWNT-annexin A5 used in combination with photothermal therapy using NIR irradiation was found for obtaining complete eradication of primary tumors (implanted subcutaneously into the flank). Having established the necessary photothermal therapy treatment conditions, we are currently undertaking an evaluation of the synergism between this photothermal therapy and the immunostimulants anti-PD1, anti-CTLA-4, anti-CD73, and cyclophosphamide to ascertain possible mechanisms of harnessing photothermal antigen production to eradicate distant metastases following primary tumor ablation.

471289 Bioproduct Design through Market Assessment: NSF Innovation Corps (I-Corps^{TM}) Participation Guides the Development of New *Escherichia coli* Strains

Abstract:

Host cell proteins (HCPs) complicate the recovery of biologically-derived products due to their presence as the target product is isolated by various unit operations that include chromatography, precipitation, and other common bioseparation techniques. While the technology exists to rewrite the genome of the host cell used for expression, few have attempted to use these technologies to reduce HCP burden on the bioseparation steps. Such a space to investigate is quite large because of the number of different techniques and conditions used for bioseparation. Nevertheless, our team set out to elucidate a product design that could gain traction with academia and industry. Participation in the I-Corps^{TM} program defined the characteristics of a minimum viable product through a combination of hypothesis-driven discovery and potential end-user comments.

To wit, a target bioseparation technique (diethylaminoethyl or DEAE ion exchange chromatography) and improvement threshold was identified based on extensive discussions with potential end-users, and a minimum viable product was constructed. An engineered *Escherichia coli* cell line (Lotus®) was built via homologous recombination that displays excellent fermentation and expression characteristics. Moreover, DEAE column capacity was improved by 37% based on chromatography breakthrough data. This presentation will describe the fundamental research efforts, I-Corps^{TM} participation, minimum viable product development, and finally the assessment of the effort.

451290 Recombinant Snake Antivenom Peptide

Abstract:

Tens of thousands of people die by snakebite every year, primarily in developing countries. Many more survivors continue to suffer permanent physical and mental disabilities from a bite. Cytotoxic snake venoms are present in many countries including from American rattlesnakes. Current Fab-based treatment for US snakes costs over $1K per vial but is covered by some insurance policies for humans but not normally for animals. A small ribosomal peptide was reported to neutralize snake venom that was identified as the N-terminus of a serum protein in the North American Opossum (*Didelphis virginiana*). We have found that the chemically synthesized peptide is able to neutralize the hemorrhagic toxins in different species of snake venoms when tested in mice. In addition to activity studies, we have explored the possibility of making the peptide recombinantly by a process that would be amenable to a large scale production. The goal of our work is to propose a large scale, low cost process for the peptide that can be implemented in developing countries where snakes are prevalent and the population cannot afford expensive horse serum-based antivenom treatments. Our process obviates the use of snake venoms that are currently required to inject horses for the production of antibodies. We have successfully produced the peptide in *E. coli* and purified sufficient quantity to demonstrate activity in mice. Process optimization is currently underway, in addition to expanded activity studies. The presentation will cover details of our process studies and the activity of the *E. coli*-produced peptide.

457484 CFD MODEL Based Near-Infrared Spectroscopy Implementation for in-LINE ASSAY Monitoring of a Pharmaceutical Suspension

Abstract:

INTRODUCTION AND PURPOSE

During bottle filling of a pharmaceutical suspension formulation, a decreasing trend in assay is observed at the end of the batch. This leads to approximately 17% of the batch content which cannot be commercialized. The root cause can most likely be found in foam present in the top layer of the suspension in the holding tank which contains less API. The aim in this study is to implement a PAT solution for establishing a batch-dependent endpoint (I.e. assay â‰¤ 95%) determination while reducing in-process sampling drastically.

MATERIALS AND METHODS

Materials

Tests were performed with a high-dose (i.e. 100 mg/mL) pharmaceutical suspension formulation. Lab scale samples were prepared in a 10L mixing vessel. Full scale batches are manufactured in a 150L compounding vessel following a one-pan process. After the transfer to a holding tank, the full scale batches are gradually discharged to a specific bottle filling system with the help of a peristaltic pump. During this process step, the suspension is stirred continuously.

Methods

NIR spectra were recorded using a Fourier-transform Near-Infrared (NIR) spectrometer (Bruker Matrix-F Duplex) equipped with a TE-InGaAs detector, a quartz halogen lamp and a fiber-optic non-contact probe. Every 30 seconds, a spectrum was acquired in the 12000-4000 cm^{-1} region with a resolution of 8 cm^{-1}, averaged over 64 scans. Ultra Performance Liquid Chromatography (UPLC) was selected as reference technique for API content determination. All UPLC measurements were performed at Janssen Pharmaceutica, Beerse, Belgium, using their developed and validated method (confidential).

RESULTS AND DISCUSSION

Samples in a range of 85-115% target concentration were analyzed with both NIR spectroscopy and UPLC. A lab-scale calibration model was developed which shows acceptable feasibility for NIR spectroscopy as potential PAT tool for the API assay determination in suspension.

For approaching the production measurement setup as good as possible, the calibration model was optimized by using a circular system with a similar flow as in production (i.e. 800 mL/min) with the help of a customized T-piece. Samples with concentrations varying between 90-103% were measured and included in the prediction model.

At production scale however, the developed model showed a weak performance with an unexpected variation in NIR assay results. It was assumed that the signal variation was created by disturbed flow patterns/turbulence in front of the lens due to possible back pulse formation at the system output. A Computational Fluid Dynamics (CFD) simulation of the flow pattern of the suspension inside the measuring interface confirmed this assumption.

CFD simulations demonstrated that an extension of the original T-piece could optimize the measurements by transferring the back pulse formation further away from the lens. This mechanical solution allowed a constant sample volume per time unit in front of the NIR laser. This was confirmed with additional lab-scale tests and the follow-up of several production scale batches.

CONCLUSIONS

Despite the dominantly present water absorption bands, NIR spectroscopy is feasible to establish good API assay predictions of this high solid-load suspension formulation.

Experiments indicated that a fine-tuning of the interfacing device is crucial to achieve adequate in-line NIR measurements. Therefore, sufficient process knowledge, including fluidum mechanics, is desired to rationalize the choice for an optimal measurement setup.

CHALLENGES AND FUTURE WORK

In the future it is aimed to investigate if NIR spectroscopy in combination with this specific interfacing device is applicable for other liquid pharmaceutical formulations.

466854 Using Optical Coherence Tomography (OCT) to Analyze in Real-Time the Tablet Coating Variability of a Continuous Drum Coating Process

Abstract:

This study demonstrates an in-line measurement system used in a continuous drum coater (Driaconti-T lab, Driam Anlagenbau GmbH). Process monitoring is performed by an optical coherence tomography (OCT) sensor (OCT Pharma 1D, Phyllon GmbH) providing cross-sectional images of film-coatings in a non-destructive and contactless manner.

The continuous coater has three chambers, each with a product volume of around 2 liters of tablets. In continuous operation mode, the tablets are inserted in chamber 1 and transported to chamber 3, while being coated in each of them. Thus, the product is handled in small batches, without exchange of product between the chambers, resulting in a defined residence time for each tablet.

OCT measurement and subsequent automatic data analysis is a method to determine the coating thickness from OCT images without the need of chemometric calibration models, based on a known refractive index of the coating materials. OCT was used to obtain the temporal evolution of the quality parameters, i.e., mean coating thickness, inter-tablet coating variability and intra-tablet coating variability during the coating process. Thereby, the influence of spray rate, air temperature and spray pattern pressure on the quality parameters was studied based on a full-factorial DoE. Additionally, the variability of single drum segments of an extended period of time was studied.

Continuous coating, here in the form of small batches, is an important enabler for continuous manufacturing of solid dosage forms. In combination with an in-line OCT system the coating process can be monitored and actively

controlled.

467352 Advanced Flexible Control System Implementation into Direct Compaction Continuous Pharmaceutical Manufacturing Pilot-Plant

Abstract:

Currently, pharmaceutical companies are going under paradigm shift from batch to continuous manufacturing. Real time monitoring and supervisory control system are necessary part of efficient continuous manufacturing. However, the minimum level of control necessary to produce the product with desired critical quality attributes depends on several factors including drug content (low dose, high dose), formulation properties (e.g. flowability, segregation), and product properties. Nevertheless, different control strategies such as feedforward (FF), feedback (FB), or combined FF/FB[1]as well as different control algorithms such as PID, MPC, or hybrid MPC-PID can be potentially used. Selection of a right control strategy and algorithm is mandatory task both for efficient manufacturing as well as regulatory point of view. However, because of different levels of complexities and unavailability of systematic method and tools, the identification of required level of control, type of control strategy and control algorithm are still a challenging task.

In this work, a flexible control system has been implemented into a direct compaction continuous tablet manufacturing pilot-plant. The flexible control system consists of three control strategies (FF, FB, combined FF/FB[1]) and three control algorithms (PID, MPC, hybrid MPC-PID) giving options for multiple control schemes. The flexible control system is designed in such a way that any control scheme can be easily selected. The control system is also modular in nature that can be easily transferred to any other control platform and manufacturing plant. The continuous tablet manufacturing pilot-plant runs under each control scheme and their performance are evaluated using standardized performance indicators. Based on the results obtained, a systematic methodology is proposed to assist the selection of control strategy and algorithm for continuous pharmaceutical manufacturing process.

Direct compaction continuous tablet manufacturing process is considered to demonstrate the implementation of flexible control strategy. The flexible control strategy consist of the five supervisory feedback control loops and three supervisory forward control loops among which the required number of control loops can be selected. Feedback loops are for drug concentration, powder level in chute, main compression force, tablet weight and hardenss control while the feedforward loops are for proactively mitigate the effects of process and raw materials variabilities. For example, measured powder bulk density is considered in real time in tablet press to proactively mitigate the effect of powder bulk density on tablet weight and hardness via changing the fill depth. Similarly, the feed forward loop to mitigate the effects of lubricant concentration variability on tablet hardness and thereby on tablet dissolution is added. As well as a feed forward control loop to mitigate the effects variation on blend composition on tablet potency. All feedforward control loops can not be used simultaneously because of limitations of number of actuators and therefore the methodology plays an important role to identify the appropriate control loops for a particular pharmaceutical product. The flexible control system are implemented using industry standard control platform (e.g. DeltaV), PAT data management tool (synTQ), OPC communication protocol, fieldbus devises, chemometric tools and real time monitoring tools.

Reference

1. Singh, R., Muzzio, F., Ierapetritou, M., Ramachandran, R. **(2015)**. A combined feed-forward/feed-back control system for a QbD based continuous tablet manufacturing process. PROCESSES Journal, 3, 339-356.

468862 Raman Spectroscopy- Towards the Prediction of Quality Attributes and Application in Cell Culture Process Development

Abstract:

In recent years, the United States Food and Drug Administration (FDA) and the European Medicines Agency (EMA) have stressed the need of improved process understanding and control in the area of pharmaceutical development, manufacturing and quality assurance, summarized in the Process Analytical Technology (PAT) initiative. Lately, Raman spectroscopy has emerged as a promising online monitoring and control tool in bioprocesses. The advantages of this technology are the invasive and nondestructive operations, the simultaneously detection of multiple process variables, as well as the online operation mode and real-time information release.

Raman spectroscopy is mainly used for monitoring purposes at production stage, where variations are small and the process is already well defined. However, the exploitation of Raman technology for process development, where different process conditions, media, products or cell lines are tested resulting in a large variability and few experiments at similar conditions, could bring very important benefits, in terms of process understanding, speed and cost reduction. This work presents several ideas, experimental results and recommendations to overcome those challenges in process development.

Different fed-batch cultures, varying in cell lines, products and platform media, were monitored online. In addition, diverse spiking strategies were performed to artificially increase the calibration range as well as the quality of the prediction models. The evaluation of advanced modeling techniques to derive optimal results from the obtained spectral data plays a central role in this work. Those techniques included an in-depth analysis of the pretreatment procedure, wavelength selection and outlier elimination with regards to robust calibration and prediction of the components of interest.

Besides classical parameters such as glucose, lactate, titer and viable cell density, all 20 amino acids could be very well predicted. More than that, even the prediction of quality attributes like aggregates and glycans showed promising results. In addition, the potential of generic models, which can be applied independently of the cell line, process platform and medium, could be proven. This enables the Raman spectroscopy as an online monitoring tool of multiple process variables and its implementation into a supervisory control of a continuous integrated bioprocess to ensure consistent process efficiency and shaping the product qualities.

470108 Hydrodynamics and Mass Transfer Effects in Mini Vessel Dissolution Testing Systems

Abstract:

In the pharmaceutical industry, dissolution testing is routinely conducted to evaluate the *in vitro*performance of solid dosage forms in pharmaceutical development, determine the dissolution behavior of formulations and provide an optimization of drug release dosage forms, and as a quality control tool to insure that solid dosage forms have consistent dissolution properties. The USP Dissolution Testing Apparatus 2 is the device most commonly used for this purpose. In a typical test, a tablet is added to the vessel and samples are manually withdrawn from the vessels over time and analyzed for their drug content.

The USP 2 drug dissolution testing method is not always appropriate, especially for dissolution testing involving very small tablets or when not enough drug substance may be available for proper dissolution testing in the USP 2 Apparatus, especially during the initial stage of drug development. Small-volume dissolution testing apparatuses (mini vessel systems) can be advantageously used instead of the standard USP 2 system to overcome these limitations. The most common of these mini vessel systems is similar in shape and operation to the USP 2 apparatus, but it only requires a small volume vessel (200 mL) and a mini paddle impeller.

Before mini vessel systems can gain wide acceptance in industry, their hydrodynamics needs to be studied, and their operating conditions must be defined in order to compare the velocity distribution and mass transfer coefficient in a mini vessel system with those in the standard USP2 system. Very few studies on mini vessel systems are available. Therefore, the objective of this work was to quantify the hydrodynamics and drug mass transfer rates in mini vessel systems, and determine the operating conditions under which mini vessel systems can be reliably used instead of USP 2 system for dissolution testing purposes.

The velocity distribution in a minivessel was obtained here using Computational Fluid Dynamics (CFD) at different agitation speeds. The predictions were compared to the velocity vector maps obtained using Particle Image Velocimetry (PIV). Four horizontal iso-surfaces were selected to be studied in detail. The CFC predicted and PIV measured flow patterns were found to be in good agreement. As expected, the 3D flow velocity magnitude increased as the agitation speed increased. However, the velocity in the inner core region below the impeller always remained very low.

The tablet-medium mass transfer process was additionally modeled, and the mass transfer coefficients at four different agitation speeds were obtained and compared to the standard system. The mini vessel system has a flow pattern similar to that of the standard USP 2 system. It was found that similar velocities at the tablet surface (on the iso-surface at y=7mm, i.e., at the vessel bottom) in the mini vessel system and in the USP 2 system were obtained when the mini vessel was operated a 100 mL, 125 rpm and the USP 2 system was operated at 900rpm and 100rpm, indicating that these two systems should result in similar dissolution profiles when operated under these conditions.

471592 3D Raman Spectroscopy – a Method for Mapping Ingredient Distribution throughout a Tablet Volume

Abstract:

Blending of powders can often result in agglomeration when one or more raw materials in a blend are cohesive, hydrophilic or have the tendency of acquiring electrostatic charges. Agglomeration often leads to failure modes in pharmaceutical tablets, particularly "low dose" formulation tablets where the active pharmaceutical ingredient (API) is present in small amounts. When powders agglomerate, the uniformity of the final tablet is adversely affected, leading to variation in the tablet potency and dissolution profiles, which are usually undesirable. New technology enables the quantitative examination of the microstructure within tablets to determine the extent of agglomeration and to assess its effects on product performance. This can be useful as a means for understanding how the process causes agglomeration. If process parameters can be correlated to the tablet microstructure, this knowledge can assist in designing tablets of desirable quality thereby reducing product failure drastically.

Raman imaging is a powerful tool for studying tablet structure. It has higher chemical specificity than NIR and is also less affected by physical characteristics of the sample. The main disadvantage of Raman, its low quantum yield, requiring long integration times, is vanishing quickly, due to vast advances made recently in laser and detector technology. This is causing a significant renewed interest in using chemical mapping to examine the microstructure of pharmaceutical tablets.

This talk will focus on the use of a new Raman imaging platform, the mPAT system from H_2Optx, which is equipped with a 10x10 micron 2D spatial resolution along with the ability to shave layers off a tablet for depth analysis. This level of spatial resolution enables the user to extract detailed structural data that reveals the relationships between process, product structure, and product performance. This technique has the ability to do a quantitative content determination within a sample and, combining this for each pixel, generate 3D maps w.r.t. the individual constituents.

Tablets manufactured from traditional batch processing methods and continuous manufacturing methods were examined, including tablets made via Direct Compaction (DC), Wet Granulation (WG) and Roller Compaction (RC). Tablets using several formulations for each of the three methods were made with various levels of induced agglomeration and were analyzed using the mPAT system. The effects of the different processing conditions on agglomeration were characterized and their effect on tablet microstructure and dissolution were quantified. These observations will be discussed in this presentation.

463044 Real-Time Monitoring and Modeling of Milling in a Continuous Tableting Line Via Dry Granulation

Abstract:

In the Pharmaceutical Industry, batch operation is the traditional mode of manufacture. During batch production, the quality of the batch usually is subject to batch-to-batch variability and thus quality is controlled by statistical sampling and if necessary rejection of the entire batch. Conversion to the continuous processing mode offers the potential for quality improvement as well as potential cost savings. However, continuous processing requires on-line measurements and active process control and thus the development of Process Analytical Technology (PAT) tools is crucial [1].

In the continuous dry granulation tableting line there are a number of variables that relate to the final product quality. For instance, composition and tablet hardness are among the CQAs in this process. In the dry granulation line, the roller compactor serves as a granulation method. In this unit, the powders are fed into a hopper and then conveyed by the feed screw to two counter rotating rolls. The powder is gripped by the rolls and pulled into the nip area of the compacting rolls before being compacted into a ribbon. Then the ribbon is milled into granules by a hammer mill [2]; generating a granular feed for the tablet press, which has favorable flow and compaction properties.

The milling step is an important unit operation because the granules particle size distribution (PSD) will affect compaction properties and subsequently the tablet hardness. Therefore, the granules PSD is also a CQA. In the roller compaction process, an increase in roll pressure will translate to an increase in ribbon density. The ribbon density has a relation to granule PSD [3]. As a consequence, the ribbon density can be considered an intermediate CQA. The relationship between ribbon density and granule PSD is difficult to establish from first principles, but a correlation can be established by conducting a series of experiments. In this design of experiments, the roll pressure is varied, the ribbon is milled at different speeds and the PSD measured through either off-line means such as sieve analysis or else on-line using laser based methods.

In continuous operations, it is desirable to measure all of the CQAs online, so as to permit the implementation of appropriate process control logic. In the tableting line installed at Purdue University, the roller compacted (RC) ribbon properties such as density and composition can be measured online using either a microwave (MW) and/or a Near-Infrared (NIR) sensor [4]. The PAT tool for online PSD measurement that we have adopted in our work is a high speed imaging camera (Innopharma Eyecon Camera), with which our team has had some previously experience in wet granulation applications [5]. This equipment counts and images particles with a sampling time of 2s and has a scanning rate of $10 m.s^{-1}$.

The milling unit operation is commonly modeled using a population balance model [6]. The key model component of the population balance model (PBM) is the breakage kernel. This component represents the process of creation of smaller particles, as larger particles are impacted by the mill hammers or pins which causes the fracture. A wide range of breakage kernels have been proposed, most of purely empirical or semi-empirical nature. Each of these models requires the estimation of model specific parameters which will depend on feed material properties as well as mill operating variables [7].

This project has three objectives. The first one is to calibrate the Innopharma Eyecon camera. The second objective is to establish the parameters of the breakage kernel and thus the PBM model and the third objective is to develop the empirical relation between the ribbon density and granules PSD both measured using the on-line PAT tools. The first objective is achieved by generating samples of granules prepared under different conditions then relating the Innopharma Eyecon Camera output to the sieve analysis results. Since the sieve analysis measures fractions mass while the camera measures particle size, there is a conversion required which depends on particle density. The data for meeting the needs of the second and third objectives is determined through a series of experiments in which the mill speed, the composition of the powder and the density of the ribbon in the roller compactor and PSD are measured on-line. This data is gathered in runs when the process reaches steady state.

The results of this three-part study enable the capability for on-line monitoring of the combined compactor-mill subsystem. The PBM model and ribbon density-PSD relationship are important components of a data reconciliation model for the entire line. In this paper we report the results of these studies and demonstrate their utility in data reconciliation.

References

[1] S. L. Lee, T. F. O. Connor, X. Yang, C. N. Cruz, L. X. Yu, and J. Woodcock, "Modernizing Pharmaceutical Manufacturing : from Batch to Continuous Production," no. 1, pp. 1–25, 2015.

[2] R. W. Miller and P. J. Sheskey, "Roller Compaction Technology for Pharmaceutical Industry." Marcel Dekker Inc., 2003.

[3] S. Yu, "Roll Compaction Of Pharmaceutical Excipients," University of Birmingham, 2012.

[4] A. Gupta, J. Austin, S. Davis, M. Harris, and G. Reklaitis, "A novel microwave sensor for real-time online monitoring of roll compacts of pharmaceutical powders online - A comparative case study with NIR," *J. Pharm. Sci.*, vol. 104, no. 5, pp. 1787–1794, 2015.

[5] A. S. El Hagrasy, P. Cruise, I. Jones, and J. D. Litster, "In-line size monitoring of a twin screw granulation process using high-speed imaging," *J. Pharm. Innov.*, vol. 8, no. 2, pp. 90–98, 2013.

[6] E. B. Litster, J. D., *The Science and Engineering of Granulation Processes*, vol. 53, no. 9. 2013.

[7] J. K. Pettersen and K. L. Sandvik, "Estimating the breakage and selection functions for a continuous mill," *Int. J. Miner. Process.*, vol. 35, no. 3–4, pp. 149–158, Aug. 1992.

Chevron
human energy

Sustainable Development and Footprint Optimization in Shale Development

Janet Peargin, Footprint Program Manager
Hong Jin, Lead HES Engineer

Chevron Energy Technology Company

November 17, 2016

The challenge
footprint of shale/tight rock development

Many wells & moving parts

Footprint changes over time

Jobs
Community GDP
Desired energy

Land/habitat
Emissions
Trucks
Noise
Waste
Resource competition

Chevron's environmental principles

Include the environment in decision making

Reduce our environmental footprint

Operate responsibly

Steward our sites

Footprint optimization: everybody plays a role

We can't solve problems by using the same kind of thinking we used when we created them.

– Albert Einstein

Shifting your perspective is so much more powerful than being smart.

– Astro Teller, Google "X" 2016 TED Talk

The opportunity: sustainable development

SEE across the asset life to create value and protect people and the environment

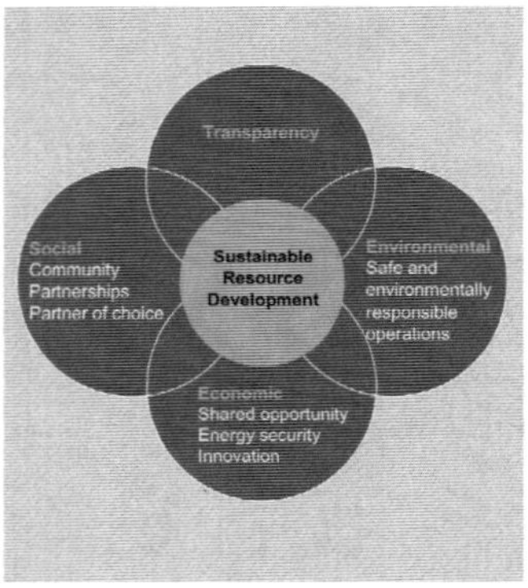

SEE triple bottom line value

SEEk win-win (all-win) ideas
Social, Environmental, Economic

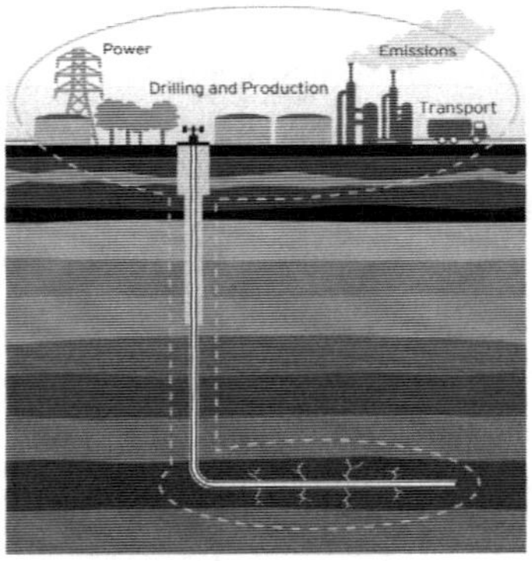

Optimize footprint

Integrated system of materials and energy that changes over time

Design with the end in mind

Decommission, restore, reuse

The win-win (all-win): optimize footprint to enhance value and reduce cost/risk

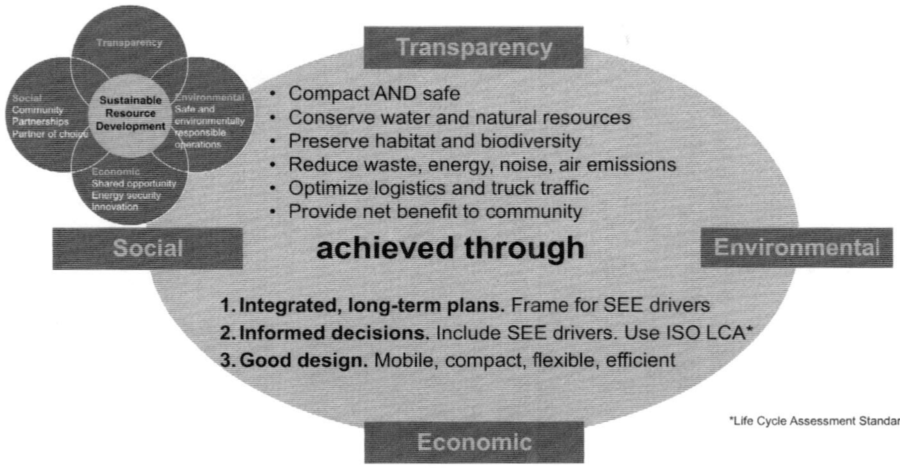

*Life Cycle Assessment Standard 14040-14044

Improve planning: SEE through the lens of sustainable development and SEE win-win (all-win) opportunities

Look at the big picture (asset, region)
Look at the long-term and change over time
Include diverse stakeholder perspectives

Take a systems view. Integrate across functions, activities, hydrocarbon/water/power
Plan and design for social, environmental, and economic (SEE) value drivers

Consider:

- Pace and scale of development
- Sequencing of events
- Degree of centralization
- Logistics and supply chain
- Ability to share infrastructure
- Power/fuel options
- Integration points

Improve decision quality: SEE more broadly and longer-term

Compliance is expected.

A traditional financial analysis is narrow and short-term.

SEE more broadly and longer term to identify risks and opportunities … and enhance decision quality.

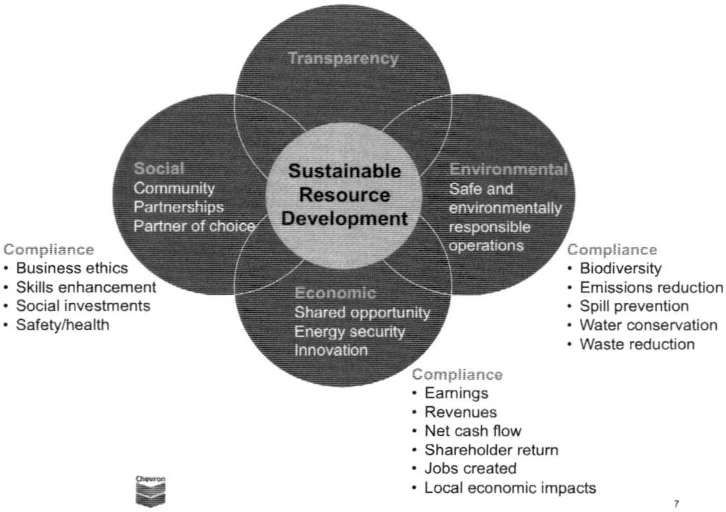

Improve decision quality: quantify environmental tradeoffs with ISO life cycle assessment (LCA)

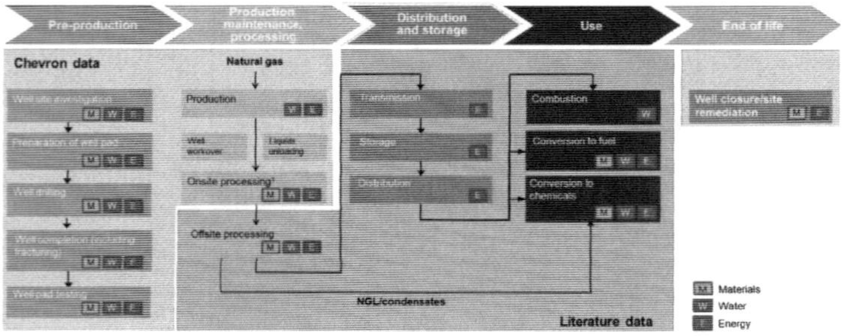

- What is our environmental footprint?
- Where are our big sources?
- What is the "best" fuel for power?
- What if I use this technology vs. that?
- What are the on and offsite impacts?

Improve design: apply win-win (all-win) principles

Plan/design for sustainable development	• Develop long-term integrated plans • Frame for SEE (social, environmental, economic)
Integrate and innovate	• Look across the asset life at hydrocarbons/water/power & different activities (drilling, operations, etc.)
Question business as usual	• What do stakeholders want? Is there a better way?
Do *more* with *less*	• Reduce – reuse – recycle – be (energy/materials) efficient • Eliminate no-value steps & equipment
Disturb less land	• Fewer pads; share infrastructure/right-of-ways; preserve habitat corridors
Conserve water	• Use less & lower quality; treat to quality needed for use; consider reuse
Reduce waste	• Flares/emissions, discharges, energy use, no-value steps/equipment
Green the process	• Consider non-toxic chemistry, power/fuel efficiencies, fewer steps
Leverage technology	• Look to downstream, offshore, suppliers, competitors, others • Think mobile, modular, compact, flexible, prefab, energy efficient, automated, standard
Leverage supply chain	• How can suppliers help? Can we hire and source locally?
Make informed decisions	• SEE the big picture. Consider short- *and* long-term costs and value. • Use life cycle assessment to quantify environmental aspects of alternatives

Chevron

Win-win example: electricity from field gas

Better emissions/safety/reliability/production/cash flow

Problem

Need cost-effective, low-emissions power source at remote site

Solution

Micro-turbines make electricity from field gas*

Win-Win **Do *more* with *less***

MORE

Safety, flexibility, schedule certainty, reliability (8–10%), production/cash flow (sooner)

LESS

Equipment, permitting, O&M, trucks, emissions, material transfer/spill potential

** Great for ESPs. Rod pumps doable with more systems engineering*

Chevron

Everyone plays a role: ask questions

- What are our stakeholder concerns?
- What perspectives are we missing?
- What are our footprint boundaries?
- What are the priorities and tradeoffs?
- What does good (enough) look like?
- How can we minimize environmental and community impacts?
- How do we do more (product) with less (material, equipment, waste, etc.)?
- How do we design for change over time to operations, region, community?
- What can we learn from other processes?
  - LeanSigma®
  - Environmental, social, health impact assessment
  - Energy Efficiency and Waste Minimization assessments

LeanSigma is a federally registered service mark of TBM Consulting, Inc.

Key takeaways

- We all play a role in creating our company footprint – a system of materials and energy that changes over time
- Think beyond compliance. Include SEE (social, environmental, economic) drivers in planning, designs, and decisions.
- SEE through the lens of sustainable development to SEE opportunities. SEE the difference.
  - Cost savings from less equipment, optimized layout/network
  - Win-win ideas to save $ AND protect people & the environment
  - Decisions that are more robust
- A financial analysis is only part of the risk/uncertainty. We <u>can</u> quantify and include SEE drivers in decision-analysis.
- Sustainable development is good business. SEE it reduce cost and risk, add value, and improve performance.
- Everyone plays a role. You can contribute. Just change your perspective to SEE the win-win (all-win) opportunities to do more (product, energy) with less (energy, materials, waste, emissions, trucks, etc.)

References

- Hardisty, P.E. *Environmental and Economic Sustainability.* CRC Press, New York. ISBN 978-14200594891-56679-643-2, 378 pp. 2010.

- Sivapalan, M.; WorleyParsons. *Sustainable Development of Unconventional Oil & Gas Resources.* Prepared for presentation at *GasTech* 2012 Conference, London, 2012.

- Tanzil, D.; Dragan, M.; Bourque, B.; Seel, K.; Taylor, T.; Golder Associates. *Evaluating Social and Environmental Risk and Opportunities in Site Selection and Routing.* Copyright 2015, Society of Petroleum Engineers. SPE-173488-MS. Prepared for presentation at the SPE E&P Health, Safety, Security, & Environmental Conference – Americas held in Denver, Colorado, USA, 16–18 March 2015.

- Thomas, Charlotte. *Industrial Ecology: Moving Beyond Traditional Engineering Borders.* Society of Women Engineers magazine, spring edition, pp. 24-30. 2010

Chevron

461852 Green Net Value Added As a Sustainability Metric Based on Life Cycle Assessment: An Application to Bounty ® Paper Towel

Abstract:

Sustainability measurement in economics involves evaluation of environmental and economic impact in an integrated manner. In this study, we combine system level economic data with environmental impact from a life cycle assessment (LCA) of a common product. We are exploring a costing approach that captures traditional costs and incorporates externality costs to provide a convenient, easily interpretable sustainability metric. Green Net Value Added (GNVA) is a type of full cost accounting that incorporates total revenue, the cost of materials and services, depreciation, and environmental externalities. It adjusts the traditional economic value added for environmental externality costs. Two, but not all, of the potential environmental impacts calculated by the standard LCIA method (TRACI) could be converted to externality cost values. We compute externality costs disaggregated by upstream sectors in order to accurately reflect the externality costs associated with different industries along the supply chain. We preset full cost, green profit and GNVA to evaluate the relative sustainability of Bounty® paper towels manufactured at two production facilities. We found that the longer running, more established line had a higher GNVA than the newer line. The dominant factors contributing to externality costs are to come from the stationary sources in the supply chain: electricity generation (27-35%), refineries (20-21%), pulp, and paper making (15-23%). Health related externalities from Particulate Matter (PM2.5) and Carbon Dioxide equivalent (CO_2e) emissions appear largely driven by electricity usage and emissions by the facilities, followed by pulp processing and transport. Supply chains for other facility purchases, product distribution, and end-of-life management (landfilling/incineration) contribute the least to the environmental health-related costs.

474653 Sustainability Metrics for Solar Powered Charging Infrastructure for Electric Vehicles

Abstract:

This presentation will include an analysis of sustainability metrics associated with the installation of solar panels in parking lots designed to provide electricity for commercial use and shade for those who park under the solar panels. The installation will also include electric vehicle supply equipment to charge batteries in electric vehicles. The system is connected to the electrical grid so that the electricity generated can be used efficiently and electric vehicle batteries can be charged at all times. Important sustainability metrics are cost of electricity generated, cost of vehicle charging, value of shaded parking, convenience of vehicle charging, land area, water, air quality, greenhouse gas emissions, and materials for the installation and equipment. Social metrics include impact on health status of the community, psychological well being, harmony, values, agency, and environmental justice.

458893 Evaluation of Low-Energy Recovery of N-Methyl-2-Pyrrolidone Solvent from Waste Streams in Specialty Resin Manufacture

Abstract:

N-methyl-2-pyrrolidone (NMP) is a common solvent used as a reaction medium in the manufacture of specialty polymeric resins and their precursors. NMP is used primarily because of its thermal stability, dipolar aprotic properties, and ability to produce a high molecular weight resin precursor. Since NMP is not consumed by the reaction, an aqueous hazardous waste stream is generated. A previous case study at a DuPont specialty chemical plant by the authors, shows that a distillation recovery system that can effectively purify NMP from the waste stream. The distillation recovery process reduces operating cost by 83%, but only reduces the life cycle carbon footprint by 40%. This is due to the high energy demand of the distillation column. Additionally, high NMP purity levels in the resin manufacture process and the complex multicomponent nature of the waste stream, required a more extensive design for the recovery operation.

Alternative distillation designs were explored, but none reduced energy consumption and life cycle carbon emissions as significantly as the economic life cycle impact. More sustainable low energy separations were investigated. Various membrane, adsorption, and extraction separation strategies have the potential to reduce energy demand, but each has its own limitations. An extraction operation was selected as the best candidate for separation, and the proposed process considers alternative uses for the recovered NMP waste in other manufacturing scenarios.

Results with a chloroform extraction operation to produce NMP of suitable quality for alternative use, provides the best option for solvent recovery. The new process can not only yield environmental life cycle improvements, based on various impact factors, but provide economic benefits as well. Yearly projections of waste reductions, life cycle improvements, energy utilization, and cost savings are presented. Life cycle carbon emissions are reduced by over 60% compared to the base case. Since NMP is a prominent solvent for several commercial sectors, methods aimed at reducing waste help promote broader sustainability goals. This study shows how sustainability metrics can be used to guide decision making when evaluating other processing alternatives and uses of the final product. (This work is partially supported by a grant from the EPA Pollution Prevention program).

468272 Thermoelectric Power Technology Choices Based on Water Availability

Abstract:

Thermoelectric Power Technology Choices Based on Water Availability National Energy Technology Laboratory (NETL) has developed a prototype water-energy model to investigate issues surrounding the energy-water nexus as it relates to the effect of power generation plant design choices on water availability and the overall costs of supplying water for thermoelectric power generation.Â The model utilizes the database developed by a team of researchers with the Water Security Program at Sandia National Laboratory (SNL). The database includes water availability, cost, and water use for electric power, agriculture, municipal and industrial for 38 conterminous states representing various Interconnects within the U.S. Data on current and future water use for thermoelectric power are derived from NETL databases on water use by technology and projections from the Energy Information Administrationâ€™s National Energy Modeling System (NEMS). Current and forecasted water availability data from the World Resources Institute are used to develop three water availability scenarios. The prototype water-energy model is used to project the amount of water available for power generation, to project water demand for future power generation based on a NEMS forecast, to identify regions that experience potential water shortfalls, and to estimate the cost of meeting water demand where water shortfalls exist for each water availability scenario. A case study of the Brazos River Basin in Texas is used to illustrate a specific application and to highlight the capabilities of the prototype water-energy model.

AUTHOR INDEX

AUTHOR INDEX

AUTHOR INDEX

AUTHOR INDEX

AUTHOR INDEX

AUTHOR INDEX